Student's Solutions Ma

to accompany

CALCULUS
Early Transcendental Functions
Third Edition

Robert T. Smith
Millersville University of Pennsylvania

Roland B. Minton
Roanoke College

Prepared by
Rebecca Torrey
Brandeis University
James Scibert
Regis University
Blake Thornton
Washington University at St. Louis

with contributions by
Ji Li
Brandeis University

 **Higher Education**

Boston   Burr Ridge, IL   Dubuque, IA   Madison, WI   New York   San Francisco   St. Louis
Bangkok   Bogotá   Caracas   Kuala Lumpur   Lisbon   London   Madrid   Mexico City
Milan   Montreal   New Delhi   Santiago   Seoul   Singapore   Sydney   Taipei   Toronto

The **McGraw·Hill** Companies

Student's Solutions Manual to accompany
CALCULUS: EARLY TRANSCENDENTAL FUNCTIONS, THIRD EDITION
ROBERT T. SMITH AND ROLAND B. MINTON

Published by McGraw-Hill Higher Education, an imprint of The McGraw-Hill Companies, Inc., 1221 Avenue of the Americas, New York, NY 10020. Copyright © 2007 by The McGraw-Hill Companies, Inc. All rights reserved.

This book is printed on acid-free paper.

5 6 7 8 9 0 QPD / QPD 0 9 8

ISBN 978-0-07-286957-6
MHID 0-07-286957-7

www.mhhe.com

Contents

Chapter 0

Preliminaries

0.1 Polynomials and Rational Functions

1. Yes. The slope of the line joining the points $(2, 1)$ and $(0, 2)$ is $-\frac{1}{2}$, which is also the slope of the line joining the points $(0, 2)$ and $(4, 0)$.

3. No. The slope of the line joining the points $(4, 1)$ and $(3, 2)$ is -1, while the slope of the line joining the points $(3, 2)$ and $(1, 3)$ is $-\frac{1}{2}$.

5. Slope is $\dfrac{6 - 2}{3 - 1} = \dfrac{4}{2} = 2$.

7. Slope is $\dfrac{-1 - (-6)}{1 - 3} = \dfrac{5}{-2} = -\dfrac{5}{2}$.

9. Slope is

$$\dfrac{-0.4 - (-1.4)}{-1.1 - 0.3} = \dfrac{1.0}{-1.4} = -\dfrac{5}{7}.$$

In exercises 11-15, the equation of the line is given along with the graph. Any point on the given line will suffice for a second point on the line.

11. $y = 2(x - 1) + 3 = 2x + 1$

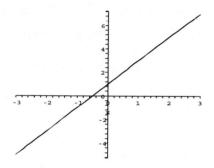

13. $y = 1$

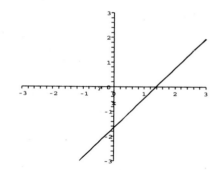

15. $y = 1.2(x - 2.3) + 1.1 = 1.2x - 1.66$

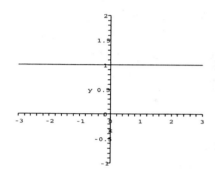

17. Parallel. Both have slope 3.

19. Perpendicular. Slopes are -2 and $\frac{1}{2}$.

21. Perpendicular. Slopes are 3 and $-\frac{1}{3}$.

23. (a) $y = 2(x - 2) + 1$

    (b) $y = -\frac{1}{2}(x - 2) + 1$

25. (a) $y = 2(x - 3) + 1$

    (b) $y = -\frac{1}{2}(x - 3) + 1$

27. Slope $\dfrac{3-1}{2-1} = \dfrac{2}{1} = 2$ through the given points. One possibility:

$$y = 2(x-1) + 1 = 2x - 1.$$

When $x = 4$, $y = 7$.

29. Slope $\dfrac{3-4}{1-0.5} = \dfrac{-1}{0.5} = -2$ through the given points. One possibility:

$$y = -2(x-1) + 3 = -2x + 5.$$

When $x = 4$, $y = -3$.

31. Yes, passes vertical line test.

33. No. The vertical line $x = 0$ meets the curve twice; nearby vertical lines meet it three times.

35. Both: This is clearly a cubic polynomial, and also a rational function because it can be written as

$$f(x) = \frac{x^3 - 4x + 1}{1}.$$

(This shows that all polynomials are rational.)

37. Rational.

39. Neither: Contains square root.

41. We need the function under the square root to be non-negative. $x + 2 \geq 0$ when $x \geq -2$. The domain is $\{x \in \mathbb{R} | x \geq -2\} = [-2, \infty)$.

43. Negatives are permitted inside the cube root. There are no restrictions, so the domain is $(-\infty, \infty)$ or all real numbers.

45. The denominator cannot be zero. $x^2 - 1 = 0$ when $x = \pm 1$. The domain is $\{x \in \mathbb{R} | x \neq \pm 1\}$ $= (-\infty, -1) \cup (-1, 1) \cup (1, \infty)$

47.

$$f(0) = 0^2 - 0 - 1 = -1$$
$$f(2) = 2^2 - 2 - 1 = 1$$
$$f(-3) = (-3)^2 - (-3) - 1 = 11$$
$$f\left(\frac{1}{2}\right) = \left(\frac{1}{2}\right)^2 - \frac{1}{2} - 1 = -\frac{5}{4}$$

49.

$$f(0) = \sqrt{0+1} = 1$$
$$f(3) = \sqrt{3+1} = 2$$
$$f(-1) = \sqrt{-1+1} = 0$$
$$f\left(\frac{1}{2}\right) = \sqrt{\frac{1}{2} + 1} = \sqrt{\frac{3}{2}} = \frac{\sqrt{6}}{2}$$

51. The only constraint we know is that the width should not be negative, so a reasonable domain would be $\{x | x > 0\}$.

53. Again, the only constraint we know for sure is that x should not be negative, i.e., a reasonable domain would be $\{x | x > 0\}$.

55. Answers vary. There may well be a positive correlation (more study hours = better grade), but not necessarily a functional relation.

57. Answers vary. While not denying a negative correlation (more exercise = less weight), there are too many other factors (metabolic rate, diet) to be able to quantify a person's weight as a function just of the amount of exercise.

59. A flat interval corresponds to an interval of constant speed; going up means that the speed is increasing while the graph going down means that the speed is decreasing. It is likely that the bicyclist is going uphill when the graph is going down and going downhill when the graph is going up.

61. The x-intercept occurs where
$$0 = x^2 - 2x - 8 = (x-4)(x+2),$$
so $x = 4$ or $x = -2$; y-intercept at $y = 0^2 - 2(0) - 8 = -8$.

63. The x-intercept occurs where
$$0 = x^3 - 8 = (x-2)(x^2 + 2x + 4),$$
so $x = 2$ (using the quadratic formula on the quadratic factor gives the solutions $x = -1 \pm \sqrt{-3}$, neither of which is real so neither contributes a solution); y-intercept at $y = 0^3 - 8 = -8$.

65. The x-intercept occurs where the numerator is zero, at
$$0 = x^2 - 4 = (x-2)(x+2),$$
so $x = \pm 2$; y-intercept at
$$y = \frac{0^2 - 4}{0 + 1} = -4.$$

67. $x^2 - 4x + 3 = (x-3)(x-1)$, so the zeros are $x = 1$ and $x = 3$.

69. Quadratic formula gives
$$x = \frac{4 \pm \sqrt{16 - 8)}}{2} = 2 \pm \sqrt{2}$$

71. $\begin{aligned} x^3 - 3x^2 + 2x &= x(x^2 - 3x + 2) \\ &= x(x-2)(x-1), \end{aligned}$
so the zeros are $x = 0, 1,$ and 2.

73. With $t = x^3$, $x^6 + x^3 - 2$ becomes $t^2 + t - 2$ and factors as $(t+2)(t-1)$. The expression is zero only if one of the factors is zero, i.e., if $t = 1$ or $t = -2$. With $x = t^{1/3}$, the first occurs only if $x = (1)^{1/3} = 1$. The latter occurs only if $x = (-2)^{1/3}$, about -1.2599.

75. If $B(h) = -1.8h + 212$, then we can solve $B(h) = 98.6$ for h as follows:
$$98.6 = -1.8h + 212$$
$$1.8h = 113.4$$
$$h = \frac{113.4}{1.8} = 63$$

This altitude (63,000 feet above sealevel, more than double the height of Mt. Everest) would be the elevation at which we humans boil alive in our skins. Of course the cold of space and the near-total lack of external pressure create additional complications which we shall not try to analyze.

77. This is a two-point line-fitting problem. If a point is interpreted as $(x, y) =$ (temperature , chirp rate), then the two given points are $(79, 160)$ and $(64, 100)$. The slope being $\frac{160 - 100}{79 - 64} = \frac{60}{15} = 4$, we could write $y - 100 = 4(x - 64)$ or $y = 4x - 156$.

79. Her winning percentage is calculated by the formula $P = \frac{100w}{t}$, where P is the winning percentage, w is the number of games won and t is the total number of games. Plugging in $w = 415$ and $t = 415 + 120 = 535$, we find her winning percentage is approximately $P \approx 77.57$, so we see that the percentage displayed is rounded up from the actual percentage. Let x be the number of games won in a row. If she doesn't lose any games, her new winning percentage will be given by the formula $P = \frac{100(415 + x)}{535 + x}$. In order to have her winning percentage displayed as 80%, she only needs a winning percentage of 79.5 or greater. Thus, we must solve the inequality

$$79.5 \leq \frac{100(415 + x)}{535 + x}:$$

$$79.5 \leq \frac{100(415 + x)}{535 + x}$$
$$79.5(535 + x) \leq 41500 + 100x$$
$$42532.5 + 79.5x \leq 41500 + 100x$$
$$1032.5 \leq 20.5x$$
$$50.4 \leq x$$

(In the above, we are allowed to multiply both sides of the inequality by $535 + x$ because we assume x (the number of wins in a row) is positive.) Thus she must win at least 50.4 times in a row to get her winning percentage to display as 80%. Since she can't win a fraction of a game, she must win at least 51 games in a row.

0.2   Graphing Calculators and Computer Algebra Systems

1. Intercepts $x = \pm 1$, $y = -1$. Minimum at $(0, -1)$. No asymptotes.

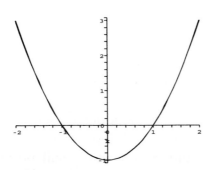

3. Intercepts $y = 8$ (no x-intercepts). Minimum at $(-1, 7)$. No asymptotes.

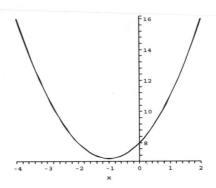

5. Intercepts $x = -1$, $y = 1$. No extrema or asymptotes.

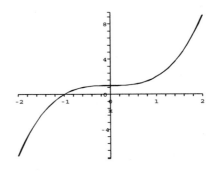

7. Intercepts $x \approx 0.453$, $y = -1$. No extrema or asymptotes.

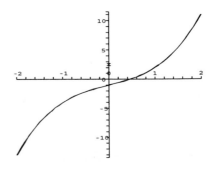

9. Intercepts $x \pm 1$, $y = -1$. Minimum at $(0, -1)$. No asymptotes.

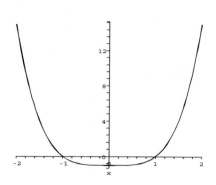

11. Intercepts $x \approx 0.475$, $x \approx -1.395$, $y = -1$. Minimum at (approximately) $(-1/\sqrt[3]{2}, -2.191)$. No asymptotes.

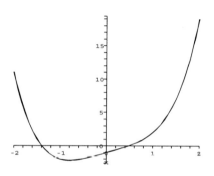

13. Intercepts $x \approx -1.149$, $y = 2$. No extrema or asymptotes.

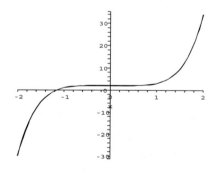

15. Intercepts $x \approx 0.050$, $y = -1$. The two local maxima occur at

$$x = \sqrt{\frac{24-\sqrt{176}}{10}} \text{ and } x = -\sqrt{\frac{24+\sqrt{176}}{10}},$$

while the two local minima occur at

$$x = \sqrt{\frac{24+\sqrt{176}}{10}} \text{ and } x = -\sqrt{\frac{24-\sqrt{176}}{10}}.$$

No asymptotes.

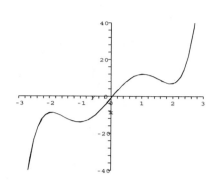

17. Intercepts $y = -3$ (no x-intercepts). No extrema. Horizontal asymptote $y = 0$. Vertical asymptote $x = 1$.

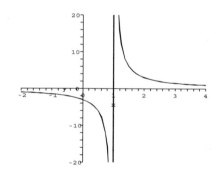

19. Intercepts $y = 0$ (and $x = 0$). No extrema. Horizontal asymptote $y = 3$. Vertical asymptote $x = 1$.

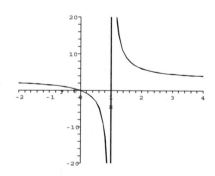

21. Intercepts $y = 0$ (and $x = 0$). Local maximum at $(0,0)$. Local minimum at $(2, 12)$. Vertical asymptote $x = 1$. Slant asymptote $y = 3x + 3$.

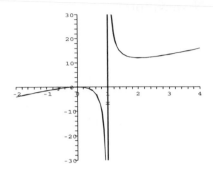

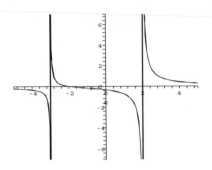

23. Intercepts $y = -1/2$ (no x-intercepts). Local maximum at $(0, -1/2)$. Vertical asymptotes $x = \pm 2$. Horizontal asymptote $y = 0$.

29. Intercepts $y = 0$ (and $x = 0$). No extrema. Horizontal asymptotes $y = \pm 3$.

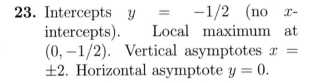

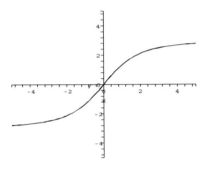

25. Intercepts $y = 3/4$ (no x-intercepts). Local maximum at $(0, 3/4)$. Horizontal asymptote $y = 0$.

31. Vertical asymptotes where $x^2 - 4 = 0 \Rightarrow x = \pm 2$.

33. Vertical asymptotes where
$x^2 + 3x - 10 = 0$
$\Rightarrow (x + 5)(x - 2) = 0$
$\Rightarrow x = -5$ or $x = 2$.

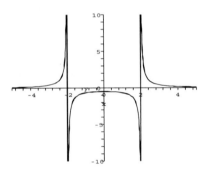

35. A vertical asymptote may occur when the denominator is zero. This denominator however is never zero, so there are no vertical asymptotes.

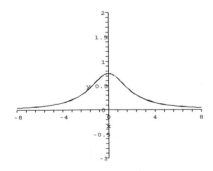

27. Intercepts $x = -2$, $y = -1/3$. No extrema. Horizontal asymptote $y = 0$. Vertical asymptotes at $x = -3$ and $x = 2$.

37. Vertical asymptotes where
$x^3 + 3x^2 + 2x = 0$
$\Rightarrow x(x^2 + 3x + 2) = 0$
$\Rightarrow x(x + 2)(x + 1) = 0$
$\Rightarrow x = 0$, $x = -2$, or $x = -1$.
Since none of these x values make the numerator zero, they are all vertical asymptotes.

39. A window with $-0.1 \leq x \leq 0.1$ and $-0.0001 \leq y \leq 0.0001$ shows all details.

41. A window with $-15 \leq x \leq 15$ and $-80 \leq y \leq 80$ shows all details.

43. From graph $x = 1$.

45. From graph $x = \pm 1$.

47. From graph $x = 0$.

49. The blow-up makes it appear that there are two intersection points. Solving algebraically, $\sqrt{x-1} = x^2 - 1$ (for $x \geq 1$) when

$$x - 1 = (x^2 - 1)^2 = ((x-1)(x+1))^2$$
$$= (x-1)^2(x+1)^2.$$

We see that $x = 1$ is one solution (obvious from the start), while for any other, we can cancel one factor of $x-1$ and find

$$1 = (x-1)(x+1)^2 = (x^2-1)(x+1)$$
$$= x^3 + x^2 - x - 1.$$

Hence $x^3 + x^2 - x - 2 = 0$.

By solver or spreadsheet, this equation has only the one solution $x \approx 1.206$.

51. The graph does not clearly show the number of intersection points. Solving algebraically,

$$x^3 - 3x^2 = 1 - 3x$$
$$\Rightarrow x^3 - 3x^2 + 3x - 1 = 0$$
$$\Rightarrow (x-1)^3 = 0 \Rightarrow x = 1.$$

So there is only one solution: $x = 1$.

53. After zooming out, the graph shows that there are two solutions: one near zero, and one around ten. Algebraically,

$$(x^2 - 1)^{2/3} = 2x + 1$$
$$\Rightarrow (x^2 - 1)^2 = (2x+1)^3$$
$$\Rightarrow x^4 - 2x^2 + 1 = 8x^3 + 12x^2 + 6x + 1$$
$$\Rightarrow x^4 - 8x^3 - 14x^2 - 6x = 0$$
$$\Rightarrow x(x^3 - 8x^2 - 14x - 6) = 0.$$

We thus confirm the obvious solution $x = 0$, and by solver or spreadsheet, find the second solution $x \approx 9.534$.

55. The graph shows that there are two solutions: $x \approx \pm 1.177$ by calculator or spreadsheet.

57. Calculator shows zeros at approximately -1.879, 0.347, and 1.532.

59. Calculator shows zeros at approximately $.5637$ and 3.0715.

61. Calculator shows zeros at approximately -5.248 and 10.006.

63. The graph of $y = x^2$ on the window $-10 \leq x \leq 10$, $-10 \leq y \leq 10$ appears identical (except for labels) to the graph of $y = 2(x-1)^2 + 3$ if the latter is drawn on a graphing window centered at the point $(1, 3)$ with

$$1 - 5\sqrt{2} \leq x \leq 1 + 5\sqrt{2}$$
$$-7 \leq y \leq 13.$$

65. $\sqrt{y^2}$ is the distance from (x, y) to the x-axis. $\sqrt{x^2 + (y-2)^2}$ is the distance from (x, y) to the point $(0, 2)$. If we require that these be the same, and we square both quantities, we have

$$y^2 = x^2 + (y-2)^2$$
$$y^2 = x^2 + y^2 - 4y + 4$$
$$4y = x^2 + 4$$
$$y = \frac{1}{4}x^2 + 1$$

In this relation, we see that y is a quadratic function of x. The graph is commonly known as a parabola.

0.3   Inverse Functions

1. $f(x) = x^5$ and $g(x) = x^{1/5}$

$$f(g(x)) = f(x^{1/5})$$
$$= (x^{1/5})^5 = x^{(5/5)} = x$$

$$g(f(x)) = g(x^5) = (x^5)^{1/5}$$
$$= x^{(5/5)} = x$$

3.

$$f(g(x)) = 2\left(\sqrt[3]{\frac{x-1}{2}}\right)^3 + 1$$

$$= 2\left(\frac{x-1}{2}\right) + 1 = x$$

$$g(f(x)) = \sqrt[3]{\frac{f(x)-1}{2}}$$

$$= \sqrt[3]{\frac{2x^3 + 1 - 1}{2}}$$

$$= \sqrt[3]{x^3} = x$$

5. The function is one-to-one since $f(x) = x^3$ is one-to-one. To find the inverse function, write

$$y = x^3 - 2$$
$$y + 2 = x^3$$
$$\sqrt[3]{y+2} = x$$

So $f^{-1}(x) = \sqrt[3]{x+2}$.

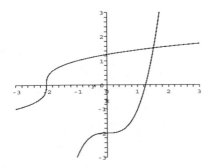

7. The graph of $y = x^5$ is one-to-one and hence so is $f(x) = x^5 - 1$. To find a

formula for the inverse, write

$$y = x^5 - 1$$
$$y + 1 = x^5$$
$$\sqrt[5]{y+1} = x$$

So $f^{-1}(x) = \sqrt[5]{x+1}$.

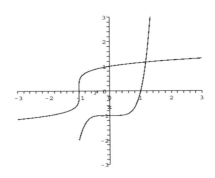

9. The function is not one-to-one since it is an even function ($f(-x) = f(x)$). In particular, $f(2) = 18 = f(-2)$.

11. Here, the natural domain requires that the radicand (the object inside the radical) be nonnegative. Hence $x \geq -1$ is required, while all function-values are nonnegative. Therefore the inverse, if defined at all, will be defined only for nonnegative numbers. Sometimes one can determine the existence of an inverse in the process of trying to find its formula. This is an example: Write

$$y = \sqrt{x^3 + 1}$$
$$y^2 = x^3 + 1$$
$$y^2 - 1 = x^3$$
$$\sqrt[3]{y^2 - 1} = x$$

The left side is a formula for $f^{-1}(y)$, good for $y \geq 0$. Therefore, $f^{-1}(x) = \sqrt[3]{x^2 - 1}$ whenever $x \geq 0$.

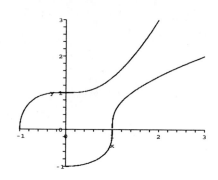

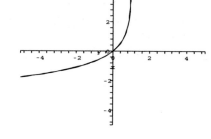

13. (a) Since $f(0) = -1$, we know $f^{-1}(-1) = 0$.

    (b) Since $f(1) = 4$, we know $f^{-1}(4) = 1$.

15. (a) Since $f(-1) = -5$, we know $f^{-1}(-5) = -1$.

    (b) Since $f(1) = 5$, we know $f^{-1}(5) = 1$.

17. (a) Since $f(2) = 4$, we know $f^{-1}(4) = 2$.

    (b) Since $f(0) = 2$, we know $f^{-1}(2) = 0$.

19. Reflect the graph across the line $y = x$.

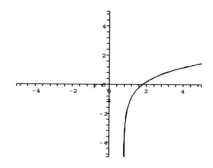

21. Reflect the graph across the line $y = x$.

23. Since 23 is halfway between 20 $(= f(2))$ and 26 $(= f(3))$, the x-value for $y = 23$ should be halfway between 2 and 3, i.e., $f^{-1}(23)$ is estimated linearly by 2.5.

Since the lines between the points fall to the right of the apparent true curve of the graph, this estimate is too high.

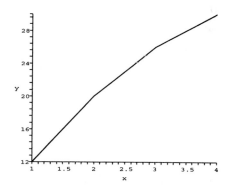

25. Since 5 is three-quarters of the way from 2 $(= f(3))$ to 6 $(= f(2))$, the x-value should be three-quarters of the way from 3 to 2, i.e., $f^{-1}(5)$ is estimated linearly by 2.25.

Since the lines between the points fall to the right of the apparent true curve of the graph, this estimate is too high.

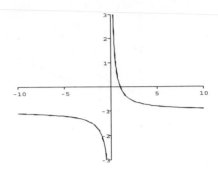

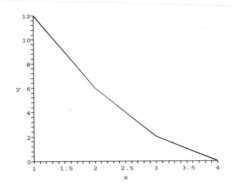

27. If $f(x) = x^3 - 5$, then the horizontal line test is passed, so $f(x)$ is one-to-one.

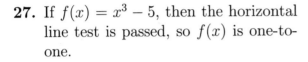

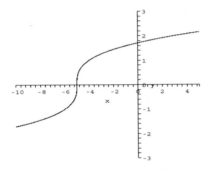

29. The function $f(x) = x^3 + 2x - 1$ easily passes the horizontal line test and is invertible.

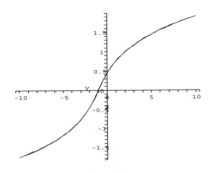

31. Not one-to-one. Fails horizontal line test.

33. If $f(x) = \dfrac{1}{x+1}$, then the horizontal line test is passed, so $f(x)$ is one-to-one.

35. If $f(x) = \dfrac{x}{x+4}$, then the horizontal line test is passed so $f(x)$ is one-to-one.

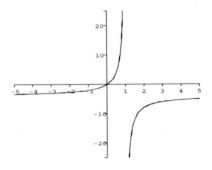

37. $f(g(x)) = (g(x))^2 = (\sqrt{x})^2 = x$
$g(f(x)) = \sqrt{f(x)} = \sqrt{x^2} = |x|.$

Because $x \geq 0$, the absolute value is the same as x. Thus these functions (both defined only when $x \geq 0$) are inverses.

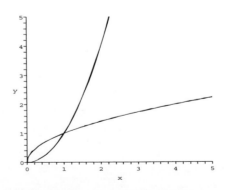

39. With $f(x) = x^2$ defined only for $x \leq 0$, (shown below as dotted) the horizontal line test is easily passed. The formula for the inverse function g is $g(x) = -\sqrt{x}$ shown below as solid and defined only for $x \geq 0$.

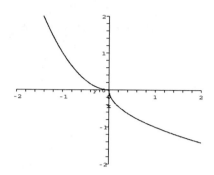

41. The graph of $y = (x - 2)^2$ is a simple parabola with vertex at $(2, 0)$. If we take only the right half $\{x \geq 2\}$ (shown below as the lower right graph) the horizontal line test is easily passed, and the formula for the inverse function g is $g(x) = 2 + \sqrt{x}$ defined only for $x \geq 0$ and shown below as the upper left graph.

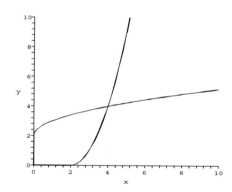

43. In the first place, for $f(x)$ to be defined, the radicand must be nonnegative, i.e., $0 \leq x^2 - 2x = x(x - 2)$ which entails either $x \leq 0$ or $x \geq 2$. One can restrict the domain to either of these intervals and have an invertible function. Taking the latter for convenience, the inverse will be found

as follows:

$$y = \sqrt{x^2 - 2x}$$
$$y^2 = x^2 - 2x = x^2 - 2x + 1 - 1$$
$$= (x - 1)^2 - 1$$
$$y^2 + 1 = (x - 1)^2$$
$$\sqrt{y^2 + 1} = \pm(x - 1)$$

With $x \geq 2$ and the left side nonnegative, we must choose the plus sign. We can then write $x = 1 + \sqrt{y^2 + 1}$.

The right side is now a formula for $f^{-1}(y)$, seemingly good for any y, but we recall from the original formula (as a radical) that y must be nonnegative. We summarize the conclusion:

$$f^{-1}(x) = 1 + \sqrt{x^2 + 1}, \ (x \geq 0).$$

This is the dotted graph below. The solid graph is the original $f(x) = \sqrt{x^2 - 2x}$.

Had we chosen $\{x \leq 0\}$, the "other half of the domain," and called the new function h, (same formula as f but a different domain, not shown) we would have come by choosing the minus sign, to the formula

$$h^{-1}(x) = 1 - \sqrt{x^2 + 1}, (x \geq 0).$$

The two inverse formulae, if graphed together, fill out the right half of the hyperbola $-x^2 + (y - 1)^2 = 1$

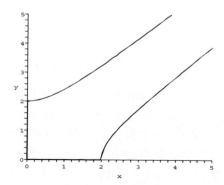

45. The function $\sin(x)$ (solid below) is increasing and one-to-one on the interval

$$-\frac{\pi}{2} \le x \le \frac{\pi}{2}.$$

One does not "find" the inverse in the sense of solving the equation $y = \sin(x)$ and obtaining a formula. It is done only in theory or as a graph. The name of the inverse is the "arc-sin" function ($y = \arcsin(x)$ shown dotted), and some of its properties are developed in the next section.

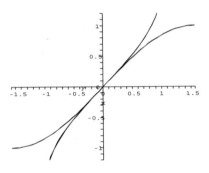

47. A company's income is not in fact a function of time, but a function of a time *interval* (income is defined as the change in *net worth*). When income *is* viewed as a function of time, it is usually after picking a fixed time interval (week, month, quarter, or year) and assigning the income for the period in a consistent manner to either the beginning or the ending date as in "... income for the quarter beginning... ." This much said, income more often than not rises and falls over time, so the function is unlikely to be one-to-one. In short, income functions *usually* do not have inverses.

49. During an interval of free fall following a drop, the height is decreasing with time and (barring a powerful up-draft, as with hail) an inverse exists. After impact, if there is a bounce then some of the heights are repeated and the function is no longer one-to-one on the expanded time interval.

51. Two three-dimensional shapes with congruent profiles will cast identical shadows if the congruent profiles face the light source. Such objects need not be fully identical in shape. (For an example, think of a sphere and a hemisphere with the flat side of the latter facing the light.) The shadow as a function of shape is not one-to-one and does not have an inverse.

53. The usual meaning of a "ten percent cut in salary" is that the new salary is 90% of the old. Thus after a ten percent raise the salary is 1.1 times the original, and after a subsequent ten percent cut, the salary is 90% of the raised salary, or .9 times 1.1 times the original salary. The combined effect is 99% of the original, and therefore the ten percent raise and the ten percent cut are not inverse operations.

The 10%-raise function is $y = f(x) = (1.1)x$, and the inverse relation is $x = y/(1.1) = (0.90909\ldots)y$. Thus $f^{-1}(x) = (0.90909)x$ and in the language of cuts, this is a pay cut of fractional value $1 - 0.90909\ldots = 0.090909$ ... or $9.0909\ldots$ percent.

0.4   Trigonometric and Inverse Trigonometric Functions

1. (a) $\left(\dfrac{\pi}{4}\right)\left(\dfrac{180°}{\pi}\right) = 45°$

    (b) $\left(\dfrac{\pi}{3}\right)\left(\dfrac{180°}{\pi}\right) = 60°$

(c) $\left(\dfrac{\pi}{6}\right)\left(\dfrac{180°}{\pi}\right) = 30°$

(d) $\left(\dfrac{4\pi}{3}\right)\left(\dfrac{180°}{\pi}\right) = 240°$

3. (a) $(180°)\left(\dfrac{\pi}{180°}\right) = \pi$

(b) $(270°)\left(\dfrac{\pi}{180°}\right) = \dfrac{3\pi}{2}$

(c) $(120°)\left(\dfrac{\pi}{180°}\right) = \dfrac{2\pi}{3}$

(d) $(30°)\left(\dfrac{\pi}{180°}\right) = \dfrac{\pi}{6}$

5. $2\cos(x) - 1 = 0$ when $\cos(x) = 1/2$. This occurs whenever $x = \frac{\pi}{3} + 2k\pi$ or $x = -\frac{\pi}{3} + 2k\pi$ for any integer k.

7. $\sqrt{2}\cos(x) - 1 = 0$ when $\cos(x) = 1/\sqrt{2}$. This occurs whenever $x = \frac{\pi}{4} + 2k\pi$ or $x = -\frac{\pi}{4} + 2k\pi$ for any integer k.

9. $\sin^2 x - 4\sin x + 3 = (\sin x - 1)(\sin x - 3) = 0$ when $\sin x = 1$ ($\sin x \neq 3$ for any x). This occurs whenever $x = \frac{\pi}{2} + 2k\pi$ for any integer k.

11. $\sin^2 x + \cos x - 1 = (1 - \cos^2 x) + \cos x - 1 = (\cos x)(\cos x - 1) = 0$ when $\cos x = 0$ or $\cos x = 1$. This occurs whenever $x = \frac{\pi}{2} + k\pi$ or $x = 2k\pi$ for any integer k.

13. $\cos^2 x + \cos x = (\cos x)(\cos x + 1) = 0$ when $\cos x = 0$ or $\cos x = -1$. This occurs whenever $x = \frac{\pi}{2} + k\pi$ or $x = \pi + 2k\pi$ for any integer k.

15. The graph of $f(x) = \sin 2x$.

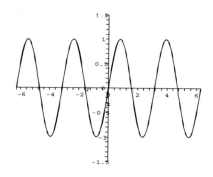

17. The graph of $f(x) = \tan 2x$.

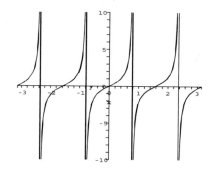

19. The graph of $f(x) = 3\cos(x - \pi/2)$.

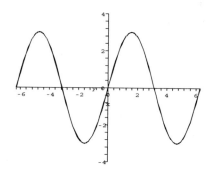

21. The graph of $f(x) = \sin 2x - 2\cos 2x$.

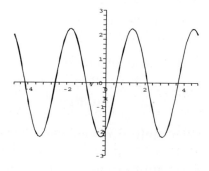

23. The graph of $f(x) = \sin x \sin 12x$.

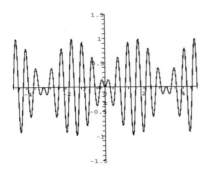

25. Amplitude is 3, period is $\frac{2\pi}{2} = \pi$, frequency is $\frac{1}{\pi}$.

27. Amplitude is 5, period is $\frac{2\pi}{3}$, frequency is $\frac{3}{2\pi}$.

29. Amplitude is 3, period is $\frac{2\pi}{2} = \pi$, frequency is $\frac{1}{\pi}$.

We are completely ignoring the presence of $-\pi/2$. This has an influence on the so-called "phase shift" which will be studied in Chapter 6.

31. Amplitude is 4 (the graph oscillates between -4 and 4, so we may ignore the minus sign), period is 2π, frequency is $\frac{1}{2\pi}$.

33. $\sin(\alpha - \beta) = \sin(\alpha + (-\beta))$
$= \sin \alpha \cos(-\beta) + \sin(-\beta) \cos \alpha$
$= \sin \alpha \cos \beta - \sin \beta \cos \alpha$

35. (a) $\cos(2\theta) = \cos(\theta + \theta)$
$= \cos(\theta) \cos(\theta) - \sin(\theta) \sin(\theta)$
$\cos^2 \theta - \sin^2 \theta = \cos^2 \theta - (1 - \cos^2 \theta)$
$= 2\cos^2 \theta - 1$

   (b) Just continue on, writing
$\cos(2\theta) = 2\cos^2 \theta - 1$
$= 2(1 - \sin^2 \theta) - 1 = 1 - 2\sin^2 \theta$

37. From unit circle $\cos^{-1} 0 = \frac{\pi}{2}$.

39. From unit circle $\sin^{-1}(-1) = -\frac{\pi}{2}$.

41. From unit circle $\sec^{-1} 1 = 0$.

43. From unit circle $\sec^{-1} 2 = \frac{\pi}{3}$.

45. From unit circle $\cot^{-1}(1) = \frac{\pi}{4}$.

47. Use the formula
$\cos(x + \beta) = \cos x \cos \beta - \sin \beta \sin x$.
Now we see that $\cos \beta$ must equal $4/5$ and $\sin \beta$ must equal $3/5$. Since $(4/5)^2 + (3/5)^2 = 1$, this is possible. We see that $\beta = \sin^{-1}(3/5) \approx 0.6435$ radians, or $-36.87°$.

49. $\cos(2x)$ has period $\frac{2\pi}{2} = \pi$ and $\sin(\pi x)$ has period $\frac{2\pi}{\pi} = 2$. There are no common integer multiples of the periods, so the function $f(x) = \cos(2x) + 3\sin(\pi x)$ is not periodic.

51. $\sin(2x)$ has period $\frac{2\pi}{2} = \pi$ and $\cos(5x)$ has period $\frac{2\pi}{5}$. The smallest integer multiple of both of these is the fundamental period, and it is 2π.

53. $\cos^2 \theta = 1 - \sin^2 \theta$
$= 1 - \left(\frac{1}{3}\right)^2 = 1 - \frac{1}{9} = \frac{8}{9}$.

Because θ is in the first quadrant, its cosine is nonnegative. Hence
$$\cos \theta = \sqrt{\frac{8}{9}} = \frac{2\sqrt{2}}{3} = 0.9428.$$

55. Second quadrant, 1-$\sqrt{3}$-2 right triangle, so $\cos \theta = -\frac{\sqrt{3}}{2}$.

57. Assume $0 < x < 1$ and give the temporary name θ to $\sin^{-1}(x)$. In a right triangle with hypotenuse 1 and one leg of length x, the angle θ will show up opposite the x-side, and the adjacent side will have length $\sqrt{1 - x^2}$. Write
$\cos(\sin^{-1}(x)) = \cos(\theta)$
$$= \frac{\sqrt{1 - x^2}}{1} = \sqrt{1 - x^2}.$$
The formula is numerically correct in the cases $x = 0$ and $x = 1$, and

both sides are even functions of x, i.e. $f(-x) = f(x)$, so the formula is good for $-1 \le x \le 1$.

59. Assume $1 < x$ and give the temporary name θ to $\sec^{-1}(x)$. In a right triangle with hypotenuse x and one leg of length 1, the angle θ will show up adjacent to the side of length 1, and the opposite side will have length $\sqrt{x^2 - 1}$. Write
$$\tan(\sec^{-1}(x)) = \tan(\theta)$$
$$= \frac{\sqrt{x^2 - 1}}{1} = \sqrt{x^2 - 1}.$$
The formula is numerically correct in the case $x \ge 1$.

Dealing with negative x is trickier: assume $x > 1$ for the moment. The key identity is $\sec^{-1}(-x) = \pi - \sec^{-1}(x)$. Taking tangents on both sides and applying the identity
$$\tan(a - b) = \frac{\tan(a) - \tan(b)}{1 + \tan(a)\tan(b)}$$
with $a = \pi$, $\tan(a) = 0$, $b = \sec^{-1}(x)$, we find
$$\tan(\sec^{-1}(-x)) = \frac{0 - \tan(\sec^{-1}(x))}{1 + 0}$$
$$= -\sqrt{x^2 - 1}$$
$$= -\sqrt{(-x)^2 - 1}$$
In this identity, $-x$ (on both sides) plays the role of an arbitrary number < -1. Consequently, the final formula is $\tan(\sec^{-1}(x)) = -\sqrt{x^2 - 1}$ whenever $x \le -1$.

61. One can use the formula $\sin(\cos^{-1} x) = \sqrt{1 - x^2}$ derived in the text:
$$\sin(\cos^{-1}(\tfrac{1}{2})) = \sqrt{1 - (\tfrac{1}{2})^2} = \tfrac{\sqrt{3}}{2}.$$

63. $\cos^{-1}(\tfrac{3}{5})$ relates to a triangle in the first quadrant with adjacent side 3

and hypotenuse 5, so the opposite side must be 4 and then
$$\tan\left(\cos^{-1}\left(\frac{3}{5}\right)\right) = \frac{4}{3}.$$

65. From graph the three solutions are 0, 1.109, and 3.698.

67. From graph the two solutions are ± 1.455.

69. Let h be the height of the rocket. Then $\frac{h}{2} = \tan 20°$
$h = 2\tan 20° \approx 0.73$ (miles)

71. Let h be the height of the steeple. Then $\dfrac{h}{80 + 20} = \tan 50°$
$h = 100 \tan 50° \approx 119.2$ (feet).

73. Using feet as the measuring standard, we find
$$\tan A = \frac{20/12}{x} = \frac{5}{3x}$$
$$A(x) = \tan^{-1}\left(\frac{5}{3x}\right)$$

The graph of $y = A(x)$ (of course, one has to choose an appropriate range to make this a function):

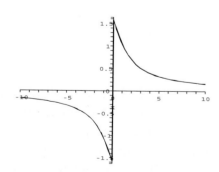

75. Presumably, the given amplitude (170) is the same as the "peak voltage" (v_p). Recalling an earlier discussion (#25 this section): the role of ω there is played by $2\pi f$ here, the frequency in cycles per second (Hz) was $\omega/2\pi$, which is now the f-parameter $(2\pi f/2\pi)$. The period was $2\pi/\omega$ (which is now $1/f$), given in this case to be $\pi/30$ (seconds). So, apparently, the frequency is $f = 30/\pi$ (cycles per second) and the meter voltage is $\frac{170}{\sqrt{2}} \approx 120.2$.

77. There seems to be a certain slowly increasing base for sales $(110 + 2t)$, and given that the sine function has period $\frac{2\pi}{\pi/6} = 12$ months, the sine term apparently represents some sort of seasonally cyclic pattern. If we assume that travel peaks at Thanksgiving, the effect is that time zero would correspond to a time one quarter-period (3 months) prior to Thanksgiving, or very late August.

The annual increase for the year beginning at time t is given by $s(t + 12) - s(t)$ and automatically ignores both the seasonal factor and the basic 110, and indeed it is the constant $2 \times 12 = 24$ (in thousands of dollars per year and independent of the reference point t).

79. As luck would have it, the trig functions csc and cot, being reciprocals respectively of sine and tangent, have inverses almost exactly where the other two do, both on the interval $\left[-\frac{\pi}{2}, \frac{\pi}{2}\right]$ but excluding the origin where neither is defined, and excluding the lower endpoint in the case of the cotangent. The range for the sine is $[-1, 1]$, hence the range for the csc is $\{|x| \geq 1\}$ and this is the domain for $\csc^{-1}$. The tangent assumes all values, and so does the cot (zero included as a value by convention when $x = \pi/2$ or $-\pi/2$), so the domain for $\cot^{-1}$ is universal. Finally, we simply

copy the language of the others:

$y = \csc^{-1}(x)$ if $|x| \geq 1$,
y lies in $\left[-\frac{\pi}{2}, \frac{\pi}{2}\right]$ and $x = \csc(y)$.
$y = \cot^{-1}(x)$ if y lies in $\left(-\frac{\pi}{2}, \frac{\pi}{2}\right]$, and $x = \cot(y)$.

0.5 Exponential and Logarithmic Functions

1. $2^{-3} = \dfrac{1}{2^3} = \dfrac{1}{8}$

3. $3^{1/2} = \sqrt{3}$

5. $5^{2/3} = \sqrt[3]{5^2} = \sqrt[3]{25}$

7. $\dfrac{1}{x^2} = x^{-2}$

9. $\dfrac{2}{x^3} = 2x^{-3}$

11. $\dfrac{1}{2\sqrt{x}} = \dfrac{1}{2x^{1/2}} = \dfrac{1}{2}x^{-1/2}$

13. $4^{3/2} = \left(\sqrt{4}\right)^3 = 2^3 = 8$

15. $\dfrac{\sqrt{8}}{2^{1/2}} = \dfrac{\sqrt{8}}{\sqrt{2}} = \sqrt{4} = 2$

17. $2e^{-1/2} \approx 1.213$

19. $\dfrac{12}{e} \approx 4.415$

21. The graph of $f(x) = e^{2x}$:

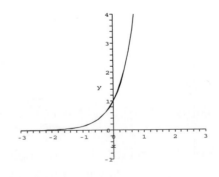

23. The graph of $f(x) - 2e^{x/4}$:

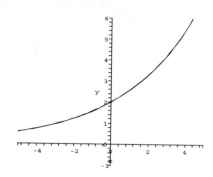

25. The graph of $f(x) = 3e^{-2x}$:

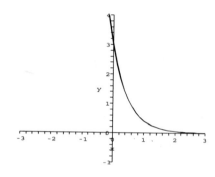

27. The graph of $f(x) = \ln 2x$:

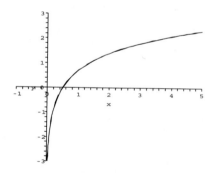

29. The graph of $f(x) = e^{2\ln x}$:

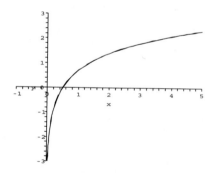

31. $e^{2x} = 2$
$\Rightarrow \ln e^{2x} = \ln 2$
$\Rightarrow 2x = \ln 2$
$\Rightarrow x = \frac{\ln 2}{2} \approx 0.3466$

33. $e^x(x^2-1) = 0$ implies either $x^2-1 = 0$ (hence $x = 1$ or $x = -1$), or $e^x = 0$ which has no solution.

35. $\ln 2x = 4$
$\Rightarrow 2x = e^4$
$\Rightarrow x = \frac{e^4}{2} \approx 27.299$

37. $4 \ln x = -8$
$\Rightarrow \ln x = -2$
$\Rightarrow x = e^{-2} = \frac{1}{e^2} \approx 0.13533$

39. $e^{2\ln x} = 4$
$\Rightarrow 2 \ln x = \ln 4$
$\Rightarrow \ln x^2 = \ln 4$
$\Rightarrow x^2 = 4$
$\Rightarrow x = \pm 2$,
but in the original equation we had the expression $e^{2\ln x}$ so $x \neq -2$ and thus the only solution is $x = 2$.

41. (a) $\log_3 9 = \log_3(3^2) = 2$

(b) $\log_4 64 = \log_4(4^3) = 3$

(c) $\log_3 \frac{1}{27} = \log_3(3^{-3}) = -3$

43. (a) $\log_3 7 = \frac{\ln 7}{\ln 3} \approx 1.771$

(b) $\log_4 60 = \frac{\ln 60}{\ln 4} \approx 2.953$

(c) $\log_3 \frac{1}{24} = \frac{\ln(1/24)}{\ln 3} \approx -2.893$

45. $\ln 3 - \ln 4 = \ln \frac{3}{4}$

47. $\frac{1}{2} \ln 4 - \ln 2 = \ln 4^{1/2} - \ln 2$
$= \ln 2 - \ln 2 = 0$

49. $\ln \frac{3}{4} + 4\ln 2 = \ln \frac{3}{2^2} + \ln 2^4$
$= \ln \left(\frac{3}{2^2} \cdot 2^4 \right)$
$= \ln(3 \cdot 2^2) = \ln(12)$

51. $f(0) = 2 \Rightarrow a = 2$.
Then $f(2) = 6$ gives $2e^{2b} = 6$, so $2b = \ln 3$ and $b = \frac{1}{2}\ln 3$. So $f(x) = 2e^{(\frac{1}{2}\ln 3)x} = 2[e^{\ln(3)}]^{x/2} = 2 \cdot 3^{x/2}$.

53. $f(0) = 4 \Rightarrow a = 4$.
Then $f(2) = 2$ gives $4e^{2b} = 2$, so $2b = \ln\frac{1}{2}$ and $b = \frac{1}{2}\ln\frac{1}{2}$. So $f(x) = 4e^{(\frac{1}{2}\ln\frac{1}{2})x}$.

55. $1 - \left(\dfrac{9}{10}\right)^{10} \approx 0.651$

57. We take on faith, whatever it may mean, that

$$\lim_{n\to\infty}\left(1 + \frac{1}{n}\right)^n = e$$

Just to take a sample starting with $n = 25$, the numbers are

$$\left(\frac{26}{25}\right)^{25}, \left(\frac{27}{26}\right)^{26}, \left(\frac{28}{27}\right)^{27},$$

and so on. If we were to try taking a similar look at the numbers in $\lim_{n\to\infty}\left(1 - \frac{1}{n}\right)^n$, the numbers starting at $n = 26$ would be

$$\left(\frac{25}{26}\right)^{26}, \left(\frac{26}{27}\right)^{27}, \left(\frac{27}{28}\right)^{28},$$

and so on.

We could rewrite these as

$$\left[\left(\frac{25}{26}\right)^{25}\right]^{\frac{26}{25}}, \left[\left(\frac{26}{27}\right)^{26}\right]^{\frac{27}{26}}, \left[\left(\frac{27}{28}\right)^{27}\right]^{\frac{28}{27}}.$$

Here, the numbers inside the square brackets are the reciprocals of the numbers in the original list, which were all pretty close to e. Therefore these must all be pretty close to $1/e$. As to the external powers, they are all close to 1 and getting closer. This limit must be $1/e$. The expression in question must approach $1 - \frac{1}{e} \approx .632$

59.

$u = \ln x$	.78846	.87547	.95551
$v = \ln y$	2.6755	2.8495	3.0096

$u = \ln x$	1.0296	1.0986	1.1632
$v = \ln y$	3.1579	3.2958	3.4249

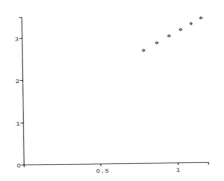

$m = \frac{3.4249 - 2.6775}{1.1632 - .78846} \approx 2$.
Then we solve $2.6755 = 2.(.78846) + b$ to find $b \approx 1.099$. Now $b = \ln a$, so $a = e^b \approx 3.001$, and the function is $y = 3.001x^2$.

61. We compute $u = \ln x$ and $v = \ln y$ for x values in number of decades since 1780 and y values in millions.

$u = \ln x$	0	0.693	1.099	1.386
$v = \ln y$	1.36	1,668	1.974	2.262

$u = \ln x$	1.609	1.792	1.946	2.079
$v = \ln y$	2.549	2.839	3.14	3.447

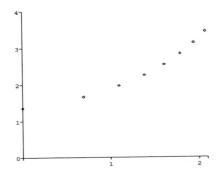

This plot does not look linear, which makes it clear that the population is *not* modeled by a power of x. The discussion in the Chapter has already strongly indicated that an exponential model is fairly good.

63. (a) $7 = -\log[H^+] \Rightarrow [H^+] = 10^{-7}$

(b) $[H^+] = 10^{-8}$

(c) $[H^+] = 10^{-9}$

For each increase in pH of one, $[H^+]$ is reduced to one tenth of its previous value.

65. (a) $\log E = 4.4 + 1.5(4) = 10.4 \Rightarrow E = 10^{10.4}$

(b) $\log E = 4.4 + 1.5(5) = 11.9 \Rightarrow E = 10^{11.9}$

(c) $\log E = 4.4 + 1.5(6) = 13.4 \Rightarrow E = 10^{13.4}$

For each increase in M of one, E is increased by a factor of $10^{1.5} \approx 31.6$.

67. (a) $80 = 10 \log \left(\dfrac{I}{10^{-12}} \right) \Rightarrow$

$8 = \log \left(\dfrac{I}{10^{-12}} \right) \Rightarrow$

$10^8 = \dfrac{I}{10^{-12}} \Rightarrow$

$I = 10^8 10^{-12} = 10^{-4}$

(b) $I = 10^{-3}$

(c) $I = 10^{-2}$

For each increase in dB of ten, I increases by a factor of 10.

69. From the graphs, we estimate:

$y = xe^{-x}$ has a max value of $1/e$ at $x = 1$

$y = xe^{-2x}$ has a max value of $1/2e$ at $x = 1/2$

$y = xe^{-3x}$ has a max value of $1/3e$ at $x = 1/3$

So we guess that $y = xe^{-kx}$ has a max value of $1/ke$ at $x = 1/k$. If one believes the first of these, then the general case follows by writing

$$xe^{-kx} = \frac{kxe^{-kx}}{k} = \frac{ue^{-u}}{k}$$

if we let $u = kx$. The numerator has max value $1/e$ when $u = 1$, i.e., $x = 1/k$. Therefore the whole expression has max value $(1/k)(1/e)$ when $x = 1/k$.

71. We know that $\cosh x = \dfrac{e^x + e^{-x}}{2}$. To show that $\cosh x \geq 1$ for all x is the same as showing that $\cosh x - 1 \geq 0$ for all x. So we ask when is the expression

$$\cosh x - 1 = \frac{e^x + e^{-x}}{2} - 1$$

greater than or equal to 0? We have:

$\dfrac{e^x + e^{-x}}{2} - 1 \geq 0$ if and only if

$\dfrac{e^x + e^{-x} - 2}{2} \geq 0$ if and only if

$e^x + e^{-x} - 2 \geq 0$ if and only if

$e^{2x} + 1 - 2e^{-x} \geq 0$ if and only if *

$e^{2x} - 2e^{-x} + 1 \geq 0$ if and only if

$(e^x - 1)^2 \geq 0$

But $(e^x - 1)^2$ is always greater than or equal to 0 since it is squared. It is actually equal to 0 at $x = 0$ (i.e., $\cosh 0 = 1$), so the range of $y = \cosh x$ is $y \geq 1$.

* In the * step (above), we have multiplied on both sides by e^x, which we are allowed to do since $e^x > 0$ for all x.

To show that the range of the hyperbolic sine is all real numbers, let a be any real number and solve the equation $\sinh(x) = a$. Let $u = e^x$. Then

$\dfrac{u - \frac{1}{u}}{2} = a$ if and only if

$u^2 - 1 = 2au$ if and only if

$u^2 - 2au - 1 = 0$ if and only if

$u = \dfrac{2a \pm \sqrt{4a^2 + 4}}{2} = a + \sqrt{a^2 + 1}.$

We simplified and chose the positive square root because $u > 0$. Because we found a unique solution no matter what a we had started with, we have shown that the range of $y = \sinh x$ is the whole real line.

73. The issue is purely whether or not $y = 0$ when $x = 315$, i.e., whether or not $\cosh(315/127.7) = \cosh(2.4667\ldots) = 5.9343\ldots$ is the same as $(757.7)/(127.7) = 5.9334\ldots$ We see that it's pretty close, and these numbers would be considered equal according to the level of accuracy reported in the original measurements.

75. Since $\sinh^{-1}(0) = 0$, the equation is solved only by $x^2 - 1 = 0$, hence $x = 1$ or $x = -1$.

77. $f = f(x) = 220e^{x\ln(2)}$
$= 220e^{\ln(2^x)} = 220 \cdot 2^x$

0.6   Transformations of Functions

1. $(f \circ g)(x) = f(g(x))$
$\qquad = g(x) + 1 = \sqrt{x - 3} + 1$
with domain $\{x | x \geq 3\}$.

$(g \circ f)(x) = g(f(x))$
$\qquad = \sqrt{f(x) - 1}$
$\qquad = \sqrt{(x + 1) - 3} = \sqrt{x - 2}$
with domain $\{x | x \geq 2 \}$.

3. $(f \circ g)(x) = f(\ln x) = e^{\ln x} = x$
with domain $\{x | x > 0\}$

$(g \circ f)(x) = g(e^x) = \ln e^x = x$
with domain $(-\infty, \infty)$ or all real numbers.

5. $(f \circ g)(x) = f(\sin x) = \sin^2 x + 1$ with domain $(-\infty, \infty)$ or all real numbers.

$(g \circ f)(x) = g(x^2 + 1) = \sin(x^2 + 1)$ with domain $(-\infty, \infty)$ or all real numbers.

7. $\sqrt{x^4 + 1} = f(g(x))$ when $f(x) = \sqrt{x}$ and $g(x) = x^4 + 1$, for example.

9. $\dfrac{1}{x^2 + 1} = f(g(x))$ when $f(x) = 1/x$ and $g(x) = x^2 + 1$, for example.

11. $(4x + 1)^2 + 3 = f(g(x))$ when $f(x) = x^2 + 3$ and $g(x) = 4x + 1$, for example.

13. $\sin^3 x = f(g(x))$ when $f(x) = x^3$ and $g(x) = \sin x$, for example.

15. $e^{x^2 + 1} = f(g(x))$ when $f(x) = e^x$ and $g(x) = x^2 + 1$, for example.

17. $\dfrac{3}{\sqrt{\sin x + 2}} = f(g(h(x)))$ when $f(x) = 3/x$, $g(x) = \sqrt{x}$, and $h(x) = \sin x + 2$, for example.

19. $\cos^3(4x - 2) = f(g(h(x)))$ when $f(x) = x^3$, $g(x) = \cos x$, and $h(x) = 4x - 2$, for example.

21. $4e^{x^2} - 5 = f(g(h(x)))$ when $f(x) = 4x - 5$, $g(x) = e^x$, and $h(x) = x^2$, for example.

23. Graph of $f(x) - 3$:

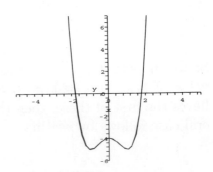

25. Graph of $f(x-3)$:

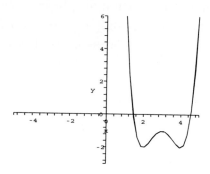

27. Graph of $f(2x)$:

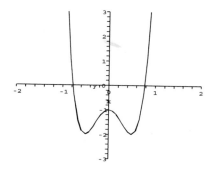

29. Graph of $4f(x) - 1$:

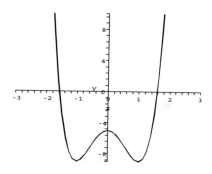

31. Graph of $f(x-4)$:

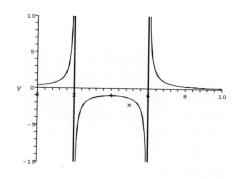

33. Graph of $f(2x)$:

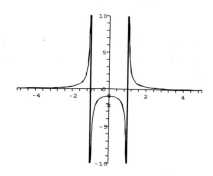

35. Graph of $f(3x+3)$:

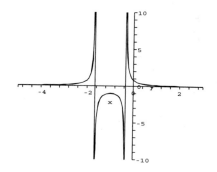

37. Graph of $2f(x) - 4$:

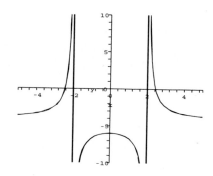

39. $f(x) = x^2 + 2x + 1 = (x+1)^2$.
Shift $y = x^2$ to the left 1 unit.

41. $f(x) = x^2 + 2x + 4 = (x^2 + 2x + 1) + 4 - 1$
$\quad = (x+1)^2 + 3$.
Shift $y = x^2$ to the left 1 unit and up 3 units.

43. $f(x) = 2x^2 + 4x + 4$
$\quad = 2(x^2 + 2x + 1) + 4 - 2$
$\quad = 2(x+1)^2 + 2$.
Shift $y = x^2$ to the left 1 unit, then

multiply the scale on the y-axis by 2, then shift up 2 units.

45. Graph is reflected across the x-axis and the scale on the y-axis is multiplied by 2.

47. Graph is reflected across the x-axis, the scale on the y-axis is multiplied by 3, and the graph is shifted up 2 units.

49. Graph is reflected across the y-axis.

51. Graph is reflected across the y-axis and shifted up 1 unit.

53. The graph is reflected across the x-axis and the scale on the y-axis is multiplied by $|c|$.

55. The graph of $y = |x|^3$ is identical to that of $y = x^3$ to the right of the y-axis because for $x > 0$ we have $|x|^3 = x^3$. For $y = |x|^3$ the graph to the left of the y-axis is the reflection through the y-axis of the graph to the right of the y-axis. In general to graph $y = f(|x|)$ based on the graph of $y = f(x)$, the procedure is to discard the part of the graph to the left of the y-axis, and replace it by a reflection in the y-axis of the part to the right of the y-axis.

57. The rest of the first 10 iterates of $f(x) = \cos x$ with $x_0 = 1$ are:

$$x_4 = \cos .65 \approx .796$$
$$x_5 = \cos .796 \approx .70$$
$$x_6 = \cos .70 \approx .765$$
$$x_7 = \cos .765 \approx .721$$
$$x_8 = \cos .721 \approx .751$$
$$x_9 = \cos .751 \approx .731$$
$$x_{10} = \cos .731 \approx .744$$

Continuing in this fashion and retaining more decimal places, one finds that x_{36} through x_{40} are all 0.739085. The same process is used with a different x_0.

59. They converge to 0. One of the problems in Chapter 2 asks the student to prove that $|\sin(x)| < |x|$ for all but $x = 0$. This would show that 0 is the only solution to the equation $\sin(x) = x$ and offers a partial explanation (see the comments for #61) of the phenomena which the student observes.

61. If the iterates of a function f (starting from some point x_0) are going to go toward (and remain arbitrarily close to) a certain number L, this number L must be a solution of the equation $f(x) = x$. For the list of iterates $x_0, x_1, x_2, x_3, \dots$ is, apart from the first term, the same list as the list of numbers $f(x_0), f(x_1), f(x_2), f(x_3), \dots$. (Remember that x_{n+1} is $f(x_n)$.) If any of the numbers in the first list are close to L, then the f-values (in the second list) are close to $f(L)$. But since the lists are *identical* (apart from the first term x_0 which is not in the second list), it must be true that L and $f(L)$ are the same number.

If conditions are right (and they are in the two cases $f(x) = \cos(x)$ (#57) and $f(x) = \sin(x)$ (#59)), this "convergence" will indeed occur, and since there is in these cases only *one* solution (x about 0.739085 in #57 and $x = 0$ in #59) it won't matter where you started.

Ch. 0 Review Exercises

1. $m = \dfrac{7-3}{0-2} = \dfrac{4}{-2} = -2$

3. These lines both have slope 3. They are parallel unless they are coincident. But the first line includes the point $(0,1)$ which does not satisfy the equation of the second line. The lines are not coincident.

5. Let $P = (1,2)$, $Q = (2,4)$, $R = (0,6)$.

Then PQ has slope $\dfrac{4-2}{2-1} = 2$

QR has slope $\dfrac{6-4}{0-2} = -1$

RP has slope $\dfrac{2-6}{1-0} = -4$

Since no two of these slopes are negative reciprocals, none of the angles are right angles. The triangle is not a right triangle.

7. The line apparently goes through $(1,1)$ and $(3,2)$. If so the slope would be $m = \frac{2-1}{3-1} = \frac{1}{2}$. The equation would be
$y = \frac{1}{2}(x-1) + 1$ or $y = \frac{1}{2}x + \frac{1}{2}$.
Using the equation with $x = 4$, we find $y = \frac{1}{2}(4) + \frac{1}{2} = \frac{5}{2}$.

9. Using the point-slope method, we find $y = -\frac{1}{3}(x+1) - 1$

11. The graph passes the vertical line test, so it is a function.

13. The radicand cannot be negative, hence we require $4 - x^2 \geq 0 \Rightarrow 4 \geq x^2$. Therefore the natural domain is $\{x \mid -2 \leq x \leq 2\}$ or, in "interval-language": $[-2, 2]$.

15. Intercepts at $x = -4$ and 2, and $y = -8$. Local minimum at $x = -1$.

No asymptotes.

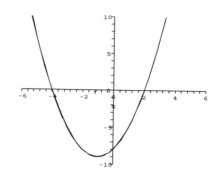

17. Intercepts at $x = -1$ and 1, and $y = 1$. Local minimum at $x = 1$ and at $x = -1$. Local maximum at $x = 0$. No asymptotes.

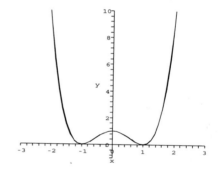

19. Intercept at $y = 0$ and at $x = 0$. No extrema. Horizontal asymptote $y = 4$. Vertical asymptote $x = -2$.

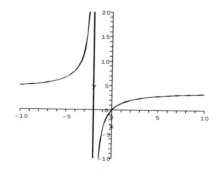

21. Intercept at $y = 0$ and $x = \frac{k\pi}{3}$ for integers k. Extrema: y takes maximum 1 and minimum -1 with great predictability and regularity. No asymptotes.

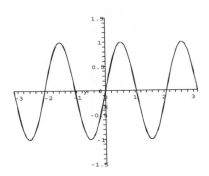

0.

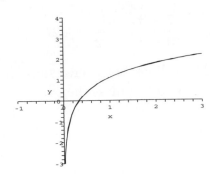

23. Intercept at $y = 2$ and from the amplitude/phase shift form $f(x) = \sqrt{5}\sin\left(x + \sin^{-1}(2/\sqrt{5})\right)$, we could write down all the intercepts only at considerable inconvenience. Extrema: y takes maximum $\sqrt{5}$ and minimum $-\sqrt{5}$ with great predictability and regularity. No asymptotes.

29. Intercepts at $x = -4$ and 2, and $y = -8$.

31. Vertical asymptote $x = -2$.

33. $x^2 - 3x - 10 = (x - 5)(x + 2)$. The zeros are when $x = 5$ and $x = -2$.

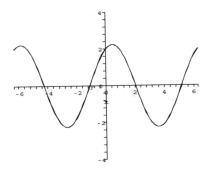

35. Guess a root: $x = 1$. Factor the left side: $(x - 1)(x^2 - 2x - 2)$. Solve the quadratic by formula:
$$x = \frac{2 \pm \sqrt{2^2 - 4(1)(-2)}}{2} = 1 \pm \sqrt{3}.$$
Complete list of three roots: $x = 1$, $x = 1 - \sqrt{3} \approx -.732$, $x = 1 + \sqrt{3} \approx 2.732$.

25. Intercept $y = 4$ (no x-intercepts). No extrema. Left horizontal asymptote $y = 0$.

37. There are 3 solutions, one at $x = 0$ and the other two negatives of one another. The value in question is $.928632\ldots$, found using the function "Goal Seek" in Excel. The result can be checked, and a graphing calculator can find them by graphing $y = x^3$ and $y = \sin x$ on the same axes and finding the intersection points.

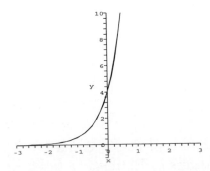

39. Let h be the height of the telephone pole. Then $\frac{h}{50} = \tan 34° \Rightarrow h = 50\tan 34° \approx 33.7$ feet.

27. Intercept $x = 1/3$ (no y-intercepts). No extrema. Vertical asymptote $x =$

41. (a) $5^{-1/2} = \frac{1}{5^{1/2}} = \frac{1}{\sqrt{5}} = \frac{\sqrt{5}}{5}$

   (b) $3^{-2} = \frac{1}{3^2} = \frac{1}{9}$

43. $\ln 8 - 2\ln 2 = \ln 8 - \ln 2^2$
$= \ln 8 - \ln 4 = \ln\left(\frac{8}{4}\right) = \ln 2$

45. $3e^{2x} = 8 \Rightarrow e^{2x} = \frac{8}{3}$
$\Rightarrow \ln e^{2x} = \ln\left(\frac{8}{3}\right)$
$\Rightarrow 2x = \ln\left(\frac{8}{3}\right)$
$\Rightarrow x = \frac{1}{2}\ln\frac{8}{3}$

47. The natural domain for f is the full real line. The natural domain for g is $\{x|1 \le x\}$. Because f has a universal domain, the natural domain for $f \circ g$ is the same as the domain for g, namely $\{x|1 \le x\}$. Because g requires its inputs be not less than 1, the domain for $g \circ f$ is the set of x for which $1 \le f(x)$, i.e., $\{x|1 \le x^2\} = \{x|1 \le |x|\}$, or in interval language $(-\infty, -1] \cup [1, \infty)$.

The formulae are easier:
$$(f \circ g)(x) = f(\sqrt{x-1})$$
$$= (\sqrt{x-1})^2 = x - 1$$
$$(g \circ f)(x) = g(x^2) = \sqrt{x^2 - 1}$$

Caution: the *formula* for $f \circ g$ is defined for any x, but the *domain* for $f \circ g$ is restricted as stated earlier. The formula must be viewed as irrelevant outside the domain.

49. $e^{3x^2+2} = f(g(x))$ for $f(x) = e^x$ and $g(x) = 3x^2 + 2$.

51. $x^2 - 4x + 1 = x^2 - 4x + 4 - 4 + 1$, so $f(x) = (x-2)^2 - 3$. The graph of $f(x)$ is the graph of x^2 shifted two units to the right and three units down.

53. Like x^3, the function $f(x) = x^3 - 1$ passes the horizontal line test and is one-to-one. To find a formula for the inverse, solve for x to find $(y+1)^{1/3} = x$ then switch x and y to get $f^{-1}(x) = (x+1)^{1/3}$ for all x.

55. The function is *even* ($f(-x) = f(x)$). Every horizontal line (except $y = 0$)

which meets the curve at all automatically meets it at least twice. The function is not one-to-one. There is no inverse.

57. The inverse of $x^5 + 2x^3 - 1$:

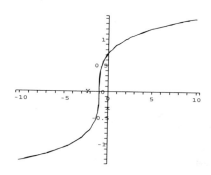

59. The inverse of $\sqrt{x^3 + 4x}$:

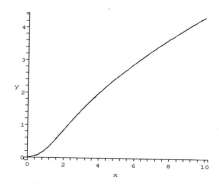

61. On the unit circle, $y = \sin\theta = 1$ when $\theta = \frac{\pi}{2}$. Hence, $\sin^{-1} 1 = \frac{\pi}{2}$.

63. Since $\tan\theta = \frac{\sin\theta}{\cos\theta}$ we want $y = \cos\theta$ to be equal to $-x = -\sin\theta$ on the unit circle. This happens when $\theta = -\pi/4$ and $\theta = 3\pi/4$. Hence, $\tan^{-1}(-1) = -\frac{\pi}{4}$ or $\tan^{-1}(-1) = \frac{3\pi}{4}$.

65. If an angle θ has $\sec(\theta) = 2$, then it has $\cos(\theta) = 1/2$. Its sine could be $\pm\frac{\sqrt{3}}{2}$. But if $\theta = \sec^{-1}(2)$, then in addition to all that has been stated, it is in the first quadrant, and the choice of sign (for its sine) is positive. In summary, $\sin(\sec^{-1} 2) = \sin\theta = \frac{\sqrt{3}}{2}$.

67. $\sin^{-1}\left(\sin\left(\frac{3\pi}{4}\right)\right) = \sin^{-1}\left(\frac{\sqrt{2}}{2}\right) = \frac{\pi}{4}$

69. $\sin 2x = 1 \Rightarrow$
$2x = \frac{\pi}{2} + 2k\pi$ for any integer k so
$x = \frac{\pi}{4} + k\pi$ for any integer k.

Chapter 1

Limits and Continuity

1.1   A Brief Preview of Calculus

1. The slope appears to be 2.

Second point	m_{sec}
$(2, 5)$	3
$(1.1, 2.21)$	2.1
$(1.01, 2.0201)$	2.01
$(0, 1)$	1
$(0.9, 1.81)$	1.9
$(0.99, 1.9801)$	1.99

3. The slope appears to be 0.

Second point	m_{sec}
$(1, 0.5403)$	-0.4597
$(0.1, 0.995)$	-0.05
$(0.01, 0.99995)$	-0.005
$(-1, 0.5403)$	0.4597
$(-0.1, 0.995)$	0.05
$(-0.01, 0.99995)$	0.005

5. The slope appears to be 3.

Second point	m_{sec}
$(2, 10)$	7
$(1.1, 3.331)$	3.31
$(1.01, 3.030301)$	3.0301
$(0, 2)$	1
$(0.9, 2.729)$	2.71
$(0.99, 2.970299)$	2.9701

7. The slope appears to be $\frac{1}{2}$.

Second point	m_{sec}
$(1, \sqrt{2})$	0.4142
$(0.1, 1.0488)$	0.488
$(0.01, 1.004988)$	0.4988
$(-1, 0)$	1
$(-0.1, 0.9487)$	0.513
$(-0.01, 0.99499)$	0.501

9. The slope appears to be 1.

Second point	m_{sec}
$(1, e)$	1.718282
$(0.1, 1.1052)$	1.051709
$(0.01, 1.0101)$	1.005017
$(-1, 0.3679)$	0.632121
$(-0.1, 0.9048)$	0.951626
$(-0.01, 0.9901)$	0.995017

11. The slope appears to be 1.

Second point	m_{sec}
$(0.1, -2.3026)$	2.5584
$(0.9, -0.1054)$	1.054
$(0.99, -0.01005034)$	1.005034
$(2, 0.6931)$	0.6931
$(1.1, 0.09531)$	0.9531
$(1.01, 0.00995)$	0.995

Note that we used 0.1 rather than 0 as an evaluation point because $\ln x$ is not defined at 0.

13. (a)

Left	Right	Length
$(0, 1)$	$(0.5, 1.25)$	0.559
$(0.5, 1.25)$	$(1, 2)$	0.901
$(1, 2)$	$(1.5, 3.25)$	1.346
$(1.5, 3.25)$	$(2, 5)$	1.820
	Total	4.6267

(b)

Left	Right	Length
(0, 1)	(0.25, 1.063)	0.258
(0.25, 1.063)	(0.5, 1.25)	0.313
(0.5, 1.25)	(0.75, 1.563)	0.400
(0.75, 1.563)	(1, 2)	0.504
(1, 2)	(1.25, 2.563)	0.616
(1.25, 2.563)	(1.5, 3.25)	0.732
(1.5, 3.25)	(1.75, 4.063)	0.850
(1.75, 4.063)	(2, 5)	0.970
	Total	4.6417

(c) Actual length approximately 4.6468.

15. (a) For the x-values of our points here we use (approximations of) 0, $\frac{\pi}{8}$, $\frac{\pi}{4}$, $\frac{3\pi}{8}$, and $\frac{\pi}{2}$.

Left	Right	Length
(0, 1)	(0.393, 0.92)	0.400
(0.393, 0.92)	(0.785, 0.71)	0.449
(0.785, 0.71)	(1.18, 0.383)	0.509
(1.18, 0.383)	(1.571, 0)	0.548
	Total	1.906

(b) For the x-values of our points here we use (approximations of) 0, $\frac{\pi}{16}$, $\frac{\pi}{8}$, $\frac{3\pi}{16}$, $\frac{\pi}{4}$, $\frac{5\pi}{16}$, $\frac{3\pi}{8}$, $\frac{7\pi}{16}$, and $\frac{\pi}{2}$.

Left	Right	Length
(0, 1)	(0.196, 0.98)	0.197
(0.196, 0.98)	(0.393, 0.92)	0.204
(0.393, 0.92)	(0.589, 0.83)	0.217
(0.589, 0.83)	(0.785, 0.71)	0.232
(0.785, 0.71)	(0.982, 0.56)	0.248
(0.982, 0.56)	(1.178, 0.38)	0.262
(1.178, 0.38)	(1.37, 0.195)	0.272
(1.37, 0.195)	(1.571, 0)	0.277
	Total	1.909

(c) Actual length approximately 1.9101.

17. (a)

Left	Right	Length
(0, 1)	(0.75, 1.323)	0.817
(0.75, 1.323)	(1.5, 1.581)	0.793
(1.5, 1.581)	(2.25, 1.803)	0.782
(2.25, 1.803)	(3, 2)	0.776
	Total	3.167

(b)

Left	Right	Length
(0, 1)	(0.375, 1.17)	0.413
(0.375, 1.17)	(0.75, 1.323)	0.404
(0.75, 1.323)	(1.125, 1.46)	0.399
(1.125, 1.46)	(1.5, 1.58)	0.395
(1.5, 1.58)	(1.88, 1.696)	0.392
(1.88, 1.696)	(2.25, 1.80)	0.390
(2.25, 1.80)	(2.63, 1.904)	0.388
(2.63, 1.904)	(3, 2)	0.387
	Total	3.168

(c) Actual length approximately 3.168.

19. (a)

Left	Right	Length
(-2, 5)	(-1, 2)	3.162
(-1, 2)	(0, 1)	1.414
(0, 1)	(1, 2)	1.414
(1, 2)	(2, 5)	3.162
	Total	9.153

(b)

Left	Right	Length
(-2, 5)	(-1.5, 3.25)	1.820
(-1.5, 3.25)	(-1, 2)	1.346
(-1, 2)	(-0.5, 1.25)	0.901
(-0.5, 1.25)	(0, 1)	0.559
(0, 1)	(0.5, 1.25)	0.559
(0.5, 1.25)	(1, 2)	0.901
(1, 2)	(1.5, 3.25)	1.346
(1.5, 3.25)	(2, 5)	1.820
	Total	9.253

(c) Actual length approximately 9.2936.

21. The sum of the areas of the rectangles is $11/8 = 1.375$.

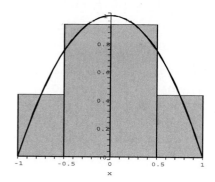

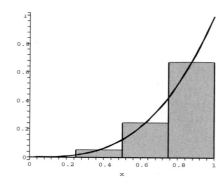

23. (a) The width of the entire region $(-1 \leq x \leq 1)$ is 2, so the width of each rectangle is $2/16 = 0.125$. The left endpoints of the rectangles are
$$-1, -1+\tfrac{2}{16}, \ldots, -1+\tfrac{28}{16}, -1+\tfrac{30}{16}$$
so the midpoints of the rectangles are
$$-1+\tfrac{1}{16}, -1+\tfrac{3}{16}, \ldots, -1+\tfrac{31}{16}.$$
The heights of the rectangles are then given by the function $f(x) = 1 - x^2$ evaluated at those midpoints. We multiply each height by the width (0.125) and add them all to obtain the approximation 1.3359375 for the area.

(b) Using the same method as in (a), the width of the rectangles is now $2/32 = 0.0625$, and the midpoints are
$$-1+\tfrac{1}{32}, -1+\tfrac{3}{32}, \ldots, -1+\tfrac{63}{32}.$$
The approximation is 1.333984375.

(c) Using the same method as in (a), the width of the rectangles is now $2/64 = 0.03125$, and the midpoints are
$$-1+\tfrac{1}{64}, -1+\tfrac{3}{64}, \ldots, -1+\tfrac{127}{64}.$$
The approximation is 1.333496094.

The actual area is $4/3$.

25. The following is a graph with 4 rectangles:

(a) Using the same method as in exercise 23, the width of the rectangles is 1/16, and the midpoints are
$$\tfrac{1}{16}, \tfrac{3}{16}, \ldots, \tfrac{15}{16}.$$
The approximation is 0.249511719.

(b) Using the same method as in exercise 23, the width of the rectangles is now 1/32, and the midpoints are
$$\tfrac{1}{32}, \tfrac{3}{32}, \ldots, \tfrac{31}{32}.$$
The approximation is 0.24987793.

(c) Using the same method as in exercise 23, the width of the rectangles is now 1/64, and the midpoints are
$$\tfrac{1}{64}, \tfrac{3}{64}, \ldots, \tfrac{63}{64}.$$
The approximation is 0.249969482.

The actual area is $1/4$.

1.2 The Concept of Limit

1. (a) $\lim\limits_{x \to 0^-} f(x) = -2$

(b) $\lim\limits_{x \to 0^+} f(x) = 2$

(c) Does not exist.

(d) $\lim\limits_{x \to 1^-} f(x) = 1$

(e) $\lim\limits_{x \to -1} f(x) \approx 0.1$

(f) $\lim\limits_{x \to 2^-} f(x) = -1$

(g) $\lim\limits_{x \to 2^+} f(x) = 3$

(h) Does not exist.

(i) $\lim\limits_{x \to -2} f(x) \approx 1.8$

(j) $\lim\limits_{x \to 3} f(x) \approx 2.5$

3. (a) $\lim\limits_{x \to 2^-} f(x) = \lim\limits_{x \to 2^-} 2x = 4$

    (b) $\lim\limits_{x \to 2^+} f(x) = \lim\limits_{x \to 2^+} x^2 = 4$

    (c) $\lim\limits_{x \to 2} f(x) = 4$

    (d) $\lim\limits_{x \to 1} f(x) = \lim\limits_{x \to 1} 2x = 2$

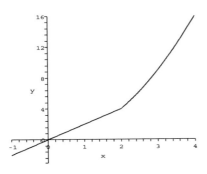

5. (a) $\lim\limits_{x \to -1^-} f(x) = \lim\limits_{x \to 1^-} x^2 + 1 = 2$

    (b) $\lim\limits_{x \to -1^+} f(x) = \lim\limits_{x \to 1^+} 3x + 1 = -2$

    (c) $\lim\limits_{x \to -1} f(x)$ does not exist

    (d) $\lim\limits_{x \to 1} f(x) = \lim\limits_{x \to 1} 3x + 1 = 4$

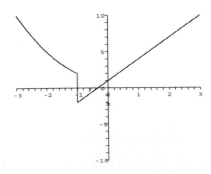

7. $f(1.5) = 2.22$, $f(1.1) = 2.05$,
    $f(1.01) = 2.01$, $f(1.001) = 2.00$.

The values of $f(x)$ seem to be approaching 2 as x approaches 1 from the right.

$f(0.5) = 1.71$, $f(0.9) = 1.95$,
$f(0.99) = 1.99$, $f(0.999) = 2.00$.

The values of $f(x)$ seem to be approaching 2 as x approaches 1 from the left. Since the limits from the left and right exist and are the same, the limit exists.

9. By inspecting the graph, and using a sequence of values (as in exercises 7 and 8), we see that the limit is approximately 2.

11. By inspecting the graph, and using a sequence of values (as in exercises 7 and 8), we see that the limit is approximately 1.

13. By inspecting the graph, and using a sequence of values (as in exercises 7 and 8), we see that the limit is approximately 1.

15. The numerical evidence suggests that the function the function blows up at $x = 1$. From the graph we see that the function has a vertical asymptote at $x = 1$.

17. By inspecting the graph, and using a sequence of values (as in exercises 7 and 8), we see that the limit is approximately 3/2.

19. The limit does not exist because the graph oscillates wildly near $x = 0$.

21. The numerical evidence suggests that

$\lim\limits_{x \to 2^-} \frac{x-2}{|x-2|} = -1$ while $\lim\limits_{x \to 2^+} \frac{x-2}{|x-2|} = 1$

so $\lim\limits_{x \to 2} \frac{x-2}{|x-2|}$ does not exist. There is a break in the graph at $x = 2$.

23. The function $\ln x$ is not defined for $x \leq 0$ so the limit does not exist. The numerical evidence suggests that the function blows up as x approaches 0 from the right. From the graph we see that the function has a one-sided vertical asymptote at $x = 0$.

25. The limit exists and equals 1.

27. Numerical and graphical evidence show that the limits

$$\lim_{x \to 1} \frac{x^2 + 1}{x - 1} \text{ and } \lim_{x \to 2} \frac{x + 1}{x^2 - 4}$$

do not exist (both have vertical asymptotes). Our conjecture is that if $g(a) = 0$ and $f(a) \neq 0$, $\lim\limits_{x \to a} \frac{f(x)}{g(x)}$ does not exist.

29. One possibility:

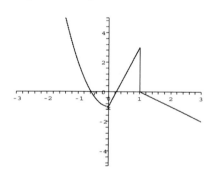

31. One possibility:

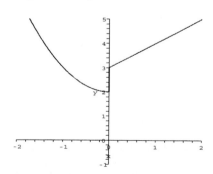

33. By inspecting the graph, and using a sequence of values (as in exercises 7

and 8), we see that the limit is approximately $1/2$.

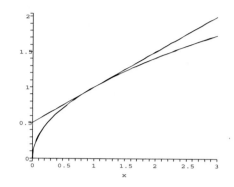

35. The first argument gives the correct value; the second argument is not valid because it looks only at certain values of x.

37.

x	$(1 + x)^{\frac{1}{x}}$	x	$(1 + x)^{\frac{1}{x}}$
0.1	2.59	-0.1	2.87
0.01	2.70	-0.01	2.73
0.001	2.7169	-0.001	2.7196

$$\lim_{x \to 0}(1 + x)^{1/x} \approx 2.7182818$$

39.

x	$x^{\sec x}$
0.1	0.099
0.01	0.010
0.001	0.001

$$\lim_{x \to 0^+} x^{\sec x} = 0$$

For negative x the values of $x^{\sec x}$ are usually not real numbers, so $\lim\limits_{x \to 0^-} x^{\sec x} = 0$ does not exist.

41. Possible answers:

$$f(x) = \frac{x^2}{x}$$

$$g(x) = \begin{cases} 1 & \text{if } x \leq 0 \\ -1 & \text{if } x > 0 \end{cases}$$

43. As x gets arbitrarily close to a, $f(x)$ gets arbitrarily close to L.

45. For $3 \le t \le 4$, $f(t) = 8$, so $\lim_{t \to 3.5} f(t) = 8$. Also $\lim_{t \to 4^-} f(t) = 8$. On the other hand, for $4 \le t \le 5$, $f(t) = 10$, so $\lim_{t \to 4^+} f(t) = 10$. Hence $\lim_{t \to 4} f(t)$ does not exist.

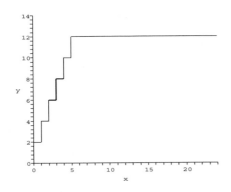

1.3 Computation of Limits

1. $\lim_{x \to 0}(x^2 - 3x + 1) = 0^2 - 3(0) + 1 = 1$

3. $\lim_{x \to 0} \cos^{-1}(x^2) = \cos^{-1} 0 = \dfrac{\pi}{2}$.

5. $\lim_{x \to 3} \dfrac{x^2 - x - 6}{x - 3}$
$= \lim_{x \to 3} \dfrac{(x - 3)(x + 2)}{x - 3}$
$= \lim_{x \to 3}(x + 2) = 3 + 2 = 5$

7. $\lim_{x \to 2} \dfrac{x^2 - x - 2}{x^2 - 4}$
$= \lim_{x \to 2} \dfrac{(x - 2)(x + 1)}{(x + 2)(x - 2)}$
$= \lim_{x \to 2} \dfrac{x + 1}{x + 2} = \dfrac{2 + 1}{2 + 2} = \dfrac{3}{4}$

9. $\lim_{x \to 0} \dfrac{\sin x}{\tan x} = \lim_{x \to 0} \dfrac{\sin x}{\frac{\sin x}{\cos x}}$
$= \lim_{x \to 0} \cos x = \cos 0 = 1$

11. $\lim_{x \to 0} \dfrac{xe^{-2x+1}}{x^2 + x}$
$= \lim_{x \to 0} \dfrac{x(e^{-2x+1})}{x(x + 1)}$

$= \lim_{x \to 0} \dfrac{e^{-2x+1}}{x + 1} = \dfrac{e^{-2(0)+1}}{0 + 1} = e$

13. $\lim_{x \to 0} \dfrac{\sqrt{x + 4} - 2}{x}$
$= \lim_{x \to 0} \dfrac{\sqrt{x + 4} - 2}{x} \left(\dfrac{\sqrt{x + 4} + 2}{\sqrt{x + 4} + 2} \right)$
$= \lim_{x \to 0} \dfrac{x + 4 - 4}{x(\sqrt{x + 4} + 2)}$
$= \lim_{x \to 0} \dfrac{x}{x(\sqrt{x + 4} + 2)}$
$= \lim_{x \to 0} \dfrac{1}{\sqrt{x + 4} + 2}$
$= \dfrac{1}{\sqrt{4} + 2} = \dfrac{1}{2 + 2} = \dfrac{1}{4}$

15. $\lim_{x \to 1} \dfrac{x - 1}{\sqrt{x} - 1}$
$= \lim_{x \to 1} \dfrac{(\sqrt{x} + 1)(\sqrt{x} - 1)}{\sqrt{x} - 1}$
$= \lim_{x \to 1}(\sqrt{x} + 1) = \sqrt{1} + 1 = 2$

17. $\lim_{x \to 1} \left(\dfrac{1}{x - 1} - \dfrac{2}{x^2 - 1} \right)$
$= \lim_{x \to 1} \left(\dfrac{1}{x - 1} - \dfrac{2}{(x - 1)(x + 1)} \right)$
$= \lim_{x \to 1} \left(\dfrac{x + 1}{(x - 1)(x + 1)} - \dfrac{2}{(x - 1)(x + 1)} \right)$
$= \lim_{x \to 1} \left(\dfrac{x - 1}{(x - 1)(x + 1)} \right)$
$= \lim_{x \to 1} \left(\dfrac{1}{x + 1} \right) = \dfrac{1}{2}$

19. $\lim_{x \to 0} \dfrac{1 - e^{2x}}{1 - e^x}$
$= \lim_{x \to 0} \dfrac{(1 - e^x)(1 + e^x)}{1 - e^x}$
$= \lim_{x \to 0}(1 + e^x) = 2$

21. $\lim_{x \to 0^+} \dfrac{\sin(|x|)}{x} = \lim_{x \to 0^+} \dfrac{\sin(x)}{x} = 1$
$\lim_{x \to 0^-} \dfrac{\sin(|x|)}{x}$
$= \lim_{x \to 0^-} \dfrac{\sin(-x)}{x}$
$= \lim_{x \to 0^-} \dfrac{-\sin(x)}{x} = -1$

Since the limit from the left does not equal the limit from the right, we see that $\lim_{x \to 0} \frac{\sin(|x|)}{x}$ does not exist.

23. $\lim_{x \to 2^-} f(x) = \lim_{x \to 2^-} 2x = 2(2) = 4$

$\lim_{x \to 2^+} f(x) = \lim_{x \to 2^+} x^2 = 2^2 = 4$

$\lim_{x \to 2} f(x) = 4$

25. $\lim_{x \to 0} f(x) = \lim_{x \to 0}(3x+1) = 3(0)+1 = 1$

27. $\lim_{x \to -1^-} f(x) = \lim_{x \to -1^-}(2x+1)$
$= 2(-1)+1 = -1$

$\lim_{x \to -1^+} f(x) = \lim_{x \to -1^+} 3 = 3$
Therefore $\lim_{x \to -1} f(x)$ does not exist.

29. $\lim_{h \to 0} \frac{(2+h)^2 - 4}{h}$

$= \lim_{h \to 0} \frac{(4 + 4h + h^2) - 4}{h}$

$= \lim_{h \to 0} \frac{4h + h^2}{h} = \lim_{h \to 0} 4 + h = 4$

31. $\lim_{h \to 0} \frac{h^2}{\sqrt{h^2 + h + 3} - \sqrt{h+3}}$

$= \lim_{h \to 0} \frac{h^2(\sqrt{h^2 + h + 3} + \sqrt{h+3})}{(h^2 + h + 3) - (h+3)}$

$= \lim_{h \to 0} \frac{h^2(\sqrt{h^2 + h + 3} + \sqrt{h+3})}{h^2}$

$= \lim_{h \to 0} \sqrt{h^2 + h + 3} + \sqrt{h+3} = 2\sqrt{3}$

To get from the first line to the second, we have multiplied by

$$\frac{\sqrt{h^2 + h + 3} + \sqrt{h+3}}{\sqrt{h^2 + h + 3} + \sqrt{h+3}}.$$

33. $\lim_{t \to -2} \frac{\frac{1}{2} + \frac{1}{t}}{2 + t}$

$= \lim_{t \to -2} \frac{\frac{t+2}{2t}}{2 + t}$

$= \lim_{t \to -2} \frac{1}{2t} = -\frac{1}{4}$

35.

x^2	$x^2 \sin(1/x)$
-0.1	0.0054
-0.01	5×10^{-5}
-0.001	-8×10^{-7}
0.1	-0.005
0.01	-5×10^{-5}
0.001	8×10^{-7}

Conjecture: $\lim_{x \to 0} x^2 \sin(1/x) = 0$.
Let $f(x) = -x^2, h(x) = x^2$. Then

$$f(x) \le x^2 \sin\left(\frac{1}{x}\right) \le h(x)$$

$$\lim_{x \to 0}(-x^2) = 0, \lim_{x \to 0}(x^2) = 0$$

Therefore, by the Squeeze Theorem,

$$\lim_{x \to 0} x^2 \sin\left(\frac{1}{x}\right) = 0.$$

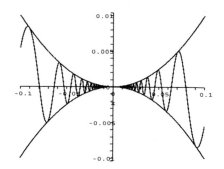

37. Let $f(x) = 0$, $h(x) = \sqrt{x}$. We see that

$$f(x) \le \sqrt{x} \cos^2(1/x) \le h(x),$$

$$\lim_{x \to 0^+} 0 = 0, \lim_{x \to 0^+} \sqrt{x} = 0$$

Therefore, by the Squeeze Theorem,

$$\lim_{x \to 0^+} \sqrt{x} \cos^2\left(\frac{1}{x}\right) = 0.$$

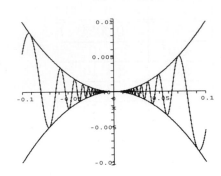

39. $\lim\limits_{x\to 4^+} \sqrt{16 - x^2}$ does not exist because the domain of the function is $[-4, 4]$.

41. $\lim\limits_{x\to -2^-} \sqrt{x^2 + 3x + 2} = 0.$

43. $\lim\limits_{x\to 0^+} \dfrac{\sqrt{1 - \cos x}}{x} = \sqrt{\dfrac{1}{2}} = \dfrac{\sqrt{2}}{2}$

45. $\lim\limits_{x\to a^-} f(x) = \lim\limits_{x\to a^-} g(x) = g(a)$ because $g(x)$ is a polynomial. Similarly,

$$\lim\limits_{x\to a^+} f(x) = \lim\limits_{x\to a^+} h(x) = h(a).$$

47. (a) $\lim\limits_{x\to 2}(x^2 - 3x + 1)$
$= 2^2 - 3(2) + 1$ (Theorem 3.2)
$= -1$

(b) $\lim\limits_{x\to 0} \dfrac{x - 2}{x^2 + 1}$
$= \dfrac{\lim\limits_{x\to 0}(x - 2)}{\lim\limits_{x\to 0}(x^2 + 1)}$
(Theorem 3.1(iv))
$= \dfrac{\lim\limits_{x\to 0} x - \lim\limits_{x\to 0} 2}{\lim\limits_{x\to 0} x^2 + \lim\limits_{x\to 0} 1}$
(Theorem 3.1(ii))
$= \dfrac{0 - 2}{0 + 1}$
(Equations 3.1, 3.2, and 3.5)
$= -2$

49. Velocity is given by the limit
$$\lim\limits_{h\to 0} \dfrac{f(2 + h) - f(2)}{h}$$

$$= \lim\limits_{h\to 0} \dfrac{(2 + h)^2 + 2 - (2^2 + 2)}{h}$$
$$= \lim\limits_{h\to 0} \dfrac{4h + h^2}{h}$$
$$= \lim\limits_{h\to 0} 4 + h = 4.$$

51. Velocity is given by the limit
$$\lim\limits_{h\to 0} \dfrac{f(0 + h) - f(0)}{h}$$
$$= \lim\limits_{h\to 0} \dfrac{(0 + h)^3 - (0)^3}{h}$$
$$= \lim\limits_{h\to 0} \dfrac{h^3}{h}$$
$$= \lim\limits_{h\to 0} h^2 = 0.$$

53. $m = \lim\limits_{h\to 0} \dfrac{\sqrt{1 + h} - 1}{h} \dfrac{\sqrt{1 + h} + 1}{\sqrt{1 + h} + 1}$
$= \lim\limits_{h\to 0} \dfrac{1 + h - 1}{h(\sqrt{1 + h} + 1)}$
$= \lim\limits_{h\to 0} \dfrac{h}{h(\sqrt{1 + h} + 1)}$
$= \lim\limits_{h\to 0} \dfrac{1}{\sqrt{1 + h} + 1}$
$= \dfrac{1}{\sqrt{1 + 0} + 1} = \dfrac{1}{2}.$

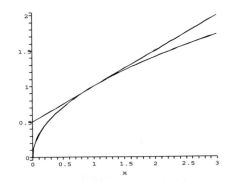

55. $\lim\limits_{x\to 0^+} (1 + x)^{1/x} = e \approx 2.71828$

57. $\lim\limits_{x\to 0^+} x^{-x^2} = 1$

59. As x gets close to 0, $1/x$ gets larger and larger in absolute value, so $\sin(1/x)$ oscillates more and more rapidly between 1 and -1, so the limit does not exist.

61. When x is small and positive, $1/x$ is large and positive, so $\tan^{-1}(1/x)$ approaches $\pi/2$. But when x is small and negative, $1/x$ is large and negative, so $\tan^{-1}(1/x)$ approaches $-\pi/2$. So the limit does not exist.

63. $\lim_{x \to a} [2f(x) - 3g(x)]$
$= 2 \lim_{x \to a} f(x) - 3 \lim_{x \to a} g(x)$
$= 2(2) - 3(-3) = 13$

65. $\lim_{x \to a} \left[\dfrac{f(x) + g(x)}{h(x)} \right]$
does not exist, because

$\lim_{x \to a} [f(x) + g(x)]$

$= \lim_{x \to a} f(x) + \lim_{x \to a} g(x)$

$= 2 - 3 = -1$

and $\lim_{x \to a} h(x) = 0$.

67. $\lim_{x \to a} [f(x)]^3$

$= \left[\lim_{x \to a} f(x) \right] \left[\lim_{x \to a} f(x) \right] \left[\lim_{x \to a} f(x) \right]$

$= L \cdot L \cdot L = L^3$

$\lim_{x \to a} [f(x)]^4 = \left[\lim_{x \to a} f(x) \right] \left[\lim_{x \to a} [f(x)]^3 \right]$

$= L \cdot L^3 = L^4$

69. We can't split the limit of a product into a product of limits unless we know that both limits exist; the limit of the product of a term tending toward 0 and a term with an unknown limit is not necessarily 0 but instead is unknown.

71. One possibility is

$$f(x) = \frac{1}{x}, g(x) = -\frac{1}{x}.$$

73. Yes. If $\lim_{x \to a} [f(x) + g(x)]$ exists, then, it would also be true that

$$\lim_{x \to a} [f(x) + g(x)] - \lim_{x \to a} f(x)$$

exists. But by Theorem 3.1 (ii)

$\lim_{x \to a} [f(x) + g(x)] - \lim_{x \to a} f(x)$
$= \lim_{x \to a} [[f(x) + g(x)] - [f(x)]]$
$= \lim_{x \to a} g(x)$

so $\lim_{x \to a} g(x)$ would exist, but we are given that $\lim_{x \to a} g(x)$ does not exist.

75. $\lim_{x \to 0^+} T(x) = \lim_{x \to 0^+} (0.14x) = 0 = T(0)$.

$\lim_{x \to 10,000^-} T(x) = 0.14(10,000) = 1400$

$\lim_{x \to 10,000^+} T(x)$
$= 1500 + 0.21(10,000) = 3600$

Therefore $\lim_{x \to 10,000} T(x)$ does not exist. A small change in income should result in a small change in tax liability. This is true near $x = 0$ but is not true near $x = 10,000$. As your income grows past \$10,000 your tax liability jumps enormously.

77. $\lim_{x \to 3^-} [x] = 2; \lim_{x \to 3^+} [x] = 3$
Therefore $\lim_{x \to 3} [x]$ does not exist.

1.4 Continuity and its Consequences

1. Discontinuous at $x = -2$ (limit does not exist), and at $x = 2$ (function undefined).

3. Discontinuous at $x = -2$ (function undefined), at $x = 1$ (function undefined), and at $x = 4$ (limit does not exist).

5. Discontinuous at $x = -2$ (limit does not exist), at $x = 2$ (function undefined), and at $x = 4$ (limit does not exist).

7. $f(1)$ is not defined and $\lim_{x \to 1} f(x)$ does not exist.

9. $f(0)$ is not defined and $\lim\limits_{x\to 0} f(x)$ does not exist.

11. $\lim\limits_{x\to 2^-} f(x) = \lim\limits_{x\to 2^-} (x^2) = 4$
$\lim\limits_{x\to 2^+} f(x) = \lim\limits_{x\to 2^+} (3x - 2) = 4$
$\lim\limits_{x\to 2} f(x) = 4; f(2) = 3$
$\lim\limits_{x\to 2} f(x) \neq f(2)$

13. $f(x) = \dfrac{x-1}{(x+1)(x-1)}$ has a removable discontinuity at $x = 1$ and a non-removable discontinuity at $x = -1$; the removable discontinuity is removed by
$g(x) = \dfrac{1}{x+1}$.

15. No discontinuities.

17. $f(x) = \dfrac{x^2 \sin x}{\cos x}$ has non-removable discontinuities at $x = \frac{\pi}{2} + k\pi$ for any integer k.

19. By sketching the graph, or numerically, one can see that $\lim\limits_{x\to 0} x \ln x^2 = 0$. Thus, one can remove the discontinuity at $x = 0$ by defining

$$g(x) = \begin{cases} x \ln x^2 & \text{if} \quad x \neq 0 \\ 0 & \text{if} \quad x = 0 \end{cases}$$

21. $f(x)$ has a non-removable discontinuity at $x = 1$.

23. $f(x)$ has a non-removable discontinuity at $x = 1$:
$\lim\limits_{x\to -1^-} f(x) = \lim\limits_{x\to -1^-} (3x - 1) = -4$
$\lim\limits_{x\to -1^+} f(x) = \lim\limits_{x\to -1^+} (x^2 + 5x) = -4$
$\lim\limits_{x\to 1^-} f(x) = \lim\limits_{x\to 1^-} (x^2 + 5x) = 6$
$\lim\limits_{x\to 1^+} f(x) = \lim\limits_{x\to 1^+} (3x^3) = 3$

25. Continuous where $x + 3 > 0$, i.e. on $(-3, \infty)$

27. Continuous everywhere, i.e. on $(-\infty, \infty)$.

29. Continuous everywhere, i.e. on $(-\infty, \infty)$.

31. Continuous where $x + 1 > 0$, i.e. on $(-1, \infty)$.

33.

$$\lim_{x\to 0^-} f(x) = \lim_{x\to 0^-} 2\frac{\sin x}{x}$$
$$= 2 \lim_{x\to 0^-} \frac{\sin x}{x} = 2$$

Hence a must equal 2 if f is continuous.

$$\lim_{x\to 0^-} f(x) = \lim_{x\to 0^-} b\cos x$$
$$= b \lim_{x\to 0^-} \cos x = b,$$

so b and a must equal 2 if f is continuous.

35. First note that
$$\lim_{x\to 3^+} f(x) = \lim_{x\to 3^+} \ln(x - 2) + x^2$$
$$= \ln(3 - 2) + 3^2 = 9.$$

Also $f(3) = 2e^{3b} + 1$, so if f is continuous, $2e^{3b} + 1$ must equal 9; that is $e^{3b} = 4$, so $b = \frac{\ln 4}{3}$. Then note that
$$f(0) = 2e^{(b)(0)} + 1 = 3.$$

Also,
$$\lim_{x\to 0^-} f(x) = \lim_{x\to 0^-} a(\tan^{-1} x + 2)$$
$$= a(\tan^{-1} 0 + 2)$$
$$= a(0 + 2) = 2a,$$

so a must equal 3/2 if f is continuous.

37. $\lim\limits_{x\to 10000^-} T(x) = \lim\limits_{x\to 10000^-} 0.14x$
$= 0.14(10,000) = 1400$

$\lim\limits_{x\to 10000^+} T(x) = \lim\limits_{x\to 10000^+} (c + 0.21x)$
$= c + 0.21(10,000)$
$= c + 2100$

$c + 2100 = 1400$

$c = -700$

A small change in income should not result in a big change in tax, so the tax function should be continuous.

39. For $T(x)$ to be continuous at $x = 141{,}250$ we must have

$$\lim_{x \to 141{,}250^-} T(x) = \lim_{x \to 141{,}250^+} T(x).$$

Now

$$\lim_{x \to 141{,}250^-} T(x) = \lim_{x \to 141{,}250^-} (.30)(x)a$$

$$= (.30)(141{,}250) - 5685$$

$$= 36690.$$

On the other hand,

$$\lim_{x \to 141{,}250^+} T(x) = \lim_{x \to 141{,}250^+} (.35)(x) - b$$

$$= (.35)(141{,}250) - b$$

$$= 49437.50 - b.$$

Hence

$b = 49437.50 - 36690 = 12{,}747.50.$

For $T(x)$ to be continuous at $x = 307{,}050$ we must have

$$\lim_{x \to 307{,}050^-} T(x) = \lim_{x \to 307{,}050^+} T(x).$$

Now

$$\lim_{x \to 307{,}050^-} T(x)$$

$$= \lim_{x \to 307{,}050^-} (.35)(x) - b$$

$$= (.35)(307{,}050) - 12{,}747.5$$

$$= 94{,}720.$$

On the other hand,

$$\lim_{x \to 307{,}050^+} T(x)$$

$$= \lim_{x \to 307{,}050^+} (.386)(x) - c$$

$$= (.386)(307{,}050) - c$$

$$= 118521.3 - c.$$

Hence

$c = 118{,}521.3 - 94720 = 23801.3.$

41. The first two rows of the following table (together with the Intermediate Value Theorem) show that $f(x)$ has a root in $[2, 3]$. In the following rows, we use the midpoint of the previous interval as our new x. When $f(x)$ is positive, we use the left half, and when $f(x)$ is negative, we use the right half of the interval. (Because the function goes from negative to positive. If the function went from positive to negative, the intervals would be reversed.)

x	$f(x)$
2	-3
3	2
2.5	-0.75
2.75	0.5625
2.625	-0.109375
2.6875	0.223
2.65625	0.557

The zero is in the interval $[2.625, 2.65625]$.

43. The first two rows of the following table (together with the Intermediate Value Theorem) show that $f(x)$ has a root in $[2, 3]$. In the following rows, we use the midpoint of the previous interval as our new x. When $f(x)$ is positive, we use the right half, and when $f(x)$ is negative, we use the left half of the interval.

x	$f(x)$
-1	1
0	-2
-0.5	-0.125
-0.625	0.256
-0.5625	0.072
-0.53125	-0.025

The zero is in the interval $[-0.5625, -0.53125]$.

45. The first two rows of the following table (together with the Intermediate Value Theorem) show that $f(x)$ has a root in $[-2, -1]$. In the following rows, we use the midpoint of the previous interval as our new x. When $f(x)$ is positive, we use the right half, and when $f(x)$ is negative, we use the left half of the interval.

x	$f(x)$
0	1
1	-0.46
0.5	0.378
0.75	-0.018
0.625	0.186
0.6875	0.085
0.71875	0.034

The zero is in the interval $[0.71875, 0.75]$.

47. $\displaystyle\lim_{x \to 2^+} f(x) = \lim_{x \to 2^+} (3x - 1) = 5$

$f(2) = 3(2)\text{-}1 = 5$

Thus $f(x)$ is continuous from the right at $x = 2$.

49. $\displaystyle\lim_{x \to 2^+} f(x) = \lim_{x \to 2^+} (3x - 3) = 3$

$f(2) = 2^2 = 4$

Thus $f(x)$ is not continuous from the right at $x = 2$.

51. A function is continuous from the left at $x = a$ if $\displaystyle\lim_{x \to a^-} f(x) = f(a)$.

(a) $\displaystyle\lim_{x \to 2^-} f(x) = \lim_{x \to 2^-} x^2 = 4$
$f(2) = 5$
Thus $f(x)$ is not continuous from the left at $x = 2$.

(b) $\displaystyle\lim_{x \to 2^-} f(x) = \lim_{x \to 2^-} x^2 = 4$
$f(2) = 3$
Thus $f(x)$ is not continuous from the left at $x = 2$.

(c) $\displaystyle\lim_{x \to 2^-} f(x) = \lim_{x \to 2^-} x^2 = 4$
$f(2) = 4$
Thus $f(x)$ is continuous from the left at $x = 2$.

(d) $f(x)$ is not continuous from the left at $x = 2$ because $f(2)$ is undefined.

53. Need $g(30) = 100$ and $g(34) = 0$. We may take $g(T)$ to be linear.
$m = \dfrac{0 - 100}{34 - 30} = -25$
$y = -25(x - 34)$
$g(T) = -25(T - 34)$

55.

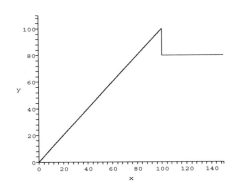

The graph is discontinuous at $x = 100$. This is when the box starts moving.

57. Let $f(t)$ be her distance from home as a function of time on Monday. Let $g(t)$ be her distance from home as a function of time on Tuesday. Let t be given in minutes, with $t = 0$ corresponding to 7:13 a.m. Then she leaves home at $t = 0$ and arrives at her destination at $t = 410$. Let $h(t) = f(t) - g(t)$. If $h(t) = 0$ for some t, then the saleswoman was at exactly the same place at the same time on both Monday and Tuesday. $h(0) = f(0) - g(0) = -g(0) < 0$ and $h(410) = f(410) - g(410) = f(410) >$

0. By the Intermediate Value Theorem, there is a t in the interval $[0, 410]$ such that $h(t) = 0$.

59.

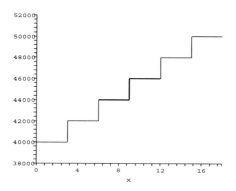

The function $s(t)$ has jump discontinuities every three months when the salary suddenly increases by \$2000. In the function $f(t)$, the \$2000 increase occurs gradually over the 3 month period, so $f(t)$ is continuous. It might be easier to do calculations with $f(t)$ because it is continuous and because it is given by a simpler formula.

61. We already know $f(x) \neq 0$ for $a < x < b$. Suppose $f(d) < 0$ for some d, $a < d < b$. Then by the Intermediate Value Theorem, there is an e in the interval $[c, d]$ such that $f(e) = 0$. But this e would also be between a and b, which is impossible. Thus, $f(x) > 0$ for all $a < x < b$.

63. $\lim_{x \to 0} x f(x) = \lim_{x \to 0} x \lim_{x \to 0} f(x)$
$$= 0 f(0) = 0$$

65. $\lim_{x \to a} g(x) = \lim_{x \to a} |f(x)| = \left| \lim_{x \to a} f(x) \right|$
$$= |f(a)| = g(a).$$

67. Let $b \geq a$. Then

$$\lim_{x \to b} h(x) = \lim_{x \to b} \left(\max_{a \leq t \leq b} f(t) \right)$$
$$= \max_{a \leq t \leq b} \left(\lim_{t \to b} f(t) \right)$$
$$= h(b)$$

since f is continuous. Thus, h is continuous for $x \geq a$.

No, the property would not be true if f were not assumed to be continuous. A counterexample is

$$f(x) = \begin{cases} 1 & \text{if } a \leq x < b \\ 2 & \text{if } b \leq x \end{cases}$$

Then $h(x) = 1$ for $a \leq x < b$, and $h(x) = 2$ for $x \geq b$. Thus, h is not continuous at $x = b$.

1.5 Limits Involving Infinity

1. (a) $\lim_{x \to 1^-} \dfrac{1 - 2x}{x^2 - 1} = \infty$.

(b) $\lim_{x \to 1^+} \dfrac{1 - 2x}{x^2 - 1} = -\infty$.

(c) Does not exist.

3. (a) $\lim_{x \to 2^-} \dfrac{x - 4}{x^2 - 4x + 4} = -\infty$

(b) $\lim_{x \to 2^+} \dfrac{x - 4}{x^2 - 4x + 4} = -\infty$

(c) $\lim_{x \to 2} \dfrac{x - 4}{x^2 - 4x + 4} = -\infty$

5. $\lim_{x \to 2^-} \dfrac{-x}{\sqrt{4 - x^2}} = -\infty$.

As x approaches 2 from below, the numerator is near -2 and the denominator is small and positive, so the faction goes to $-\infty$.

7. $\lim\limits_{x \to -\infty} \dfrac{-x}{\sqrt{4 + x^2}}$

$= \lim\limits_{x \to -\infty} \dfrac{-x}{-x\sqrt{\frac{4}{x^2} + 1}}$

$= \lim\limits_{x \to -\infty} \dfrac{1}{\sqrt{\frac{4}{x^2} + 1}}$

$= \dfrac{1}{\sqrt{1}} = 1$

9. $\lim\limits_{x \to \infty} \dfrac{x^3 - 2\cos x}{3x^2 + 4x - 1}$

$= \lim\limits_{x \to \infty} \dfrac{x^2 \left(x - \frac{2\cos x}{x^2}\right)}{x^2 \left(3 + \frac{4}{x} - \frac{1}{x^2}\right)}$

$= \lim\limits_{x \to \infty} \dfrac{\left(x - \frac{2\cos x}{x^2}\right)}{3 + \frac{4}{x} - \frac{1}{x^2}} = \infty$

11. $\lim\limits_{x \to \infty} \ln 2x = \infty$.

Note that $\ln 2x = \ln 2 + \ln x$, so it is enough to show that $\ln x$ goes to ∞ as x goes to ∞. This can be seen from the graph of the function $\ln x$ on page 51.

13. $\lim\limits_{x \to 0^+} e^{-2/x} = 0$.

When x is small and positive, $-2/x$ is large and negative, and e raised to a large negative power is very small.

15. $\lim\limits_{x \to \infty} \cot^{-1} x = 0$.

(Compare Example 5.8) We are looking for the angle that θ must approach as $\cot \theta$ goes to ∞. Look at the graph of $\cot \theta$. To define the inverse cotangent, you must pick one branch of this graph, and the standard choice is the branch immediately to the right of the y-axis. Then as $\cot \theta$ goes to ∞, the angle goes to 0.

17. $\lim\limits_{x \to \infty} e^{2x-1} = \infty$.

As x gets large, $2x - 1$ gets large, and e raised to a large positive power is large and positive.

19. $\lim\limits_{x \to \infty} \sin 2x$ does not exist. As x gets larger and larger, the values of $\sin 2x$ oscillate between 1 and -1.

21. As x goes to ∞, both e^{3x} and e^x go to ∞ as well. Furthermore, as x goes to ∞, so does $\ln x$. Thus it looks like

$$\lim\limits_{x \to \infty} \left(\dfrac{\ln(2 + e^{3x})}{\ln(1 + e^x)}\right) = \dfrac{\infty}{\infty}.$$

This is an indeterminate form, i.e., we don't know from this analysis what happens in this limit. Looking at numerical and/or graphing evidence, we guess that the limit is 3.

23. $\lim\limits_{x \to \frac{\pi}{2}^-} e^{-\tan x} = \lim\limits_{x \to \infty} e^{-x}$

$\qquad\qquad = \lim\limits_{x \to -\infty} e^x = 0$, but

$\lim\limits_{x \to \frac{\pi}{2}^+} e^{-\tan x} = \lim\limits_{x \to -\infty} e^{-x}$

$\qquad\qquad = \lim\limits_{x \to \infty} e^x = \infty$,

so the limit does not exist.

25. Since $4 + x^2$ is never 0, there are no vertical asymptotes. We have

$\lim\limits_{x \to \infty} \dfrac{x}{\sqrt{4 + x^2}}$

$= \lim\limits_{x \to \infty} \dfrac{x}{x\sqrt{\frac{4}{x^2} + 1}}$

$= \lim\limits_{x \to \infty} \dfrac{1}{\sqrt{\frac{4}{x^2} + 1}}$

$= \dfrac{1}{\sqrt{1}} = 1$

and

$\lim\limits_{x \to -\infty} \dfrac{x}{\sqrt{4 + x^2}}$

$= \lim\limits_{x \to -\infty} \dfrac{x}{-x\sqrt{\frac{4}{x^2} + 1}}$

$= \lim\limits_{x \to -\infty} \dfrac{-1}{\sqrt{\frac{4}{x^2} + 1}}$

$= \dfrac{-1}{\sqrt{1}} = -1$,

so there are horizontal asymptotes at $y = 1$ and $y = -1$.

27. $4 - x^2 = 0 \Rightarrow 4 = x^2$ so we have vertical asymptotes at $x = \pm 2$. We have

$$\lim_{x \to \pm\infty} \frac{x}{4 - x^2}$$

$$= \lim_{x \to \pm\infty} \frac{x}{x^2\left(\frac{4}{x^2} - 1\right)}$$

$$= \lim_{x \to \pm\infty} \frac{1}{x\left(\frac{4}{x^2} - 1\right)} = 0.$$

So there is a horizontal asymptote at $y = 0$.

29. The denominator factors: $x^2 - 2x - 3 = (x - 3)(x + 1)$. Since neither $x = 3$ nor $x = -1$ are zeros of the numerator, we see that $f(x)$ has vertical asymptotes at $x = 3$ and $x = -1$.

$f(x) \to -\infty$ as $x \to 3^-$,
$f(x) \to \infty$ as $x \to 3^+$,
$f(x) \to \infty$ as $x \to -1^-$, and
$f(x) \to -\infty$ as $x \to -1^+$.

We have

$$\lim_{x \to \pm\infty} \frac{3x^2 + 1}{x^2 - 2x - 3}$$

$$\lim_{x \to \pm\infty} \frac{3 + 1/x^2}{1 - 2/x - 3/x^2} = 3.$$

So there is a horizontal asymptote at $y = 3$.

31. The function $\ln x$ has a one-sided vertical asymptote at $x = 0$, so $f(x) = \ln(1 - \cos x)$ will have a vertical asymptote whenever $1 - \cos x = 0$, i.e., whenever $\cos x = 1$. This happens when $x = 2k\pi$ for any integer k. Since $1 - \cos x \geq 0$ for all x, $f(x)$ is defined at all points except for these vertical asymptotes. Thus as $f(x)$ approaches any of these asymptotes (from either side), it behaves like $\ln x$ approaching 0 from the right,

so $f(x) \to -\infty$ as x approaches any of these asymptotes from either side.

33. The function is continuous for all x, so no vertical asymptotes. We have

$$\lim_{x \to \infty} 4\tan^{-1} x - 1 = 4(\lim_{x \to \infty} \tan^{-1} x) - 1$$
$$= 4(\pi/2) - 1$$
$$= 2\pi - 1$$

and

$$\lim_{x \to -\infty} 4\tan^{-1} x - 1$$
$$= 4(\lim_{x \to -\infty} \tan^{-1} x) - 1$$
$$= 4(-\pi/2) - 1$$
$$= -2\pi - 1,$$

so there are horizontal asymptotes at $y = 2\pi - 1$ and $y = -2\pi - 1$.

35. Vertical asymptotes at $x = \pm 2$. The slant asymptote is $y = -x$.

37. Vertical asymptotes at

$$x = \frac{-1 \pm \sqrt{17}}{2}.$$

The slant asymptote is $y = x - 1$.

39. When x is large, the value of the fraction is close to 0.

41. When x is large, the value of the fraction is very close to $\frac{1}{2}$.

43. $\displaystyle\lim_{x \to \infty} \frac{x^3 + 4x + 5}{e^{x/2}} = 0.$

45. When x is close to -1, the value of the fraction is close to 1.

47. When x is close to 0, the value of the fraction is large and negative, so the limit appears to be $-\infty$.

49. We multiply by

$$\frac{\sqrt{4x^2 - 2x + 1} + 2x}{\sqrt{4x^2 - 2x + 1} + 2x}$$

to get:

$$\lim_{x \to \infty} \left(\sqrt{4x^2 - 2x + 1} - 2x \right)$$
$$= \lim_{x \to \infty} \frac{-2x + 1}{\sqrt{4x^2 - 2x + 1} + 2x} \cdot \frac{1/x}{1/x}$$
$$= \lim_{x \to \infty} \frac{-2 + 1/x}{\sqrt{4 - 2/x + 1/x^2} + 2}$$
$$= \frac{-2}{\sqrt{4} + 2} = -\frac{1}{2}.$$

51. $\lim_{x \to \infty} \left(\sqrt{5x^2 + 4x + 7} - \sqrt{5x^2 + x + 3} \right)$

If we multiply by
$$\frac{\sqrt{5x^2 + 4x + 7} + \sqrt{5x^2 + x + 3}}{\sqrt{5x^2 + 4x + 7} + \sqrt{5x^2 + x + 3}},$$
we get
$$\lim_{x \to \infty} \frac{(5x^2 + 4x + 7) - (5x^2 + x + 3)}{\sqrt{5x^2 + 4x + 7} + \sqrt{5x^2 + x + 3})}$$
$$= \lim_{x \to \infty} \frac{3x + 4}{\sqrt{5x^2 + 4x + 7} + \sqrt{5x^2 + x + 3}}$$
$$= \lim_{x \to \infty} \frac{3 + \frac{4}{x}}{\sqrt{5 + \frac{4}{x} + \frac{7}{x^2}} + \sqrt{5 + \frac{1}{x} + \frac{3}{x^2}}}$$
$$= \frac{3}{2\sqrt{5}} = \frac{3\sqrt{5}}{10}.$$

53.

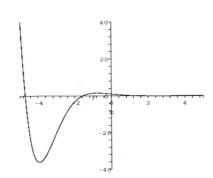

on $[-10, 10]$ by $[-100, 100]$

The horizontal asymptote is $y = 0$ approached only as $x \to \infty$. The graph crosses the horizontal asymptote an infinite number of times.

55. $\lim_{x \to \infty} \left(1 + \frac{1}{x} \right)^x = \lim_{x \to 0^+} (1 + x)^{1/x}$

$\qquad = \lim_{x \to 0^-} (1 + x)^{1/x} = \lim_{x \to -\infty} \left(1 + \frac{1}{x} \right)^x$

57. $h(0) = \dfrac{300}{1 + 9(.8^0)} = \dfrac{300}{10} = 30$ mm

$\lim_{t \to \infty} \dfrac{300}{1 + 9(.8^t)} = 300$ mm

59. $\lim_{x \to 0^+} \dfrac{80x^{-.3} + 60}{2x^{-.3} + 5} \left(\dfrac{x^{.3}}{x^{.3}} \right)$

$\qquad = \lim_{x \to 0^+} \dfrac{80 + 60x^{.3}}{2 + 5x^{.3}}$

$\qquad = \dfrac{80}{2} = 40$ mm

$\lim_{x \to \infty} \dfrac{80x^{-.3} + 60}{2x^{-.3} + 5} = \dfrac{60}{5} = 12$ mm

61. $f(x) = \dfrac{80x^{-0.3} + 60}{10x^{-0.3} + 30}$

63. $\lim_{t \to \infty} v_N = \lim_{t \to \infty} \dfrac{Ft}{m} = \infty$

$\lim_{t \to \infty} v_E = \lim_{t \to \infty} \dfrac{Fct}{\sqrt{m^2 c^2 + F^2 t^2}}$

$\qquad = \lim_{t \to \infty} \dfrac{Fct}{t \sqrt{\dfrac{m^2 c^2}{t^2} + F^2}}$

$\qquad = \lim_{t \to \infty} \dfrac{Fc}{\sqrt{\dfrac{m^2 c^2}{t^2} + F^2}}$

$\qquad = \dfrac{Fc}{\sqrt{F^2}} = c$

65. As in Example 5.10, the terminal velocity is $-\sqrt{\dfrac{32}{k}}$. When $k = 0.00064$, the terminal velocity is $-\sqrt{\dfrac{32}{.00064}} \approx -224$. When $k = 0.00128$, the terminal velocity is $-\sqrt{\dfrac{32}{.00128}} \approx -158$.

Solve $\sqrt{\dfrac{32}{ak}} = \dfrac{1}{2}\sqrt{\dfrac{32}{k}}$. Squaring both sides, $\dfrac{32}{ak} = \dfrac{1}{4} \cdot \dfrac{32}{k}$ so $a = 4$.

67. We must restrict the domain to $v_0 \geq 0$ because the formula makes sense only if the rocket is launched upward. To find v_e, set $19.6R - v_0^2 = 0$. Using $R \approx 6{,}378{,}000$ meters, we get $v_0 = \sqrt{19.6R} \approx 11{,}180$m/s. If the

Create an outline of the passage using key words. In your outline, record only major points, so ignore the hook or opening remarks and any specific details. Include approximately four to five key words for each paragraph. You should be able to consult your outline and locate information as well as check the overall main idea. Also, indicate the basic paragraph organization by separating paragraph topics and grouping details with the appropriate topic. Be brief but organized. You don't want to spend too much time on the outline, but you also want it to be useful when you start answering questions.

Look at the following passage map for the passage on cell division:

cell = sack, trillions
mitosis = division / 4 stages

#1
molecule signal → nucleus
more molecules + organelles (groups)

#2
synthesis = 2 X DNA
DNA = chain, gene order, instructions

#3
wall gone, no nucleus
contents divided

#4
2 new nucleus, new wall between → separate

READING STRATEGIES: QUESTION TYPES

Now, you can review strategies for each question type in the Reading section of the TOEFL. There are 10 unique question types in the Reading section. All of them require the general reading strategies explained above. However, each type of question is also based on particular information from the passage, so you need to review ways to find that information accurately and quickly. Finally, each question type also includes particular incorrect answers, or distracters, which manipulate or alter ideas from the passage. Therefore, the following steps review the best ways to find the correct answer(s) and avoid the incorrect ones for each question type.

The strategies for each question type below include a short reading passage and a sample question. The short passage and sample question are used to illustrate the steps necessary to answer each question type. Remember to practice the general reading strategies when you read the short passage. Try to answer the sample question before you read the strategies, and then review the strategies carefully and identify any steps you may have missed in your first attempt. The correct choice to each sample question is given in the strategies that follow it. These sample questions are not included in the Answer Key at the end of the Reading chapter. The Answer Key explains the correct answer(s) and distracters for the questions in the Practice section for each question type.

Question Type 1: Fact/Detail Questions

Detail questions test your basic comprehension of key facts in the passage. For each passage, you may receive from 3 to 6 questions about specific facts, or details. Like all questions in the reading section, detail questions are multiple-choice questions. You have four choices and either one or two choices is correct. (The question will specify whether you should choose one or two correct answers.) The other choices are distracters, or incorrect answer choices. Detail questions ask about facts related to one of the interrogative pronouns: what, who, whom, whose, which, why, how, when, or where, even though the questions might not always use these pronouns.

Once you receive a detail question, there are several actions you should perform in addition to the general strategies you've already learned. The strategies below review these recommended actions, but first read the following short excerpt from a passage on the ancient Greek god Apollo:

Apollo

One of art's earliest purposes was to serve as a marker of a holy place, either a place of worship or a place of protection. To illustrate the latter, consider ancient Greece, where statues of gods were placed on city walls to safeguard the people. These statues were often of the Greek god Apollo, whose importance to the Greeks must be understood before the artwork dedicated to him can be fully appreciated.

Apollo was an embodiment of all the virtues that Greek society upheld as worthy. Physical beauty and talent were highly valued, so every depiction of Apollo is of a physically ideal body. The first statues of Apollo showed him in a very limited pose, representing little or no activity, thus reminding onlookers that he was an authoritarian deity and not to be crossed. Some stances show Apollo with a serious, grave expression, extending an arm as a warning to those who have not followed his wishes.

However, the last representations of Apollo in ancient Greece allowed him more freedom of gesture and detailed expressions. This may be interpreted as a result of weakening cultural morals, but it was certainly meant as a tribute to the beauty and strength of Apollo's mind. The graceful poses of his body, whether he is depicted holding a bow and arrow or a musical instrument, attest to Apollo's intellectual power. His face, portraying pensiveness or determination, permitted the Greek people to identify themselves with him and celebrate his inner qualities, in addition to his attractive physique.

Unfortunately, the argument that the loosening of Apollo's strict representation corresponded to a breakdown in public values may also be true. It's unfortunate that the power wielded by the ancient religion began to decline precisely as the statues of Apollo became more lifelike.

Strategy 1: Identify a key idea or set of ideas in the question.

Read every question on the exam carefully. Students often hurt their own chances when they read questions too quickly. Each detail question involves a particular noun, action, or state, or set of nouns, actions, or states. Identify these and use them to locate the relevant information in the passage. This is necessary even if the question refers to a specific paragraph number. Regardless of any directions in the question, the body paragraphs in the reading passage could be very long. Therefore, even though the question may tell you to look at a particular paragraph, you must still locate the necessary information in that paragraph.

You must identify the idea(s), not the specific words or vocabulary. Often, the question will paraphrase, or restate, ideas from the passage. Of course, some words are not easily paraphrased, such as technical terms (DNA) and names (Apollo), and so don't assume every word is restated in the question. Usually, the restatement involves a synonym or different part of speech. Look at the following detail question about the passage above:

According to the author, the final carved depictions of Apollo

- ○ were superior to the early, serious versions.
- ○ represented a warning to the people.
- ○ possibly led to weakening cultural morals.
- ○ represented more than just his physical beauty.

The key ideas in the question are in the phrase *final carved depictions*. Of course the noun *Apollo* is important, but it is repeated too often in the passage since it is the main topic, or subject, of the passage. The name is less useful, then, for locating the necessary information in the passage.

Strategy 2: Scan the passage for the key idea(s).

Scanning is useful for locating information quickly to select the correct choice(s) as well as eliminate incorrect ones. Scanning is the opposite of reading actively or closely. It means to look over the passage without really reading the words. When you scan a passage, you don't want to understand the sentences; you want only to find some information.

You are probably looking for a restatement of the key idea(s) in the question, so you must still think about what you are seeing. However, you don't need to reread the passage; just look it over until you find a word, phrase, or clause that could be a possible paraphrase of the key idea(s). Since a detail question involves only specific facts from the passage, you have to read only a specific part of the passage closely. Scanning allows you to find that part by locating the key idea(s) quickly.

If the question directs you to a particular paragraph, then begin there. If the question does not mention a paragraph, use your outline or map of the passage to guide you. If your map is well organized, it can save you time by limiting the amount of scanning you must do. For example, the sample detail question above doesn't refer to a specific paragraph, so you should use your outline to give you a clue to the relevant information, or you can simply scan the whole passage. The following could be an outline for the Apollo passage:

P1: early art purpose=holy, protect, Apollo - safeguard
P2: first: A. perfect body + talent, authority
P3: last: A. weak morals ?, movement, exercise, music → intellect
P4. ↑ lifelike = ↓ public values

Based on an accurate outline of the passage, the third paragraph contains the necessary information related to the final depictions of Apollo.

Strategy 3: Read the relevant information closely and carefully.

Once you've found the key idea(s), read the sentence and surrounding sentences slowly. Don't skim them. Students often make avoidable mistakes by reading the passage too quickly and carelessly for detail questions. As you will learn below, detail questions contain distracters based on rearrangements of the ideas in the passage. In order to avoid confusion and error, pay close attention to the actors, actions, states, and their relationship. Identify the correct relationship of cause, effect, reason, intention, and so on. Also, a correct choice may paraphrase the information in more than one sentence.

Strategy 4: Look for accurate paraphrasing in the correct answer(s).

A correct answer for a detail question is usually paraphrased to some degree. Although certain ideas may not be restated, some of the information from the passage will certainly be in different words or sentence structure. As you will learn below, distracters often repeat the exact vocabulary from the passage. Correct answers restate some or all of the information from the passage. For instance, (D) is the correct answer to the detail question above, and it paraphrases the key point in Paragraph 3: the later statues of Apollo showed Apollo's mental superiority as well as his physical perfection.

Strategy 5: Choose an answer based on stated information, not inferences or assumptions.

Detail questions ask about stated, or written, information. Therefore, you should be able to find the correct answer choice(s) restated somewhere in the passage. Don't choose an answer based on an idea that you think is true but can't find in the passage.

Question Forms

Fact/detail questions ask about specific information related to people, places, things, times, methods, reasons, events, statements, etc. They may include interrogative pronouns such as *who, whom, what*, but they may not. You can recognize detail questions based on the following forms or some variation:

According to the passage, X was…
According to paragraph 1, X did Y because…
According to Paragraph 2, why did X do Y?

Which of the following statements about…is supported by paragraph X?

What does the author say about…?

The author's description of X mentions which of the following?

Distracters

Distracters are incorrect answers that look correct, but they are actually incorrect. To distract means to draw someone's attention away from something; in this case the distracter takes your attention away from the correct answer and encourages you to choose a wrong one. The multiple-choice questions on the TOEFL are designed to be challenging, not easy or obvious. Detail questions that require one correct answer have three distracters; detail questions that require two correct choices have two distracters.

Regardless of the type, the distracters for detail questions follow a certain pattern. First, distracters in detail questions usually repeat vocabulary exactly as it appears in the passage. Since you know that a correct answer choice paraphrases all or most of the information from the passage, repetition is a strong clue that a choice could be wrong. Repetition is also common in distracters for most other types of questions as well, so don't be distracted by repetition. However, some vocabulary is not easily paraphrased, so repetition is a clue, but not proof, of a distracter. You should always check an answer choice with the passage. Review the following strategies to become more familiar with the various types of distracter.

Distracter 1: Answer choice includes unmentioned information.
This is perhaps the easiest distracter to recognize and avoid, but it usually includes some repeated idea(s) from the passage with the unmentioned idea(s). If you always check your choice with the information in the passage, you should be able to avoid this distracter because you won't be able to find all of the idea(s) from this choice in the passage. Choice (A) is an example of this type. There is no judgment or measurement of the quality of any of the statues. However, some readers might be distracted by the repetition of *serious*, which is used to describe the earlier statues. Also, a reader could incorrectly assume that the representation of inner and outer features in the later statues indicates better quality, but this assumption is not supported by any statements in the passage. Often, this distracter is based on incorrect assumptions.

Distracter 2: Answer choice refers to ideas mentioned in the passage but related to a different key point.
This type of distracter can trick many students because it uses ideas from the passage, but the ideas don't relate to the question. This is why it is important to identify key ideas in the question, and then read the passage carefully once you locate the information in the passage. You need to make sure that you find the right information, not just any information. Choice (B) is an example of this type. The warning is mentioned in relation to the earlier statues and in relation to certain of those statues, not all of them, but the question asks about the later or last statues of Apollo.

Distracter 3: Answer choice rearranges stated ideas incorrectly.

A distracter can take ideas written in the passage and change their relationship to each other. This distracter type can reverse a cause and effect relationship, reorder the sequence, interchange details (time, place, etc.), among many other possibilities. For instance, (C) in the detail question above rearranges stated ideas from the passage. The answer choice states that the statues might have caused, or led to, weakening cultural morals, but according to the passage, the weakened morality of society was a possible cause of the new statues. The distracter reverses the correct cause and effect relationship.

Distracter 4: Answer choice reverses the positive or negative quality of an action or state.

This kind of distracter states the correct ideas from the passage, but changes them from positive to negative or from negative to positive. The incorrect choice could remove or add *no* or *not* to the ideas, or it could use an antonym. As always, you must read the relevant sentence(s) closely to avoid any distracter.

Practice

Now, practice the strategies you've just reviewed by answering several detail questions. Read actively, pay attention to main idea, purpose, and key points, and record the basic outline of the passage in organized notes.

Homeschooling

American parents today are faced with a stark choice. The country's public schools are becoming more crowded, more violent, and less effective in preparing children for employment or college. Private schools may be too expensive or unavailable. To ensure that their children receive an adequate education, an increasing number of parents are simply teaching their children at home. While homeschooling offers many benefits to both child and parent, its three most important advantages are its flexibility of curriculum, its adaptability to different learning styles and speeds, and its more positive, supportive social environment.

If, for example, the child is interested in dinosaurs, that subject could be used to teach scientific concepts in geology, biology, or even history. Moreover, in the home environment, there is plenty of room for spontaneous discussion, field trips, and other learning experiences that classroom logistics make difficult, expensive, or challenging. Homeschooling puts the child's natural curiosity to use, limited only by the imaginations of the child and parent.

Children can move through the material at a rate that challenges them positively. In the conventional classroom, most lessons are aimed at the middle level of ability. Thus, some students are rushed along much faster than is optimal, or faster than necessary for satisfactory results, while others yawn or find distractions because the pace is too slow. Nor can a teacher pay much attention to any single student in a classroom of 30 or discover how individual students learn best. But the parent at home, who knows the child better than any teacher, can readily make adjustments to content, teaching strategy, or pace, as the child requires.

The final important advantage of homeschooling lies in the socialization children are able to receive. Homeschooled children are less subject to the stresses and pressures experienced by conventional students who spend six, seven, or eight hours a day with their peers. They are less likely to become involved with gangs or drugs. On the other hand, homeschooled children spend much more time in the company of appropriate role models: parents, other adults, and older siblings. In this environment, they are better able to learn from actual life situations, and how to interact with people of all ages. In particular, homeschooling fosters healthy family relationships because both children and their parents are able to play larger and more complete roles in one another's lives.

If both parents work out of the home, care must be found for young children while the parents are away. Indeed, working parents may be unable to find the time to provide schooling for their children at all, and hiring a tutor to fill that role is an expensive proposition. Second, parents may be attacked for choosing what many people feel is an antisocial or elitist option—for thinking that their children are better than anyone else's, for refusing to participate in an important social institution, or even for trying to destroy public schools by depriving them of students and funding. Third, not all parents will be comfortable in the role of teacher. They may not have the patience required, the basic knowledge of the material, or the energy to encourage and motivate their children when necessary.

Homeschooling is not a panacea for the institutional deficiencies found in American public schools; these can only be addressed through a large-scale restructuring of public education policies nationwide. Nevertheless, homeschooling offers a number of significant advantages to parents and children. And it works. Homeschooled children, on average, place in the 87th percentile on standardized exams—the national average is the 50th percentile—and have been admitted to all major universities and military academies in the country. Clearly, homeschooling is a serious, positive alternative for motivated parents and their children.

1. According to the passage, one of the benefits of homeschooling is that the environment

   ○ enables children to develop skills that will make them good parents.

   ○ prepares students for being self-directed members of the workforce.

   ○ gives the child ready access to books and other educational materials.

   ○ provides more flexibility in the choice of learning activities

2. Which of the following is mentioned as a problem with today's public schools?

   ○ They are a frequent recruitment ground for violent gangs.

   ○ They overly emphasize the importance of standardized exams.

   ○ They have too many students and are very distracting.

   ○ They structure their lessons for children at the lowest level of ability.

3. The author's description of a parent's homeschooling role mentions the importance of parents

- ○ being trained to use appropriate teaching methods.
- ○ challenging children to work at increasingly higher levels.
- ○ modifying the instruction to suit the needs of the child
- ○ providing alternative opportunities for children to interact with their peers.

Question Type 2: NOT/EXCEPT Questions

NOT/EXCEPT questions ask about factual information from the passage; however, you must then choose the answer that is not true or the answer that is not mentioned in the passage. In other words, NOT/EXCEPT questions are the inverse, or opposite, of detail questions. Although both questions discuss detailed information from the passage, NOT/EXCEPT questions ask you to recognize and choose answers that contradict or contrast with the passage. Like all questions in the Reading section of the TOEFL, NOT/EXCEPT questions are multiple choice. There are four choices, and only one choice is correct. The other three are distracters, or wrong answers.

Since NOT/EXCEPT questions involve factual information, the strategies are very similar to those for fact/detail questions. First, read the following short excerpt about the music form called rock and roll, then review the strategies.

Rock and Roll

Rock and roll is a form of music that was invented in the United States in the 1950s. It has become popular in the United States, Europe, and many other parts of the world. African-American performers like Little Richard, Fats Domino, Ray Charles, and Big Joe Turner were among the first people to come out with true rock and roll, a combination of various elements from country and western, gospel, rhythm and blues, and jazz. The influences of early performers like bluesman Muddy Waters, gospel performer Ruth Brown, and jazz musician Louis Jordan on rock and roll are still felt today. For example, the songs of early country legend Hank Williams affected musicians from early rock star Buddy Holly to 1980s rocker Bruce Springsteen.

At first, only a small but growing segment of the population had heard the new musical form. In the segregated 1950s, African-American musical forms were not initially considered appropriate for white audiences. Therefore, much of the U.S. population had not been exposed to them. All that changed, however when, in 1953, Cleveland disc jockey Alan Freed began to play rhythm and blues to a largely non-African-American radio audience. Freed was successful and sold many records. The music spread, and the term that Freed had adopted for the music—rock and roll—began to spread as well.

Teenagers, and the money they were willing to spend on records, provided an impetus for rock and roll. On their way to becoming rock stars, many performers copied songs from the original artists. For instance, Pat Boone scored a hit with a toned-down version of Little Richard's song, "Tutti Frutti," causing Little Richard to comment, "He goes and outsells me with my song that I wrote." Elvis Presley's first television appearance in January 1956 marks rock and roll's ascension into the world of pop music.

Strategy 1: Identify the question type based on the clue words NOT or EXCEPT.

The words NOT and EXCEPT are always in capital letters to emphasize the type of question. Therefore, if you read the question carefully, you should recognize them easily. It's important to emphasize the peculiar nature of NOT/EXCEPT questions. The correct answer is actually not mentioned in the passage, or it is different from what is written in the passage. As the clue words suggest, the correct answer to a NOT/EXCEPT question is not true in relation to the passage. For example, read the following sample question based on the passage above:

> Which of the following is NOT mentioned in the passage as a factor in the commercial success of early rock and roll?
>
> ○ The use of different broadcast media
> ○ The purchasing power of young enthusiasts
> ○ The charismatic personality of disc jockey Alan Freed
> ○ The exposure of a non-African-American audience to African-American musical forms

This question type can be confusing. Three of the four answer choices are accurate restatements of ideas in the passage. If you miss or forget the significance of *NOT* or *EXCEPT*, you may connect one of these answer choices with information in the passage and choose that answer, which would be a mistake. The correct choice is the one that is unmentioned or inaccurately paraphrased.

Strategy 2: Identify a key idea or set of ideas in the question.

The word NOT or EXCEPT identifies the question type, but you must still find clues to the relevant sentence(s) in the passage. So, as you did for detail questions, you must choose key ideas (words, a phrase, or a clause) for which you will scan the passage. For example, the question mentions the key words *factor*, *commercial success*, and *early*. *Rock and roll* can't be considered too useful since it is part of the main topic of the passage. The term *rock and roll* is repeated too often, but the nouns *factor* and *success* are far more useful in identifying particular parts of the passage. However, remember that you must look for the ideas, not the exact words, since the passage will contain restatements.

Strategy 3: Identify a key idea or set of ideas in each answer choice if the question has none.

As is the case with detail questions, a NOT/EXCEPT question could include no key ideas. For example, the preceding question could be rewritten as follows:

> All of the following are mentioned in the passage EXCEPT
>
> ○ The use of different broadcast media
> ○ The purchasing power of young enthusiasts
> ○ The charismatic personality of disc jockey Alan Freed
> ○ The exposure of a non-African-American audience to African-American musical forms

As you can see, the question itself gives no clue about what idea(s) to look for in the passage or where to look. Understandably, this makes the question more difficult to answer, for you must identify key ideas in each answer choice and scan for them individually.

Strategy 4: Scan the passage for the key idea(s).

First, use your outline or passage map as a guide. Then scan the passage to locate the relevant information based on the key ideas from the question and/or the answer choices. The following could be a possible outline of the excerpt from the rock and roll passage:

> P1: rock and roll, popular, combination of styles, influences
>
> P2: not all, change = Freed R&B to white Americans
>
> P3: teens, money, stars copied songs, Elvis on TV = pop music

The necessary information might be located in one paragraph, or it could be located in several different paragraphs throughout. The outline above indicates that the information about rock and roll's success is probably in the second and third paragraphs. If you don't find the information immediately, scan the whole passage as well as use your memory of your first reading to guide you.

Strategy 5: Read the relevant sentences closely.

Read the passage closely, not quickly, when you find ideas related to the question. Don't skim the passage, because you need to pay attention to the particular details related to identity, place, time, direction, sequence, reason/cause, effect/result, intention, number, and so on. Check the information in the passage with each answer choice, and eliminate any choice that accurately paraphrases the passage.

Strategy 6: Choose the answer with missing, altered, or contradictory ideas.

The best choice for NOT/EXCEPT questions is the one that does not support or reflect information in the passage. The correct choice for this question type resembles one or several of the distracters for fact/detail questions. The correct choice could rearrange detail (actors, actions, states, objects, etc) or it could add information that isn't in the passage or that is the opposite of the information in the passage. For example, the correct answer to the sample question above is (C). Although Alan Freed is mentioned, the passage mentions nothing about his personality.

Question Forms

You can recognize NOT/EXCEPT questions based on the clue words and the following forms or some variation of them:

> Which of the following is NOT mentioned?
>
> According to paragraph 1, which of the following is NOT true of X?
>
> All of the following are mentioned in paragraph 1 …EXCEPT…
>
> The author makes all of the following statements about… EXCEPT…
>
> In Paragraph 2, the author mentions all of the following EXCEPT…

Distracters

Distracters are incorrect answers that appear correct when they are actually incorrect. Each NOT/EXCEPT question is a multiple-choice question with four choices; only one is correct and the other three are distracters. Incorrect answer choices for this question accurately reflect information from the passage (remember that the correct choice for this question type does NOT match the information in the passage). Therefore, the only type of distracter for NOT/EXCEPT questions is a challenging but correct paraphrase of the passage. All good distracters for this question type try to confuse you by making an answer choice seem like it doesn't reflect the passage when it actually does. This is one of the many reasons why careful, active reading is so vital to the test, especially for NOT/EXCEPT questions.

Distracter 1: Answer choice accurately paraphrases the passage, using correct but unmentioned vocabulary.

A paraphrase that uses correct but new vocabulary can be confusing. Choice (B) paraphrases the passage with synonyms that don't resemble the vocabulary in the passage but have the same meaning. Look at the following restatement of one factor in the success of rock and roll:

Original Statement	Paraphrased Answer Choice
Teenagers, and the money they were willing to spend on records, provided an impetus for rock and roll.	The purchasing power of young enthusiasts

In this context, the teenagers are not only young but also enthusiasts, or fans, of rock and roll. Moreover, money is the same as purchasing power since the teenagers used their money to purchase records. This example demonstrates the importance of context to vocabulary and paraphrasing.

Distracter 2: Answer choice accurately paraphrases two or more separate ideas together.

A NOT/EXCEPT question can involve ideas from throughout a passage, not just from one sentence or paragraph. This type of distracter can paraphrase these distinct ideas in one answer choice. The answer choice uses a more general term for these distinct ideas, which could confuse you since this one general idea is not discussed on its own. Look at the following paraphrase of two ideas from the rock and roll passage. Radio and television are mentioned in different parts of the passage, but both are examples of broadcast media:

Original Statements	Paraphrased Answer Choice
1. All that changed, however when, in 1953, Cleveland disc jockey Alan Freed began to play rhythm and blues to a largely non-African-American radio audience.	The use of different broadcast media
2. Elvis Presley's first television appearance in January 1956 marks rock and roll's ascension into the world of pop music.	

Practice

Now, practice the strategies you've just learned. Read the following passage and answer the NOT/EXCEPT questions that follow. They are followed by detail questions for you to review as well.

The Electoral College

Among the democracies of the world, thé United States is distinguished by the manner in which its people select the country's head of state. Neither a parliamentary system like that of the United Kingdom or Japan, nor a system of direct popular vote as in France or South Korea, the Electoral College used in the United States is complex, anachronistic, and a handicap to the democratic process. Some people argue that the elimination of the College is necessary to bring the United States into the world of modern democracy, with an energetic, involved electorate and presidents who are in touch with the needs and wants of the citizens who vote for them.

The great complexity of the current system has the unfortunate consequence of blinding most citizens to its workings. In effect, the Electoral College makes the presidential election into a two-stage process. Each of the 50 states is allotted a number of electoral votes, which are used only to elect the president. These votes correspond to the number of that state's Congressional members: two for each state's two senators, and a variable number for each state's representatives, for a total of 538. As a result, states with small populations, like Alaska and Vermont, may have only three or four electors, while large states like California, Texas, or New York may have dozens. On Election Day, each state holds its own presidential vote, making the race into 50 little mini-elections. First, within each state, one presidential candidate wins the popular vote, which is the vote by the citizens. The winner chosen by the people is usually awarded all of that state's electors. The ultimate victor is the candidate who wins the largest number of electoral votes nationwide. Therefore, it's important for a candidate to win the popular vote in states with the most electoral votes.

Why was such a complex and problematic system ever imposed in.the first place? The answer lies in the origins of the American federal system. When the country was established, there was relatively little sense of national identity. People identified themselves as citizens of their states first, as Americans second. Each state functioned a lot like an independent country, and so it made sense to make decisions that affected the entire nation at the state level. Furthermore, even in its earliest days, the United States was a very large country, stretching over 1,600 kilometers of coastline. Communication and transportation systems between disparate parts of the country were extremely poor, and so running campaigns nationally, rather than on a state-by-state basis, would have been quite difficult. So the Electoral College was provided as a solution.

But neither of these factors is any longer the case. Americans have developed a very strong sense of national identity and demand to play a direct role in the selection of their leaders. Mass media and powerful party organizations make national political campaigns easy to conduct. But there are further problems with the Electoral College system. Because presidential candidates know that they only need electoral votes, not popular votes, they avoid campaigning in small states, or states where they know their opponents are likely to win, creating a gulf between themselves and a significant fraction of the electorate. Furthermore, many members of political minorities don't bother to vote at all, because they know that the candidate they support won't win in their state anyway. Both situations have the effect of reducing citizen representation and form obstacles to a healthy democracy. The final problem with the electoral system is by far the largest one. Because of its "winner-take-all" nature, the Electoral College can actually elect a candidate who received fewer popular votes than the opposition, altogether thwarting the purpose of holding an election in the first place. This unfortunate circumstance has in fact come about several times in the nation's history, most recently in the 2000 election of George W. Bush.

Its original justifications outmoded, its operations inscrutable, and its effects at odds with the goals of a democracy, the Electoral College is an institution that some would like to abandon. In its place, the United States should adopt a modern system of electing the president, one that will promote, not discourage, the full participation of all citizens. Such a method will remind our presidential candidates that it is the peoples' voices that matter most.

4. Which of the following is NOT mentioned in the passage as being a problem with the Electoral College system?

   ○ There is a lack of connection between candidates and citizens.

   ○ Many members of political minorities do not vote.

   ○ Presidential candidates don't campaign in small states.

   ○ Larger states have more influence over national policy smaller states.

5. In Paragraph 3, the author's description of the United States at the time of the nation's founding mentions all of the following, EXCEPT

   ○ The independence of each state

   ○ The size of the nation's population

   ○ The country's transportation network

   ○ The nation's communication system

6. The author makes all the following statements about modern Americans, EXCEPT

  ○ They are loyal to their political parties.

  ○ They have a strong sense of national identity.

  ○ They want to be directly involved in choosing their leaders.

  ○ They do not understand how the electoral system functions.

7. Which of the following details supports the main idea?

  ○ Each of the 50 states is allotted a number of electoral votes corresponding to the size of that state's congressional delegation.

  ○ Even in its earliest days, the United States was a very large country, stretching over 1,600 kilometers of coastline.

  ○ Because of its "winner-take-all" nature, the Electoral College can actually elect a candidate who received fewer popular votes than the opposition.

  ○ Those who would cling to the Electoral College are motivated by self-interest or by a misguided sense of tradition.

8. What does the author state about the Electoral College?

  ○ It retains the support of the major political parties.

  ○ It has been a source of political controversy since its creation.

  ○ It has been adopted as a basis for electoral systems in other countries.

  ○ It was created to overcome the difficulties of running a national campaign.

Question Type 3: Referent Questions

Referent questions ask you to identify the correct antecedent of a referent. A referent is a pronoun, which replaces another word, a phrase, or a clause. A pronoun replaces, or represents, an antecedent. In the following sentence, the pronoun *it* replaces the noun *the sun*:

  The sun changes color as *it* sets.

It is the referent and *the sun* is the antecedent. The following chart lists the various types of pronouns:

Personal Pronouns	Relative Pronouns (Conjunctions)	Demonstrative Pronouns	*Adjectives
he, she, we, it, they him, her, us, it, them his, her, our, its, their his, hers, ours, its, theirs himself, herself, itself, ourselves, themselves	who, whom, whose, which, when, where, that	singular: this, that plural: these, those	each, every, little, few, some, any, all, most, many, much, more another, the other, others, few others, all others, most others, some others, one, ones, one other, two others, etc *These adjectives may also function as pronouns.

On the test, there are at most two referent questions for each passage. It's possible that one passage may not have one at all, but other passages will. In these questions, the referent is always highlighted in the passage, and the paragraph is also sometimes identified in the question by a number, so you don't need to scan the passage. Referent questions are multiple-choice questions. There are four choices, and you must choose one.

The strategies below explain the structure, information, and steps necessary to answer these questions correctly. Read the following excerpt on road rage. The excerpt contains a sample referent question used to explain the strategies for this question type:

Road Rage

According to some accounts, the term *road rage* was initially used in London around 1994. The word officially entered the English vocabulary in 1997, when the Oxford English Dictionary defined road rage as "a violent anger caused by the stress and frustration of driving in heavy traffic." However, according to psychologists, this is not completely accurate. Road rage is certainly an expression of anger, but it is not caused by stressful traffic alone. When drivers are frustrated in traffic, they can still decide how they will respond. Road rage comes from a choice to retaliate, and certainly some drivers in this situation need to learn how to make a more peaceful one.

First, it is clear that road rage has become a serious problem. The American Automobile Association's Foundation for Traffic Safety reported in 1996 that the average number of violent incidents had increased 51 percent since 1990. In 1996, police nationwide reported about 2,000 accidents in which road rage was cited as a factor. However, many people believe that more incidents of road rage occur than ever get reported. The problem is being studied by government agencies at many levels, and a number of states have already considered legislation to help correct it. Many experts feel that, while some of these remedies may help, reeducation and a change in attitude are also really needed. It is clear that steps should be taken to find a solution to this escalating problem.

Strategy 1: Be familiar with the grammar for pronouns and adjective clauses.
Be aware of how pronouns can and can't be used grammatically in order to guide your choice of antecedent. Referents follow specific grammar rules, which can help you choose the correct antecedent. However, you must always check the meaning as well as the grammar since the grammar alone cannot indicate the correct choice. The chart below includes some, but not all, of the relevant rules:

Personal Pronouns

Rules	Examples (referent and antecedent in *italics*)
1. The form of the personal pronoun matches the function.	Subject: he, she, it, we, they Object: him, her, it, us, them Possessive: his, her, its, our, their Reflexive: himself, herself, itself, ourselves, themselves
2. *It* and related forms must refer to a singular animal, thing, or place, not a person.	The *bear* was shot with a tranquilizer to remove *it* safely from the area. The brochure recommends *Vancouver* for *its* beauty. The *sun* can damage the skin most when *it* is directly overhead.
3. *They* and related forms can refer to all plural nouns (people, things, animals, etc.)	The rangers captured the *bears* to protect *them*. *Visitors* to Vancouver have many choices as long as *they* enjoy the outdoors. The sun's *rays* are brightest when *they* are directly overhead.
4. A self-reflexive (himself, itself, etc.) is used when the object of a verb or phrase is the same as the subject.	A *bear* looking for food in the garbage might attack someone to protect *itself*. The *ranger* killed the bear to protect *herself*. Due to the harmful UV rays, *people* should protect *themselves* in the sun. Thanks to Vancouver's many attractions, *visitors* enjoy *themselves* immensely.

Relative Pronouns

Rules	Examples (referent and antecedent in *italics*)
1. A relative pronoun has two functions: it replaces a noun, phrase, or clause, and it is the conjunction of an adjective clause.	The state of Nevada passed an antismoking law last year *that* took effect in December 2006. Studies have shown that secondhand smoke harms *children, whom* lawmakers want to protect the most. Smoking inside public buildings is banned due to *secondhand smoke, which* can harm nonsmokers. The government argues *that public smoking harms nonsmokers, which* many studies support.
2. *Who* and *whom* must modify a person.	Some people believe that U.S. voters may not approve of *candidates who* smoke cigarettes. Several previous *presidents* of the U.S., *who* were photographed and filmed less frequently in the past, smoked openly. *Bill Clinton, whom* voters elected twice, enjoys smoking cigars occasionally.
3A. When *which* modifies a noun, it cannot be a person. 3B. *Which* can also modify a whole clause (a subject and verb)	Many U.S. states have passed *laws that* ban smoking in public buildings. *The federal government refuses to pass national legislation, which* angers some antismoking activists.
4. *That* can modify a person, animal, or thing.	22 U.S. states have passed *a law that* bans smoking in public buildings. *Bears that* attack are often defending their young. Some people believe that U.S. voters may not elect a Presidential *candidate that* smokes cigarettes.
5. *A relative pronoun cannot replace/modify a noun on the opposite or far side of a verb. (*Very important)	Some people believe that U.S. voters may not approve of *candidates who* smoke cigarettes. Explanation: *Who* cannot replace/modify *voters* or *people* because both words are on the wrong side of the verb *approve*. *Who* can only replace a noun on the right side of the verb *approve*.

Demonstrative Pronouns

Rules	Examples
This and *that* alone usually don't replace a person, only an idea/opinion or entire clause. *These* and *those* can replace any plural nouns, ideas, opinions, etc. *All four demonstrative pronouns always refer backwards; they always replace preceding words, phrases, or clauses.	*Governments at many levels have passed antismoking laws,* but *this* hasn't stopped some smokers from fighting. *Many famous U.S politicians,* such as Roosevelt, *smoked cigarettes,* and *that* never hurt their careers in the past. In *the 1950s,* a majority of American adults smoked cigarettes, but *those* were different times.

Adjectives

Rules	Examples
1. Each, every, few, any, and many must modify/replace a countable noun: Each/every + singular noun Few/many + plural noun	Most *members* of Congress belong to one of the two major parties, but *each* can vote independently. Although *members* of Congress can cast votes according to their own beliefs, *few* actually oppose the consensus of their own party.
2. Little and much must modify an uncountable noun.	The sun produces huge amounts of *radiation,* but *little* actually reaches the surface thanks to the ozone layer.
3. Some, all, and most can replace countable or uncountable nouns.	Although new legislation affects the lives of all *citizens,* only *some* actually read new laws.
4A. *Another* replaces a singular, countable noun. 4B. *Others/the others, etc* replace a plural, countable noun. 4C. *The other* replaces a singular, countable noun.	When your car is stopped by the police, one *officer* will approach your car while *another* stays farther behind. 22 U.S. *states* have antismoking legislation while *many others* are considering it. The president and vice-president always fly on separate *planes; one* carries the president and *the other* is reserved for the vice-president.

Strategy 2: Carefully read the sentence with the referent as well as the surrounding sentences.
The question specifies a particular referent, which is highlighted in the passage, and possibly a paragraph as well. Look at the following sample question related to the passage on road rage:

Look at the word **this** in paragraph 1. The word this in the passage refers to

- ○ An expression of anger
- ○ The Oxford English Dictionary
- ○ The first appearance of road rage in London
- ○ The definition of the term *road rage*

Once you locate the highlighted referent in the passage, read more than just one sentence. Read also the sentence before and after the sentence containing the highlighted word. Although the four choices might not be in all surrounding sentences, you should make sure you understand the full context, not just one part.

There could be clues to meaning in the surrounding sentences even if some choices aren't there, so don't skim the sentences. Students often read too quickly and carelessly due to nervousness and pressure. Referent questions require careful attention and analysis. In the excerpt above about road rage, the highlighted pronoun *this* is the referent, and you should read at least from the beginning of the previous sentence (the word officially entered…) until the end of the next one (…by stressful traffic alone).

Strategy 3: Pay attention to the pronoun's number, gender, and type (person or thing).
Since referent questions are multiple-choice, you need to eliminate obviously wrong answers. The specific qualities of the pronoun are clear indications of which antecedents might be incorrect. Therefore, identify whether the pronoun is singular or plural, male or female, and so forth. For example, the pronoun *this* is singular and probably refers to a noun thing or idea, not a person, based on the known rules outlined above.

Strategy 4: Identify the function of the referent and its immediate context.
You need to know how the pronoun relates to its own sentence. The immediate context is the action(s), state(s), and related detail (place, time, reason, contrast, etc.) within the sentence. Determine the pronoun's relationship to those ideas. Is the pronoun a subject or object? If it is a subject, what is the action or state and what or whom is it affecting? If it is an object, what action is it experiencing or receiving? Is the sentence a continuation of previous ideas (moreover, furthermore) or is it a contradiction of them (however, although, etc.).

The pronoun *this* is a subject that is described as inaccurate, so the correct antecedent must be something that can be inaccurate. Moreover, the inaccuracy is the judgment or opinion of a specific group, psychologists. Therefore, the antecedent must be something that psychologists could discuss or refer to. Finally, the sentence begins with *but*, which expresses contrast. The contrasting word fits the negative tone of the adjective *inaccurate* and suggests disagreement or debate. All of these details (inaccurate, psychologists, but) are part of the pronoun's immediate context and they are important clues to the proper antecedent.

Strategy 5: Relate the referent to the broader context.

The broader context refers to the actions, states, actors, objects, and details (places, times, reasons, etc.) in the surrounding sentences. Ask yourself, "What is happening in the sentence before and after the pronoun?" and "Who or what is causing it?" Determine how the referent affects or is affected by these ideas around it. For instance, the pronoun *this* in the sample question comes after a description of the first use of the term *road rage* and a quoted definition from the Oxford English dictionary. The pronoun comes before a more specific explanation of what road rage involves. Therefore, the pronoun *this*, which is described as inaccurate, comes in between two sentences that try to define and then redefine road rage.

Strategy 6: Use cohesive devices to understand the context accurately.

Cohesion means union or connection, and cohesive devices are words or phrases that bind, or connect, ideas together. They help the writer (and reader) move from one set of ideas to the next. Cohesive devices include pronouns as well as other parts of speech, and they are important for many question types: referent, inference, rhetorical structure, coherence, and paraphrasing questions. The following chart reviews the basic cohesive devices in a reading passage (and a lecture).

The sample referent question above involves a transitional adverb *however*, which indicates contrast or opposition. Therefore, the referent is involved in a contradiction or refutation of an earlier statement.

Cohesive Device	Example (cohesive device in *italics*)
*Pronouns: personal, demonstrative, relative, etc. *See above for a full discussion of pronouns.	Today's teachers are faced with a stark choice. *They* must decide whether to teach in a way that helps students pass standardized exams or teach in way that actually helps students learn. OR Speaking a foreign language is an enormous asset in the modern world. *This* is why so many junior high and elementary schools are now offering language classes.
Articles: a, an, the, some. Commonly, a noun is general when it is first mentioned, and then it becomes specific every time it is repeated after. The use of general and specific articles is also combined with repetition or synonyms (see below).	Distractions in the typical classroom make it difficult for many students to focus on their studies. *The* distractions that tend to be universal are desire to socialize with classmates or to play with toys they have brought from home.

Transitional phrases/transitional adverbs Time: then, next, later, finally, etc. Cause: therefore, as a result Contrast: however, on the other hand Definition: that is, in other words Example: for instance, for example	Peers and older siblings have a major influence on how schoolchildren behave. *For instance*, young children often repeat bad words they hear from their friends or brothers and sisters.
Repetition, synonym, or slight variation of a word	To foster healthy relationships between children in a classroom, teachers should provide time for games and fun activities. Having healthy *relationships* with their peers helps children gain confidence useful in other parts of their lives. OR Before choosing a public or a private school for their children, parents should take into account the cost. The *price* of a private school is usually much higher than that of a public school. OR Many people feel that money is a panacea for our educational problems. Unfortunately, the *problems in education* we now face are too extensive to be solved by money alone.

Strategy 7: Look for an antecedent before the referent in the same sentence or in the preceding one.

As the chart above outlines, most antecedents come before the pronoun, so you should focus your attention on the choices that precede, or come before, the pronoun. The correct antecedent could be in the same sentence as the pronoun or in the preceding sentence, but it can't be two or more sentences away. In the excerpt above, (C) is in the first sentence, which is two sentences away from the pronoun *this*. That means you can eliminate (C), which is too far away.

Strategy 8: Look for an antecedent after the pronoun in specific cases.

A pronoun may precede the antecedent (a word, phrase, or clause) when both are in the same sentence and when the sentence has very specific structure. When the sentence begins with an adverb clause, a pronoun in the adverb clause can come before the antecedent in the second, independent clause:

Although *it* can be hard to explain, *road rage* is not hard to recognize.

Since *they* began keeping records on road rage, *government agencies* have seen a continual rise in the number of cases.

Almost as soon as *it* was defined, experts began to debate the root causes of *road rage*.

When *he* arrived on the scene, *the police officer* separated the two drivers who were fighting.

Note that the pronoun and antecedent are still in the same sentence in all four examples above. Antecedents do not normally come in a separate sentence after the pronoun because this would confuse the reader or listener. In the sample question above, the pronoun *this* does not fit the pattern above, so (C) cannot be chosen based on this specific rule.

Strategy 9: Use sentence structure to eliminate some choices.
As you can see in the table above, many pronouns can replace only words with specific types, gender, number, and so on. For example, *this* has two key patterns: First, it can only replace a preceding noun, phrase, or clause. In other words, *this* must refer back to something that comes in an earlier sentence. Therefore, you can eliminate any choices that come after the pronoun, such as (A) in the sample question. The expression of anger is mentioned in the following sentence and can't be replaced by *this*. Moreover, as you learned in Strategy 6, pronouns very rarely refer back more than one independent sentence, so any choices such as (C) in the first sentence of the excerpt can be omitted as well, based on grammar.

Strategy 10: Replace the referent with each remaining choice and check the meaning.
After you have eliminated as many choices as possible based on number, gender, and structure, you should place each remaining choice in the same place as the pronoun and see if the sentence makes sense based on the function of the pronoun and the surrounding ideas. For example, the only choice that can correctly be described as inaccurate by psychologists is the definition of the term road rage, so (D) is correct. Although (A) can be eliminated based on grammar, it can also be rejected based on meaning. An expression of anger is emotional, and subjective expressions of emotions cannot be described as inaccurate. Furthermore, (B) refers to an entire reference book (the Oxford English Dictionary) that is well respected and unlikely to be entirely inaccurate. Also, psychologists can debate the definition of one term that refers to psychological issues, but they can't have a professional opinion about an entire dictionary. Finally, (C) can be eliminated based on its distance from the pronoun and also because there is nothing to suggest that the incident is inaccurately described. Only the place and time are given, which are not very exact anyways.

Question Forms

A referent question mentions a specific pronoun in the question and possibly a particular paragraph as well. You can recognize a referent question based on the following forms or some variation:

Look at the word X in paragraph…The word X in the passage refers to…

Look at the word X. The word X in the passage refers to which of the following?

Distracters

A distracter is an incorrect answer that appears correct but is really incorrect. Each referent question is a multiple-choice question with four choices. Only one choice is correct, and the other three are distracters. By definition, a distracter confuses the reader by seeming correct; it is not an answer choice that is easily avoided or eliminated, so the following strategies don't review incorrect answers that confuse things and people or that have the wrong gender or number. For example, the pronoun *he* cannot replace the noun *book*, and the pronoun *this* cannot replace the noun *problems*. These errors are clear and don't distract, or confuse. Therefore, they need little explanation. Review the following explanations of the more challenging distracters for referent questions.

Distracter 1: Answer choice fits all the characteristics except the context.

This is certainly the most common type of distracter. All the traits of the choice are correct: the pronoun and choice match in number, gender, and grammar, and as noun things or people. Only the meaning of the sentence can tell you that this choice is incorrect. Of course, this kind of analysis requires a good vocabulary, which is one of the many reasons that you should read and study a wide range of topics. For example, the following sentence from the chart above contains several possible antecedents for the pronoun *some*:

> Although new legislation affects the lives of all citizens, only *some* actually read new laws.

The nouns *legislation*, *lives*, and *citizens* are all possible antecedents for the pronoun *some*, which can replace countable nouns (lives and citizens) and uncountable nouns (legislation). Since all three choices are nouns, they can all be subjects like *some*. However, *some* is the subject of the verb *read*, so the correct antecedent must also be able to read. Therefore, only *citizens* can be the antecedent. The relationship between the pronoun and the verb is part of the immediate context, to which you should always pay close attention.

Distracter 2: Answer choice is right next to the pronoun but doesn't fit the context and/or grammar.

Students often assume incorrectly that the closest possible choice is often the best one. Although the closest pronoun is the correct antecedent in the example for Distracter 1 above, this is not always true. Don't assume that proximity matters very much. While a pronoun cannot replace a noun, phrase, or clause that is more than one independent sentence away, it does not have to be directly next to the antecedent. The antecedent and referent can be separated by one or more phrases and clauses, depending on the structure and meaning. Look at the following sentence from the chart above:

> The problem is being looked at by government agencies at many levels, and a number of states have already considered legislation to help correct *it*.

The nouns *problem*, *number*, and *legislation* are all singular noun things and possible antecedents for the pronoun *it*. The noun *legislation* is closest to the pronoun, but it doesn't fit the context of the sentence. Based on the infinitive *to correct*, the pronoun *it* must require correction. All three nouns could be corrected in different circumstances, but there is nothing to indicate that something is wrong with the number or the legislation. Since the noun *problem* is a synonym for road rage in this context, that noun is the best choice.

Practice

Now, practice the strategies you've just learned. Read the sample passage below and then answer the referent questions that follow. Read actively and outline the passage in brief but organized notes.

Bridge History

The earliest bridges were simple beam bridges—something like a log or plank of wood laid across a stream—a basic and simple design that has existed since prehistory. What we call modern bridge technology began with the Romans.

The Romans perfected the art of building arch bridges and built them throughout the Roman Empire. An arch is the top half of a circle, and the Romans built an arch by piling stones on top of each other. In their design, large stone blocks were wedged against each other in two separate piles. As the two piles got higher, the stones got closer together until they met to form an arch. The final stone at the highest point of the arch was called the keystone, and it locked all the others in place. It was a fantastic design, capable of withstanding heavy weights and a great deal of wear and tear.

The Romans built many simple single arch bridges, but they also built bridges that used an array of arches, and even multitiered arches, which made for much sturdier bridges; in fact, many Roman bridges still stand today.

In the 12th century, the rulers of the Holy Roman Empire took over the construction and maintenance of bridges throughout Europe, for they wanted to help Catholic priests move easily throughout the countryside. In France, there was even an order of priests, the *Freres du Pont*, devoted entirely to the design and building of bridges. The French priests adorned their bridges with statues of saints and chapels that travelers could rest and worship in. And these bridge chapels were the precursors of the tollbooths that collect money on toll bridges even today.

In terms of materials, both the Romans and the 12th century Catholic priests built bridges out of concrete in addition to using stones. After the 12th century, the technology for making concrete was gradually lost and most bridges were built with bricks and mortar. Historians emphasize, however, that many wooden bridges were also built during those years, as they have been throughout all periods of bridge construction. The next big revolution in bridge technology came in the 18th and 19th centuries, with the introduction of iron and the expansion of the railroads. During the first decades of the 19th century, iron was the bridge building material of choice; however, these early railroad bridges experienced a number of spectacular failures, which led to efforts to strengthen them. Complicated trusses--systems of beams and bars--were used along with iron to provide the additional strength required for the heavier loads of trains.

By far the biggest revolution in bridge design came in the mid-19th century with the introduction of steel. By the end of the century, it had pretty much replaced all other materials as the prime material for bridge building. In addition, many bridges were built using complex truss systems, many of them based on the ideas of Gustav Eiffel—the designer of the Eiffel Tower in Paris. It is interesting to note that two of the engineering marvels of the 19th century, the Eiffel Tower and the Ferris Wheel, were both elaborations of the truss systems used in contemporary bridge design. Bridge technology has been instrumental in pushing engineering as a field.

The last major technological revolution in bridge design was John Roebling's suspension bridge. On a suspension bridge, the roadway is suspended by cables that are anchored by towers at either end of the bridge, and with supporting structures for the cables placed at regular intervals. Roebling pioneered the use of steel in suspension bridges and developed the technology to make wire cables. He designed the Brooklyn Bridge, which is a masterpiece of 19th century suspension bridge design. Truly, there was more development in the technology and materials of bridge building in the 19th century, than in the previous two or three thousand years combined.

Practice

9. Look at the word **others** in paragraph 1. The word others in the passage refers to

   ○ heavy weights
   ○ stones
   ○ piles
   ○ Romans

10. Look at the word **they** in Paragraph 3. The word they in the passage refers to

    ○ wooden bridges
    ○ periods of bridge construction
    ○ years
    ○ historians

11. Look at the word **them** in Paragraph 3. The word them in the passage refers to

    ○ the first decades
    ○ railroad bridges
    ○ failures
    ○ efforts

12. Look at the word **it** in Paragraph 4. The word it in the passage refers to

    ○ bridge design
    ○ revolution
    ○ the century
    ○ steel

Question Type 4: Vocabulary Questions

There are three to five Vocabulary questions for each reading passage. A vocabulary question tests your understanding of a particular word. The word could be a noun, verb, adjective, or adverb, and it is highlighted in the passage, so you don't have to scan or search for it. Vocabulary questions are multiple-choice, and you must choose the best synonym from a list of four choices.

Vocabulary questions do not test your knowledge of advanced terms or concepts from a specific field. Although the word is part of an academic passage, it won't be technical or specialized. Any advanced term will be referenced in the glossary function, and you can click on it to read the definition. Of course, the glossary function explains only the specialized terminology, not the highlighted words in vocabulary questions, so don't rely on the glossary function to answer any question. The glossary function only provides background for the main idea of the passage. The highlighted words for vocabulary questions are applicable to a wide range of fields, so they can be familiar to a general audience at the college or university level.

This question type tests your vocabulary, not your ability to analyze the passage. Therefore, the passage contains few clues to guide your answer. In fact, the passage may contain more distracters than accurate clues to the correct choice. The best situation for students is when they understand the vocabulary word right away, and they can choose the correct synonym without too much effort. When you are certain of the correct choice, answer the vocabulary question quickly. This saves time for the more difficult questions, which require you to scan the passage and reread sentences.

When you don't recognize the correct synonym immediately, you must analyze the parts of the vocabulary word, the choices, and the passage. The following strategies explain the kind of information to look for and analyze when you don't immediately know the correct answer based on your own knowledge. All the strategies have been explained in a logical order. However, depending on the word, passage, and your own knowledge, you may need to follow all or only some of the following steps. First read the short excerpt below that includes a sample vocabulary word; then review the strategies, which refer to the highlighted word in the excerpt:

Walt Disney—Turning Fantasy into Reality

Walt Disney, the man behind one of the most recognized names in the world, was an enterprising dreamer who visualized whimsical characters and fanciful worlds, and made them materialize. The creations of the Disney studios team of artists—particularly Mickey Mouse and Donald Duck—are watched in virtually every corner of the globe, delighting children whose parents and grandparents also grew up laughing at their antics. Disney dreamed of a magical park—a clean, safe, inviting place where children and their parents could enjoy spending a day together. This dream became a reality with Disneyland.

On July 21, 1954, Disney started building his dream park on 160 acres in Anaheim, California, not too far from Los Angeles. Disneyland was a new type of amusement park, and because nothing like it existed anywhere, everything had to be created from scratch.

At a cost of $17 million, Disney gave his park rivers, waterfalls, mountains, a fairy-tale castle, flying elephants, giant tea cups, moon rockets, and a Mississippi river boat.

The park included five distinct areas. The first was called Main Street, U.S.A., a replica of a small American town as it would have looked at the beginning of the 20th century. Adventureland, on the other hand, was meant to conjure up an exotic place far from civilization, while Frontierland recreated the pioneer days of the American West. Fantasyland depicted the world in children's storybooks—that of Sleeping Beauty, Peter Pan, and Alice in Wonderland. Tomorrowland, as its name suggests, represented the world of the future, and the scientific and technological wonders to come.

Strategy 1: Locate the highlighted word and read its sentence carefully.

Students sometimes don't read the passage carefully because the four choices are not in the passage. This is a mistake because the passage may contain context clues that can help you eliminate some choices. Therefore, always find the highlighted word in the passage and read the passage. Relate the vocabulary word to the surrounding actions, states, and so on. For example, look at the following sample vocabulary question:

Look at the word **antics** in paragraph 1. The word antics in the passage is closest in meaning to

- ○ dreams
- ○ toys
- ○ traditions
- ○ behaviors

Strategy 2: Identify the part of speech, function, and immediate context of the highlighted word.

All four choices will be the same part of speech as the highlighted word, so the part of speech alone will not help you eliminate any choices. It will, however, help you better understand how the word relates to the immediate context, which includes the actions, states, and details (descriptions, explanations, places) in the sentence. If the word is an adjective, then determine what it is describing. If the word is an adverb, then determine what action, adjective, or other adverb is being modified. For example, the highlighted word *antics* is a noun and it is the object of the preposition *at* in the participle phrase *laughing at their antics*.

The possessive adjective *their* is an important clue; it connects the noun to people or things that can explain the meaning of the noun. The possibilities are *creations*, *children*, *parents*, and *grandparents*. As you've already learned, determining the correct antecedent is a combination of meaning and logic. That phrase follows the verb *grew up*, and the phrase indicates the action (laughing) that occurred at the same time as the verb (grew up). Therefore, the subjects of the verb *grew up* (parents and grandparents) were engaged in an action (laughing) at or during the same time as the verb *grew up*. Based on the structure and meaning of the sentence, the highlighted word *antics* was the object, and cause, of this laughter.

Also indicates that the parents and grandparents did the same thing as the children (laughing) and the similarity between *delighting* and *laughing* is another clue that all enjoyed the antics. *Delighting* is a participle that refers to an action simultaneous to that of the previous sentence, whose subject is *creations*. Moreover, logically parents and grandparents can't grow up while laughing at the antics of their own children, so the antics must belong to the creations. Therefore, the antecedent of *their* is the noun *creations*.

Tone or attitude is an important part of both the immediate and broader context of a word. Try to determine if the vocabulary word is positive, negative, or neutral based on the related vocabulary in the sentence. Based on the vocabulary discussed in the paragraphs above (delighting, laughing, etc.), *antics* relates to positive, entertaining, and enjoyable activities.

Strategy 3: Look at the broader context.
The broader context includes the surrounding sentences, and their actions, states, and details. Ask yourself questions such as the following: What is happening in the sentences around the vocabulary word? Who or what is causing the actions or states? Who or what is affected? How are they affected? Why? Is the highlighted word affected/modified? For example, the first sentence of the excerpt contains clues about the creations whose antics are the focus of the vocabulary question. The first sentence refers to Disney's *whimsical characters* and *fanciful worlds*. You can assume that the noun *characters* in the first sentence is the same as *creations*; making something materialize is similar to creating it. Moreover, the sentence after the highlighted word mentions the safe, clean, inviting place where Disney's characters materialized. The tone of the surrounding vocabulary (safe, clean, enjoy) is positive, which could help you eliminate any answer choices that contradict this positive tone.

Strategy 4: Look for examples and/or definitions.
In your careful reading of the surrounding sentences, look for transitions or clue words of examples (for example, for instance, such as, particularly, especially). You may be able to use some known qualities of the examples as clues to the meaning of the vocabulary word. The excerpt includes two examples: Mickey Mouse and Donald Duck. Since the examples are the creations directly related to the antics in the vocabulary question, they are excellent clues to the meaning of *antics* if you are familiar with those two characters. Finally, the vocabulary word itself will not be defined in the passage, but other related vocabulary might be defined or explained. Look for verbs of definition (be, mean, include, involve, refer to, be called, be defined as, etc.).

Strategy 5: Look for key words of contrast and comparison.
When you relate the vocabulary word to the immediate and broader context of the passage, be careful that you understand whether the word is similar or different to that context. Pay attention to transitions and any words that indicate similarity (match, mirror, reflect, like, similar to, alike, equal to, etc.) or difference (unlike, different, dissimilar, not alike, oppose/opposite/opposition, etc.). Moreover, identify how the vocabulary word is affected by these comparative or contrasting words. You might be able to better understand the vocabulary word by identifying what it is like or not like, but you must be clear what the case is. For

instance, the excerpt includes the adverb *also*, which tells you that the action of the parents and grandparents (grew up) is similar to that of the children.

Strategy 6: Use your knowledge of prefixes, suffixes, and roots.
Prefixes, suffixes, and roots are specific parts of words, and they can help you understand not only the vocabulary word and the answer choices, but they could also be useful to understanding the context. The vocabulary word and choices may not be the only unfamiliar words in the passage; you may have to guess the meaning of a word in the surrounding sentences, so it is useful if you are familiar with all three parts of a word. However, don't look for a synonym with the same prefix, suffix, or root. Most likely, this is a distracter since most synonyms do not resemble each other. For example, the verbs *destroy* and *ruin* are synonymous, yet neither word resembles the other. Also, the verbs *mislead* and *misplace* both begin with the same prefix (*mis*), but they have different meanings: *to mislead* means to confuse or trick, and *to misplace* means to lose. Therefore, you have to use the clues within each word to understand its meaning, but you can't use a similarity between two words to make the right choice (see also Distracters below for more detail on incorrect answers).

Prefixes
A prefix is an addition to the beginning of a word. Most prefixes in English, especially those in scientific terms, come from Latin and ancient Greek. Some common prefixes for verbs include *be-* (belong), *in-* (involve), *pro-* (promote), and *ob-* (observe). Unfortunately, not all prefixes clearly indicate the meaning of the word: *Believe* and *behave* both begin with the prefix *be-*, but the prefix gives little clue to the meaning of either word. More important, each word has a different meaning despite the fact that they both have the same prefix: *to believe* means to think in a certain way or to accept certain ideas, and *to behave* means to act a certain way.

However, some prefixes add a specific meaning to the word, altering its definition slightly. These prefixes are more interchangeable, meaning they can be attached to different words, and they carry their specific meaning to the new word. The following list includes some, but not all, of these moveable prefixes:

The chart below lists some of the more common **negative prefixes**. When added to an action or state, these prefixes indicate that the opposite or reverse is true.

Prefixes and Meaning	Examples
un-: reverse	untie, undo, unfurl, unwind, etc.
im-: not	immobile, immature, impatient, etc.
de-: remove, lower, take away	demote, deregulate, defuse, debase, etc.
dis-: stop/end, remove	disuse, discharge, dislodge, disjointed, etc.
anti-: against, opposed to	antivirus, antidote, etc.
mis-: not correct, wrong	mislead, misinform, miscue, misshapen, etc.

The following prefixes indicate the **starting point, direction, or destination** of an action or state.

Prefixes	Examples
sub-/under-: below	submerge, substrata, subterranean, etc.
medi-: middle, between	Mediterranean

The following prefixes indicate **when an action or state occurs.**

Prefixes and Meaning	Examples
pre-: before	predate, prehistoric, prenatal, etc.
post-: after	postpone, etc.
re-: again	reelect, reform, renew, reclaim, etc.
medi-: middle, between	medieval

These prefixes indicate the **relationship** of an action or state with other verbs, or actors, places, etc.

Prefixes and Meaning	Examples
inter-: between, among	interpersonal, intercollegiate, etc.
intra-: within, inside	intramural, etc.
extra-: outside	extracurricular, extraordinary, etc.

These prefixes indicate the **severity or degree** of an action or state, or the fact that the action or state is too extreme.

Prefixes and Meaning	Examples
out-: perform/be better than others	outlast, outperform, outlive, etc.
ultra-, mega-, super-: be extreme/a high degree	superheat, ultraviolet, megalomaniac, etc.

Suffixes

A suffix is an addition to the end of a word; the suffix changes a word's part of speech (noun, verb, adjective, adverb). Like many prefixes, suffixes are found on many different words. However, unlike some prefixes, most suffixes don't indicate very much about the meaning of a word; suffixes usually indicate word form only. The adjectives *courteous* and *spontaneous* both end in the same suffix (-eous), yet the suffix indicates only the form of the words (adjective) and nothing about their meaning: *courteous* means polite and well-mannered, and *spontaneous* means unplanned and immediate. The following is a partial list of the suffixes that suggest some of a word's meaning:

Suffixes and Meaning	Examples
-ism: an ideology, a system of beliefs/ideas	capitalism, socialism, Taoism, etc.
-ic, -eur, -er, -or, -eer, -ee, -ist: a person, a trained professional	a mechanic, an entrepreneur, a welder, an editor, an engineer, an etymologist, etc.
-ology: an academic field of study, the study of a topic	biology, archeology, paleontology, anthropology, etymology, etc.
-graphy: the use of maps, pictures, or images in the study of something	oceanography, topography, geography, etc.

Roots

The root of a word is the part of the word that doesn't change or changes very little when the word changes form (noun to verb, verb to adverb, etc.), and the root is the most useful part of the word for guessing its meaning. As you've seen, many words can share a prefix or suffix, but two unrelated words rarely share the same root. The root of a word is unique to that word and its related forms. Because roots change little, they are the most powerful clues to meaning. For example, the verbs *edit* and *editorialize* and nouns *edition, editorial,* and *editor* all have the same root (edit). Of course, all have slightly different meanings since some refer to actions (edit) while others refer to a person (editor) or thing (edition). However, as long as you know the meaning of one of the forms, such as the verb *edit* (collect, correct, and prepare printed or recorded work for publication), you can make an educated guess about the meaning of its related forms.

Etymology is the academic study of the origins and evolution of different languages, and etymologists can trace the roots of English words back in time through European languages (mostly French and German as well as Spanish and Italian), Latin, ancient Greek, and even further back to the early Indo-European languages. Because English has absorbed so much from other languages—half of all words in English either come from French or share a root with a French equivalent—people who speak a language related to English can guess the meaning of English words far more easily than people who don't speak one of those languages.

Of course, you cannot and should not try to become an etymologist to prepare for the TOEFL. However, you should always pay special attention to roots, especially if your native language does not share an etymological history with English. When you learn a new word, record some of its related word forms, and identify the root. When you see another word with the same root, use your knowledge of the original word to guess the meaning of the new one. The Latin word for *land* or *ground* is *terra*. This Latin root eventually became part of the English words *territory, territorial,* and *terrain.* Notice that the root of those related words is *terr,* which comes from the Latin. You can use your knowledge of this root and the prefix *sub* (under) to guess the meaning of the adjective *subterranean* in the following excerpt:

> For much of human history, a well has simply been a deep hole in the ground. The hole must be deep enough to reach a *subterranean* source of fresh water. Often the water was discovered by accident, perhaps while digging the foundation of a building.

Based on the context (deep hole, digging), the root (*terr-*) and prefix (*sub-*), you can make an educated guess that *subterranean* means "under ground." These strategies are useful even if subterranean is not the vocabulary word. Correctly understanding that adjective could be necessary for fully understanding the context and relating it to another word.

Strategy 6A: Use your imagination when making connections between words with similar parts.

You often need to think creatively to use word clues, such as prefixes, suffixes, and roots. Essentially, you must recognize a part of the word in the passage that you've seen elsewhere in another word in another context. Then you have to remember what that previous word means and try to associate part of the other word's meaning to the new word in the passage. Usually, word clues, such as prefixes, can suggest a similarity between words, but the similarity is based on an association of ideas in different contexts. For example, the root *audi-* is a part of many words related to the production and reception of sound: audio, audit, auditor, auditory, audition, audience, audible, and auditorium. Like many words in English, these words have multiple meanings as well (an audit can be a passive observation of a class or an active investigation of a person or company), but most of their meanings involve sound. If you know that audience refers to a group of people who listen to a speech or watch a performance, you can use that understanding when you recognize the same root in one of the other words. However, if you try to guess the meaning of another word, such as *auditorium,* based on *audience,* you can't assume that *auditorium* refers to people or listening. You must pay attention to differences of word form (adjective, noun, etc) and use context to determine related ideas, such as people, places, things, actions, and so on. Moreover, you can always think about other words that contain similar parts. *Auditorium* has the same suffix as *stadium,* which is a very large theater with spectators on all sides of a field. The similar suffix is a good clue that *auditorium* is also a place. The key difference is that a stadium is a larger theater used for sports events and concerts while an auditorium is a smaller one used for lectures and plays, and an auditorium has an audience on only three sides of the stage. Of course, students with a broad vocabulary can make more of these connections, so you should always improve your vocabulary.

Strategy 7: Use word clues and context clues to eliminate answer choices.

When you decide which answer choice to select, you must combine your understanding of the context (actions, states, actors, details, tone, etc.) and the word clues (prefixes, suffixes, roots) with your knowledge of each answer choice. Ask yourself which answer choice best fits the surrounding context. Think about the differences between the following qualities: states and actions, visible and invisible actions, desirable and undesirable events, instantaneous and prolonged actions, spontaneous and nonspontaneous events, noun things and noun people, abstract and concrete nouns, human traits and animal traits, adult behavior and childish behavior, and so forth.

For example, (A) in the sample question refers to mental processes, and dreams are not normally shared by different generations of people. Moreover, since the possessive pronoun *their* refers to *creations*, it is unlikely that children and their parents could laugh at the dreams of other people, especially the dreams of fictionalized characters. (B) might at first seem correct since children play with toys, but the pronoun *their* means that the toys would belong to the creations, not the children. Also, some students might incorrectly assume that the toys are the creations, but the examples (Mickey Mouse and Donald Duck) indicate that the characters are the creations, not the toys. Finally, according to the context, the creations are watched by people all over the world, and the use of toys alone to entertain children seems very limited and specific, so the choice is unlikely and should be compared carefully against the others. (C) might seem correct based on the implied history and past time in the surrounding sentences. According to the sentence, people have been enjoying the antics for generations, and traditions are practices that have a long history. However, based on the verb *laughing*, this choice contradicts the overall positive tone of the passage. Laughing at people's traditions seems cruel and negative. Also, based on the pronoun *their*, the traditions would belong to the fictional characters, yet real people normally have traditions. (D) is the best choice. The creations have entertained people with their funny behavior, and it is acceptable for people to laugh at this behavior.

Strategy 8: Check your final answer by putting it in the original sentence.

Besides your own vocabulary, the sentence in the passage is the best guide to the correct choice. Therefore, always read the sentence again with your best choice in the place of the highlighted word. Sometimes, you can only tell if something sounds right by reading it to yourself.

Question Forms

You can identify a vocabulary question based on the identification of a particular word from the passage. Review the following forms:

> The word X in the passage is closest in meaning to...
>
> In stating X, the author means that...

Distracters

Distracters are incorrect answers. There is only one correct choice for a vocabulary question, so the other three choices are incorrect. However, the three distracters may seem correct for various reasons. The test writers hope to draw your attention away from the correct choice through several methods. Review the following explanations of the types of distracters for vocabulary questions.

Distracter 1: Answer choice is a synonym or related word form for a word in the surrounding sentences.

This distracter is basically a restatement of a word or phrase from another part of the sentence. It's a difficult distracter for many students because sentences often include paraphrases. It is common for a writer to use a variety of vocabulary in a passage. Therefore, this distracter seems familiar to many students. Choice (A) from the sample question above is an example of this type of distracter. The noun *dreams* is directly related to the noun person *dreamer* from the first sentence, the verb *dreamed* from the third sentence, and the noun thing *dream* in the fourth. Although passages often include this kind of variety in word form, don't assume that the correct choice should resemble or match any of the restatements. In fact, as you've already seen in detail questions, incorrect answers often repeat vocabulary from the passage.

Distracter 2: Answer choice only appears to fit the context, or the choice fits part of the context.

This incorrect answer choice has some connection to one or more words in the passage, but the connection doesn't fit all the vocabulary. (B) is an example of this type. The noun *toys* seems to fit the context; children use them, and toys are positive, playful things. However, children play with toys at home or in kindergarten, but not at a theme park. Also, the possessive *their* relates toys to the creations, and it seems unlikely that Disney's characters would entertain children with toys in movies or in films. This type of distracter is one reason why you must think about the context carefully and test each vocabulary word based on as many associations as possible.

Practice

Now, practice the strategies you've just learned along with the general reading strategies. Read the passage below, and answer the vocabulary questions that follow. Also, you can review pronoun questions for the same passage.

The Processes of Diffusion and Osmosis

Diffusion is defined as a type of transport phenomenon, that is, a means by which matter moves from one place to another. Diffusion results from the kinetic energy of random motion that all matter possesses. To visualize how diffusion occurs, consider an eye dropper full of red food coloring squeezed into a large glass of water. Over time, the food coloring begins to disperse through the water, until both liquids are uniformly mixed and the water becomes slightly tinted. This is because the molecules of the food coloring are in constant motion, as are the water molecules. Technically, diffusion is the movement from areas of high concentration to areas of low concentration. In our example, the food coloring moves from an area of higher concentration (the food dye in the dropper) to an area of lower concentration (the dye-free glass of water). In this case, one would say the food coloring has diffused into the water.

For diffusion to occur there needs to be a gradient between two different material fields. In the example above, the two fields are the eye dropper and the glass of water. The rate of diffusion from one field to another is proportional to the difference in concentration of molecules in those fields. So while molecules move continuously in both directions, their tendency is to move from the more concentrated to the less concentrated molecular fields—which explains why the food coloring diffuses into the water, and not the other way around. In general, the greater the difference in concentrations between two molecular fields, the faster the rate of diffusion.

Osmosis is a particular type of diffusion that refers specifically to the movement of water across a semipermeable membrane. Generally defined, osmosis is simply an operation in which water diffuses through a membrane to an area with a larger quantity of dissolved substances, or solutes. In other words, if two solutions of different concentrations of dissolved material are separated by a wall that permits the smaller water molecules to pass through, but keeps out the larger molecules of the dissolved solids, then the water diffuses across the membrane from the less concentrated to the more concentrated solution. The solution containing more solutes has less water in it, and conversely, the solution with fewer solutes has more water in it. The water therefore will flow from high concentration to low concentration.

Given two solutions, the one with fewer solutes is said to be "hypotonic," while the one with more solutes is called "hypertonic." If an equal quantity of solutes exists on both sides of a membrane, an "isotonic" environment results, and no osmosis will occur.

Osmosis is tremendously important in biology, where water is the primary solvent, carrying many kinds of nutrients, and helping to regulate cells. Plants, for example, absorb water through osmosis. Because most plants are hypertonic in relation to the soil where they live, water moves automatically from the soil, through plant cell membranes, and into the roots. Without this process, plants would be unable to survive. Osmosis is an important process in humans and animals as well. Small blood vessels called capillaries wind through our bodies, coming in close contact with many cells. The exchange of important fluids between the capillaries and cells requires osmosis. Plasma, which is an important component of blood, is hypertonic compared to the surrounding cells. Therefore, liquids naturally move towards the capillaries, nourishing the blood in the process.

A process called *reverse osmosis* is employed by industrial chemists to achieve the purification of water. Because seawater has a high concentration of salt, it is hypertonic when separated from fresh water by a thin, semipermeable membrane. The fresh water crosses the membrane to unite with the seawater and become salt water. Chemists who need to distill seawater into fresh water for drinking or other purposes have found that by exerting pressure on the seawater, the natural procedure of osmosis is reversed and the pure water component passes from the salt water into the fresh water, leaving salt behind.

One question regarding osmosis is how cells can withstand the invasion of new water without bursting. Cells have a built-in security measure called *turgor pressure*. As new liquid moves into a cell, turgor pressure increases until it reaches a point whereby it is able to block incoming water and force it back out. The concept of turgor pressure is what allows industrial chemists to conduct reverse osmosis.

13. Look at the word **kinetic** in paragraph 1. The word kinetic in the passage is closest in meaning to

   ○ heavy
   ○ inactive
   ○ dynamic
   ○ unpredictable

14. Look at the word **specifically** in paragraph 3. The word specifically in the passage is closest in meaning to

   ○ usually
   ○ certainly
   ○ accurately
   ○ exclusively

15. Look at the word **withstand** in paragraph 7. The word withstand in the passage is closest in meaning to

   ○ endure
   ○ oppose
   ○ protect
   ○ compete

16. Look at the word **this** in paragraph 1. The word this in the passage refers to

   ○ the color of the food coloring
   ○ the constant motion of molecules
   ○ the mixture of the food coloring and water
   ○ the kinetic energy of random motion

17. Look at the word **which** in paragraph 2. The word which in the passage refers to

   ○ the rate of diffusion
   ○ the tendency of molecules
   ○ the movement in both directions
   ○ the less concentrated molecular field

18. Look at the word **where** in paragraph 5. The word where in the passage refers to

   ○ water
   ○ biology
   ○ osmosis
   ○ solvent

Question Type 5: Inference Questions

Inference is a crucial skill for not only reading but also listening, speaking, and writing. To infer means to understand unwritten or unstated ideas based on logic and detail. Although an inference is based on written detail, the inference itself is not written or stated in the passage. Moreover, the key detail could be a number (a date, time, statistic, etc.), any part of speech (a noun, verb, etc.), a phrase, or a combination of ideas in one or more sentences. Inference is directly related to implication. To imply means to communicate or suggest unstated or unwritten information based on logic and stated detail. A writer implies and a reader infers. Read the following sentence:

> Jack Kerouac went to Columbia University on a football scholarship.

In order to recognize the implications of this sentence, you need to understand the noun *scholarship*. Along with your reasoning abilities, inference questions require you to apply your vocabulary skills in context. A scholarship is financial support provided to a gifted student by a school that hopes the student's performance in sports, the arts, or sciences will benefit the school. Based on the fact that scholarships are only given to skilled students, you can infer from the above sentence that Jack Kerouac was both athletic and a good football player. Otherwise, he could never have gotten a scholarship to play football for a major American school. At the same time, the writer implies the same facts (athletic, good football player) about Jack Kerouac by mentioning his football scholarship. Using your logical reasoning and some detail or combination of details, you can understand something that is unstated, but implied, in the sentence. Moreover, you can often combine ideas from more than one sentence in a passage for more complex inferences. However, on the actual test, you don't need to make an inference on your own since you have four choices.

Inference questions are multiple-choice, and you must choose the correct inference from a list of four choices. You can identify the correct inference by finding some detail(s) in the passage that can be used to infer one of the choices. Therefore, in order to answer inference questions, you must logically connect information in the passage to the correct choice. The following strategies review the clues to the right information and the kind of analysis necessary. First, read the excerpt below from a longer passage on a famous American writer. The passage is referred to in a sample inference question discussed in the strategies:

Jack Kerouac

San Francisco, America's romantic city by the bay, has always been a haven for artists. One of the great American romantics who wrote in San Francisco was Jack Kerouac. His autobiographical novels and wayward travels made him the most celebrated member of the Beat Generation. The Beats, or beatniks, were a group of writers and poets from cities across the U.S., who shared a love of jazz, experimentation, and adventure.

Born on March 12, 1922, in Lowell, Massachusetts, Kerouac came from a working-class family. Like many families of that era, his family struggled financially during the Great Depression of the 1930s. Fortunately, Kerouac attended Columbia University in New York on a football scholarship, but a leg injury kept him off the field. Intellectually gifted but undisciplined, Kerouac eventually dropped out of Columbia twice but continued to pursue a career in writing. Writing in the bars of New York's Lower East and Lower West sides, Kerouac met and worked with William S. Burroughs and Allen Ginsberg, before they all traveled west and started a literary and cultural revolution.

Kerouac first landed in San Francisco in 1947. There he joined his soul brother, Neal Cassady, whose frenetic letters and cross-country travels spurred Jack to write *On the Road,* perhaps his preeminent work, during the month of April 1951. Since the book was written as a simple personal testament "in search of his writing soul," Kerouac had no idea that it would, a decade later, encourage a generation onto the highways and into the social activism of the Vietnam era.

Almost overnight Kerouac became a national—even mythical—figure. But in the end he could not live with the myth he had created. His later years were spent drinking and living with his mother—an ironic turn on the life of freedom he had written about. When he died in 1969 from complications related to alcoholism, Kerouac had little money, but his estate is now valued at over $20 million.

Strategy 1: Think about possible inferences as you read actively.
Inference is an important part of active reading. When you read actively, you are trying to understand not only the meaning of the words but also the connections and relationships among the ideas. These relationships can imply certain facts, and you can improve your understanding and your score by inferring those facts. Therefore, inference is a skill that you can exercise, or use, even when you are reading the passage for the first time. You don't want to read the passage too slowly at first since a close, careful reading is necessary only once you are answering a specific question. However, you still want to think about possible implications or connections among the various ideas as you read.

The sample excerpt above offers many opportunities to infer connections and facts based on the ideas in the text. For example, look again at the first paragraph. The first sentence describes San Francisco as a haven, or safe place, for artists, and then the second sentence mentions that Jack Kerouac wrote there. However, since *artist* is a more general noun than *writer*, you can infer that other artists, such as musicians and painters, must have been in San Francisco as well, not just writers. In the third sentence, Kerouac's writing is described as autobiographical, meaning that it included elements of Kerouac's life even though the stories were not always about only him. Therefore, when the same sentence also refers to Kerouac's wayward travels, you can infer that he included at least some traveling in his writing. If Kerouac included elements of his own life in his work, and he traveled a lot, then logically his work must discuss traveling to some degree. You can infer this even though it isn't written explicitly in the sentence. Finally, the last sentence of the introduction describes the common interests of beatniks (jazz, experimentation, and adventure). Since Kerouac is referred to as a popular member of this group in the previous sentence, then you can infer that Kerouac also enjoyed jazz, experimentation, and adventure. The use of the verb *shared* in the last sentence of the introduction clearly supports that fact, and the verb is a good example of the importance of key words to correct inferences.

You don't need to record all possible inferences in note form. The paragraph above describes the kind of active thinking you should try to do while you read; it doesn't indicate how much you should write down. Remember that your initial notes from your first reading should be a basic outline of the passage: the main idea (the writer's thesis in the introduction), the key supporting points (the topic of each body paragraph), and the most important details (dates, names, etc.). The inferences that you make during your first reading can simply augment, or improve, your understanding of the passage, and they may eventually help you to answer some questions quickly or to avoid some distracters. Basically, the more you understand from your reading, the higher your score.

Strategy 2: Identify a key idea or ideas in the question.
Like detail questions, an inference question can include key ideas that you can use to locate the relevant information in the passage. Some questions might identify a particular paragraph, but not all questions. Moreover, even if the question specifies a paragraph by number with a phrase like *in paragraph 2*, the paragraph could be long enough that you must still use the key idea(s) to locate the relevant information within the paragraph.

It's important to identify key ideas, not specific words, in the question since the question restates information from the passage. Although some words such as names aren't easily paraphrased, any important concept from the passage will be restated. Look at the following sample Inference question for the excerpt above:

> Which of the following can be inferred about Kerouac's time as a student in New York?
>
> ○   He took some of the same classes as William Burroughs and Allen Ginsberg.
> ○   He probably failed because he spent all his time playing football.
> ○   He could afford university thanks to his athletic abilities.
> ○   He decided to become a writer only after he dropped out of school for the second time.

The key ideas in the question are *time*, *student*, and *New York*. Notice how the nouns *time* and *student* are not used in the passage. You have to scan for related ideas, not the exact words. *New York* is repeated from the passage since place names are not easily paraphrased. You should use all key words together to identify the necessary information in the passage since New York is mentioned twice and you must make sure that you focus on the correct time in New York.

Strategy 3: Identify a common idea or theme in the answer choices if the question has no clues.
Sometimes, the question does not include any key words. For example, the sample question above could be rewritten as follows: *Which of the following can be inferred about Kerouac?* Since the name (Kerouac) is repeated too often throughout the passage, you can't use it to locate any information. In this case, you have to look for a set of related ideas, or theme, in

the answer choices. The answer choices for the sample question above all relate to Kerouac's time at university (classes, failed, university, school), so this should be the idea you scan for.

It is possible that there is little or no common theme in the answer choices. Each answer choice could relate to a different paragraph in the passage. In this case, you must choose a key idea from each answer choice, and scan for each one separately. However, most inference questions will specify a paragraph by number, or contain some key idea(s) in the question or some common theme in the choices.

Strategy 4: Scan the passage for the relevant information.

To scan a passage means to look superficially through the passage without reading too closely for meaning; when you scan, you are trying to recognize related information, and then you will stop to read a particular part more closely. Scan the passage for words, phrases, or expressions that have a connection to the key words from the question or theme from the choices. If you made an outline of the passage during your initial reading of the passage, use it to help you identify a paragraph. Don't reread the passage while you scan since this could take too much time. Remember that you only have less than two minutes for each question. Based on the key words in the sample question on Jack Kerouac, you should focus on the second paragraph of the passage.

Strategy 5: Read the relevant sentences carefully; don't skim.

Once you locate the correct part of the passage and/or paragraph, you need to read the information carefully. It's important that you not skim the sentences. Many distracters, or incorrect answers, are based on one or more repeated words from the passage, so you may carelessly choose a distracter if you read the passage too quickly and superficially at this point. Also, read more than just one sentence. Usually, an accurate inference requires a combination of ideas, but it's possible that one key word could indicate the correct answer.

For example, the second paragraph in the sample excerpt begins with a description of Kerouac's working-class upbringing, which means that his family members worked in trades (plumbing, mechanics, etc.) or other manual jobs. Although this doesn't imply extreme poverty, it means that his parents were not office workers or business owners and his family was not wealthy. These facts are emphasized by the second sentence, which mentions the family's financial trouble, or lack of money and work, during the Depression. This information is important background for the third sentence, which refers to Kerouac's scholarship to Columbia University.

Strategy 6: Use cohesive devices to connect ideas.

Cohesive devices are methods of showing meaningful connections between words, clauses, sentences, or paragraphs. These devices include pronouns (he, she, this, that, etc.), adjectives, adverbs, articles, transitions (however, therefore, etc.), repetition, and varied word forms. In the excerpt on Kerouac, there are many useful cohesive devices, and the adverb *fortunately* is directly related to the inference question. In the context of the passage, the adverb suggests not only a positive outcome, or result, but also a lucky result. The difficulties mentioned in the previous sentence imply that the scholarship helped him avoid a negative or worse outcome (not going to university).

Strategy 7: Check each answer choice against the passage, and choose the most logical answer.
Once you have a clear understanding of the ideas in the relevant sentences and their relationship to each other, check to see which answer choice is the most logical. Think about facts that are associated, or related, to the words in the passage but are unmentioned, such as the facts related to *working class* (physical labor, not wealthy), *struggled financially* (little money or savings), and *fortunately* (avoided a negative future). Based on these clues, (C) is the most logical answer to the sample question.

Question Forms

Inference questions can be stated in a variety of forms. An inference question can be active or passive, and it may or may not specify a paragraph by number. Review the following forms for inference questions:

> The author implies that…
> What can be inferred about X ?
> What does the author suggest about X?
> Which of the following can be inferred about X?
> According to paragraph #, which of the following can be inferred about X?
> Based on the information in paragraph #, the author implies that…

Distracters

Each inference question has four answer choices. Only one choice is correct, so the other three choices are distracters, or incorrect choices. A distracter is an incorrect answer that appears, or looks, correct. In order to make the question more difficult, many incorrect answers on the TOEFL seem connected to the passage through a variety of tricks, or distractions. Like the distracters for many other question types, the distracters for inference questions often repeat vocabulary from the passage, and distort it. The following explanations review these distractions.

Distracter 1: Answer choice repeats vocabulary, but alters it or adds incorrect information.
Don't choose an answer simply because it includes vocabulary from the passage. Always check the ideas in the choice (actors, actions, states, reasons, etc.) with those in the passage. Often, an incorrect choice adds illogical or impossible ideas to familiar ones, or the choice distorts, or alters, the repeated ideas incorrectly. The distortion might involve an incorrect sequence, cause, effect, time, place, and so on. If you read the passage too quickly and carelessly, you might be fooled by the repetition. (A) and (D) are examples of this type of distracter.

In (A) the names William Burroughs and Allen Ginsberg are repeated from the passage, but it is unlikely that either man attended classes with Jack Kerouac. In the passage, the two names don't appear in connection with Kerouac's time at university. Instead, they are connected to the bars in New York. Based on the participle phrase *writing in the bars*, the passage clearly states that Kerouac met Burroughs and Ginsberg in or around the bars, not in class at Columbia.

In (D), the answer choice repeats the verb phrase *dropped out of school*, but it rearranges the order of the ideas. The answer choice states that Kerouac's decision to be a writer came after he left school, but the passage implies that Kerouac's choice predated, or came before, his decision to leave school based on the verb *continued*. Also, (D) uses the adverb *only*, and you should be careful about choices that use overly certain or definitive language (see Distracter 2 below).

Distracter 2: Answer choice alters the scope, degree, or certainty of the ideas in the passage.

This type of distracter is related to Distracter 1 above since the distracter alters the ideas. However, it alters the details by making them too specific, too general, less certain, too certain, too weak, or too extreme. The specificity of an idea is the degree that it is specific to a particular type, group, person, time, place, etc. For example, the sample excerpt states that Kerouac was a celebrated, or praised, person, but the passage specifies that he was the most highly regarded member of the Beat Generation of writers, who worked during a particular period of time (middle of the 20th century). This kind of distracter might incorrectly state that Kerouac was the most celebrated American artist of the 20th century. This broadens the scope of the original statement in the passage by comparing Kerouac to too many other artists (all other Americans in his century), and the writer never makes or implies that kind of comparison.

This type of distracter might also distort the details of Kerouac's sports injury. According to the passage, Kerouac was injured in the leg and couldn't continue playing football. This type of distracter could state inaccurately that Kerouac was almost killed on the football field. Although the injury was serious (it ended his football career), it was not life threatening (one cannot be killed by a leg injury that is uninfected and properly treated).

Distracter 3: Answer choice is a plausible inference in another context.

This is a difficult distracter because it seems logical or likely, but it still doesn't match the stated details in the passage. (B) is a good example of this kind of distracter. It seems possible that a football player could focus too much on his sport and neglect his studies. In fact, this is a common problem among good and even below average athletes. However, nothing in the passage supports this idea. In fact, the passage contradicts that choice based on the fact that Kerouac had to stop playing football due to an injury.

Practice: Passage

Now, practice the strategies you've just learned. Read actively, think about the main idea and purpose, record the key points in an outline of the passage, and don't read too slowly right away.

The Olympic Effect

Since their beginning in 1896, the modern Olympic Games have been the gold standard of athletic prowess. Olympic competitors are the best in the world, and no trophy compares with an Olympic gold medal. Increasingly, however, the athletic competitions have been matched, if not overshadowed, by the fierce competition between cities and nations vying to host the Olympics. The amount of money and prestige at stake is so great that the phenomenon has been given a name: the Olympic effect. Although it is primarily an economic force capable of affecting the business climate of the host community for years before, during, and after the Games, the Olympic effect can also exert a powerful influence on the environment and society of the host community.

The economic impact of the Olympic Games is complex. Preparation for the Games usually generates vast investments not only in the stadiums, tracks, and other sports venues, but also in local highways, hotels, and airports. Thousands of jobs are created and billions of dollars are spent: Australia spent $2.1 billion for the Sydney Games in 2000, and Greece spent at least $11 billion only four years later. Long-term effects are generally positive, as the Games tend to improve the international image of the host city. While tourism is generally thought to increase during the event, the region hosting the Games may actually experience a drop in tourism and retail income, as people try to avoid the crowds, traffic, and price hikes they expect to find in the host city. Some Utah ski resort operators experienced a 20–30 percent drop in visitors during the Salt Lake City Games.

The environmental repercussions of the Olympic effect are also mixed. On the one hand, in the years following the Sydney Games, also known as the "Green Games," participating communities developed recycling programs, emphasized renewable energy sources, and built or extended mass transit systems. Salt Lake City, for instance, built a 21-kilometer TRAX light rail system. Some, like Sydney and Beijing (2008), have also made notable efforts to clean up their municipal environments. Sydney actually built most of its Olympic venue over a reclaimed toxic waste dump that had once blighted the city. To clear its polluted air, Beijing has converted much of its energy production from coal to natural gas, and implemented new smokestack and vehicle emissions guidelines. Medical researchers estimate that the benefit to Beijing residents' health will amount to many billions of dollars over the coming decades. On the other hand, the Games create enormous consumption of fossil fuels and strain the waste disposal systems. The Winter Games in particular often result in accelerated development of important wildlife habitats.

Social aspects of the Olympic effect are varied and include a growth in civic pride. But the biggest social impact of the Olympic effect is a predictable one: a sharp growth in public interest in those sports highlighted during recent Games. In the United States, this can best be seen in the ice sports. The figure skating drama between Nancy Kerrigan and Tonya Harding in 1994 played a part in piquing interest, as did the gold medal of Tara Lipinski in 1998. American successes in speed skating and hockey have also inspired interest. The result has been a 50 percent increase in the number of ice arenas in the United States in the last 15 years. Dallas, Texas, built eight between 2000 and 2002 alone. Related sports such as speed skating, hockey, and even curling have seen a tremendous growth in interest. Even southern cities like Tampa Bay, Florida, and Atlanta, Georgia, now have hockey teams.

When the modern Olympic Games were first held in 1896, they were conceived as a celebration of the human body and the human spirit. This has not changed. But now the Games are something more. They profoundly shape the lives of people who never set foot inside the arenas. They are the paramount international pageant, turning the spotlight on cities, nations, and people, prompting them to find the best in themselves and hold it up for all to see.

Practice

19. The author implies that the Sydney Games

   ○ Led to the discovery of a toxic waste dump

   ○ Caused a great deal of environmental damage

   ○ Led to an increase in environmental programs

   ○ Cost less to host than most other Olympic Games

20. Based on the information in paragraph 4, what can be inferred about ice sports?

   ○ Americans have greater success in playing ice hockey than they did 15 years ago.

   ○ Figure skating has always been a very dramatic sport.

   ○ It is surprising that the southern states are now interested in ice sports.

   ○ Tampa Bay, Florida, and Atlanta, Georgia, have grown because of ice hockey.

21. Based on the information in paragraph 2, what can be inferred about the Olympic Games?

   ○ The economic effects of the Olympic Games are unpredictable.

   ○ The costs of hosting the Olympic Games increase every four years.

   ○ The Olympic Games always have a long-term positive effect on tourism.

   ○ Most cities find their investment in the Olympic Games to be worthwhile.

Question Type 6: Rhetorical Structure Questions

Rhetorical structure refers to the ways that a writer supports the key points in a passage. A writer supports the key points in a passage with detail: examples, description, definition, and explanation. Rhetorical structure questions ask about the kind of detail in a passage and its purpose or function. In order to answer this type of question, you must recognize what kind of detail is used in the passage (description, explanation, etc.) and how the detail is used (to describe X, to define Y, to emphasize Z, etc.). Moreover, you must identify the point(s) supported by the detail (X, Y, Z in the previous sentence).

Rhetorical structure questions are multiple-choice questions. Only one answer choice is correct, and the other three choices are incorrect answers, or distracters. Moreover, there are two forms of rhetorical structure question. Both types discuss the same concept (rhetorical structure), but each one asks about it in a different way.

Rhetorical Structure Question: Types	Examples
Type 1 The first type of rhetorical structure question identifies a particular idea or set of ideas, and asks about its purpose.	Why does the writer mention X? to define A to give an example of B to emphasize C to illustrate D
Type 2 The second type of rhetorical structure question mentions a rhetorical structure, and asks how the writer achieves or accomplishes it.	How does the writer explain A? by mentioning X by comparing X and Y by referring to Z by describing Y

Although there are two types of rhetorical structure question, both involve the same kind of information in the passage. Therefore, both types are discussed together. The strategies for type 1 and type 2 are very similar, so the following strategies discuss an example of each type. First, read the following excerpt that is used in the explanation of the strategies for rhetorical structure questions.

Orangutans

Orangutans, found in the rain forests of Sumatra and Borneo, differ from other species of great apes in several important ways. Though their intelligence and relatively long lifespan are traits shared by the other great primates, the social behavior and general lifestyle of orangutans are quite different.

The other two groups of great apes, chimpanzees and gorillas, live in mixed social groups consisting of one alpha male, a group of females, and children of varying levels of maturity. Orangutans, on the other hand, live semisolitary existences. That is to say, male orangutans live alone most of the time, occupying and defending their territory and the females who live there. Though females residing within the boundaries of one male orangutan's domain belong to him, they do not live with him. Their only intimate contact occurs during weeklong periods of fertility every few years, between pregnancies. Once a female is pregnant, she takes care of herself and eventually raises her baby completely alone. Even when several adult females and their babies group around the same fruit tree—a rare but possible event—they do not fight or share as other primate groups would, but instead, ignore each other entirely.

Like most primates, orangutans are comfortable in the trees. However, only orangutans live there almost constantly since they are physically more adapted to a life of swinging and hanging from trees than chimpanzees or gorillas. For instance, their arms are extremely long and muscular compared to their legs. Furthermore, their joints, especially the joints of their knees, hips, elbows, and shoulders, are incredibly mobile. Finally, when orangutans walk on their hands and feet, they use the outside edges, rather than their palms or soles. Thus, they can eat, sleep, and play in the trees, only climbing down to find sticks and branches to build their nighttime nests.

Strategy 1: Be familiar with the types of detail and rhetorical structure.

Rhetorical structure questions ask about the various ways that detail can be used to support key points in a passage. Therefore, you should be aware of the various types of detail, and how that detail can be used. Although each type of detail is discussed separately below, many details are mixed in a reading passage (and a lecture). In other words, one type of detail, such as explanation, can also involve any other type, such as examples or comparison.

Type of Detail	Examples from the Sample Passage
Description A description is a very broad category of detail that gives the characteristics of the topic. Descriptive detail can be any part of speech, and it helps the reader imagine the topic by referring to appearance, smell, touch, taste, emotions, personality, attitude, manner, speed, quality, condition, degree, frequency, cost, origin, material, time, place, sequence, name/title, etc.	Orangutans, found in the rain forests of Sumatra and Borneo… (location) The other two groups of great apes, chimpanzees and gorillas,…(name) Their only intimate contact occurs during weeklong periods of fertility every few years…(time) …a rare but possible event…(frequency) …their arms are extremely long and muscular compared to their legs. (appearance) …when they walk on their hands and feet, they use the outside edges…(manner)
Definition A definition is a specific type of description. A definition gives the meaning of a term or the characteristics that make the term unique. Definitions give essential or distinguishing information.	That is to say, male orangutans live alone most of the time… (a definition of *semi-solitary*)
Comparison/Contrast Both comparison and contrast are also types of description. A comparison is a description of similarities, and a contrast is a description of differences.	Orangutans…differ from other species of great apes in several important ways. (contrast) Like most primates, orangutans are comfortable in the trees. However, only orangutans live there almost constantly… (comparison followed by contrast)
Example An example is a specific term from a more general category or type. Possible examples include various nouns, statistics, anecdotes, or imaginary situations. If the topic is disease, AIDS and tuberculosis are examples of a communicable disease while cancer and Alzheimer's are examples of a non-communicable disease.	For instance, their arms are extremely long and muscular compared to their legs. Furthermore, their joints, especially the joints of their knees, hips, elbows, and shoulders, are incredibly mobile. (examples of physical adaptation) …several adult females and their babies group around the same fruit tree—a rare but possible event… (a hypothetical, or imaginary, example of a meeting of orangutans)
Explanation An explanation gives the reason(s) for the ideas in the passage. An explanation may involve cause and effect logic, intentions, purposes, goals, desires, or any other type of detail (sequence, time, examples, etc.).	Finally, when orangutans walk on their hands and feet, they use the outside edges, rather than their palms or soles. Thus, they can eat, sleep, and play in the trees… (description of manner + thus + result)

Rhetorical structure is the use of detail to support the key points in the passage. However, there are specific types, or methods, of support, and you need to be familiar with them. Since rhetorical structure expresses a writer's purpose in using detail, the various types are usually expressed as infinitives. Some rhetorical structures are simply the infinitive form of the nouns above since each type of detail has a related purpose in the passage (a description = to describe, a definition = to define, an explanation = to explain, etc.), but there are other types since there is more than one use for some detail.

Rhetorical Structure	Meaning/Definition
To illustrate/to demonstrate	An illustration is literally a picture or drawing, but a writer uses words instead of images; a writer illustrates an idea by using concrete examples that help the reader imagine, or picture, it. Likewise, a demonstration is a performance, but a writer can demonstrate an idea by using concrete details that show how the topic operates or is used in a practical situation. Detailed descriptions, anecdotes, and imaginary scenarios are commonly used in both illustrations and demonstrations.
To clarify	A clarification is an attempt to avoid confusion. A writer clarifies an idea by explaining a point that could be misunderstood or misinterpreted. Definitions, detailed descriptions, and comparisons are often used to clarify ideas.
To distinguish/to differentiate	This is very similar to a clarification because a difference is made clearer. However, a clarification doesn't have to contrast two topics, but a writer contrasts two topics when distinguishing them in a passage. Contrast, definitions, and explanations are often used by writers to distinguish topics.
To expand/to elaborate	A writer elaborates by adding more details. Description and examples are most often used to expand a topic.
To emphasize	A writer emphasizes an idea by elaborating and adding detail related to a specific aspect or quality of the topic. Usually, one or more details are repeated in the elaboration of the idea. Detailed description, definition, and examples are commonly used to emphasize an idea as well as key words, such as just, even, so, and only.
To prove	Proof is conclusive or convincing support for an argument. Usually, the topic is debatable or controversial, and the writer wants to convince the reader with details that are difficult or impossible to reject. Accepted facts and data, usually from real experiments or scientific studies, are the most common ways to prove something.
To refute	A refutation is an argument against an opposing plan, opinion or point of view. A writer can mention an opponent's argument and then explain a flaw or weakness in that position. Logical explanation is essential for a refutation as well as concrete examples and possibly definitions.

Strategy 2: Identify a key idea in the question.

For both types 1 and 2, you can use a key idea from the question to help you locate the relevant information in the passage. Sometimes, the question specifies a particular paragraph by number, but not always. Also, even if a particular paragraph is identified, the paragraph could be so long that you will still need the key idea to focus on the right part of the paragraph. As always, your outline of the passage can also help you find the right information.

You need to identify key ideas, not words, since they could be paraphrased from the passage rather than copied or quoted. Of course, some words could be repeated, such as the species name orangutans, but not all the words will be copied.

Look at the sample questions below about the rhetorical structure in the excerpt about orangutans:

[**Type 1**]

Why does the writer mention an infrequent gathering of orangutans at a food source?

- ◯ To emphasize their lack of social interaction
- ◯ To define the meaning of semi-solitary
- ◯ To emphasize that they eat only fruit
- ◯ To describe the difficulty of finding food in the jungle

[**Type 2**]

How does the writer explain why orangutans can live mostly above the ground?

- ◯ By pointing out that orangutans can't walk on their hands and feet properly
- ◯ By mentioning their fear of predators
- ◯ By referring to their search for sticks and branches
- ◯ By giving examples of their physical adaptations

The key ideas in type 1 are *infrequent*, *gathering*, and *food source*. The key ideas in type 2 are the words *explain*, *why*, *can live*, and *above the ground*. The key ideas for both question types may provide clues to the type of detail (*explain* and *why* = explanation/reason) and clues to the specific supporting points (infrequent gathering, live mostly above ground). Also, notice that the noun *orangutans* is not a useful word since it is repeated too often throughout the passage.

Strategy 3: Scan the passage for the key ideas.

Use the key ideas from the question to locate the relevant information in the passage. This is very similar to the procedure for Detail questions, NOT/EXCEPT questions, and inference questions. Remember to look for restatements as well as repetition. The key words from the first question (type 1) above should lead you to the discussion of the fruit tree at the end of the second paragraph based on the words *group*, *rare*, and *fruit tree* in the passage. The key words from the second question (type 2) should take you to the third paragraph based on the vocabulary in the first two sentences (*in the trees* and *only orangutans live there almost constantly*).

Strategy 4: Read the surrounding sentences carefully, and analyze the rhetorical structure.
Don't skim the passage since you need to think about how the detail supports the point(s) in the passage. Read the sentences closely and use the cohesive devices to understand the sequence of ideas. Cohesive devices are ways to relate and connect ideas within the same sentence, in different sentences, or in different paragraphs. They include pronouns, articles, adjectives, conjunctions, and transitions. For example, paragraph 3 includes the transitions *for instance* (examples), *furthermore* (continuation/similarity), and *thus* (effect/result). These can help you identify how the detail is used and what the detail supports. The correct answer for the second question above (type 2) is (D).

Strategy 5: Be prepared to infer rhetorical structure based on sequence and meaning.
You can't always rely on cohesive devices to indicate how ideas relate to each other. Writers often imply the connection between ideas based on their sequence and the vocabulary. In order to make the necessary inferences, you need to think about paragraph organization and the sequence of ideas. A paragraph is organized around a unique supporting point, mentioned in the topic sentence at the beginning of the paragraph. All the detail in the paragraph supports that unique topic. Moreover, the details usually follow a logical sequence in most paragraphs: general to specific, problem to solution, cause to effect.

For instance, at the end of the second paragraph, the reference to a group of orangutans around a fruit tree is not preceded or followed by any useful cohesive devices. However, all the detail supports the semisolitary lifestyle of orangutans: the brief meeting during fertility, the unsupported pregnancy and the raising of babies alone, and the meeting at the fruit tree. Therefore, there is a continuation and repetition of a theme: orangutans live in social isolation. This repetition and its relationship to the topic of the paragraph should tell you that choice (A) is the correct answer to the first question (type 1).

Question Forms
You can recognize both types of rhetorical structure questions based on the following forms or some variation:

> How does the author illustrate/explain/define…
> Why does the author mention/refer to X?
> The author uses X as an example of…
> The author discusses X in paragraph 1 in order to…

In Paragraph 2, the author discusses X in order to…

In Paragraph 3, why does the author state that X…

Distracters

Rhetorical structure questions are multiple-choice questions. Only one answer is correct, and the other three choices are distracters, or incorrect answers. You should review the types of distracters for rhetorical structure questions (and all question types) because distracters are wrong answers that can seem correct. The correct answer is never obvious due to the kind of distractions that are included in the choices. Therefore, review the different types of distracters for rhetorical structure questions.

Distracter 1: Answer choice refers to a rhetorical structure from another part of the passage.

Always make sure that a rhetorical structure (definition, explanation, examples, etc.) relates to the key ideas in the question. It's possible that an incorrect choice could refer to a something from the passage but unrelated to the question. (B) in the first sample question (type 1) is an example of this type. The definition of the adjective *semisolitary* occurs at the beginning of the paragraph. Although the definition relates to the same paragraph topic as the description of the tree, the meeting at the tree emphasizes the idea of a solitary lifestyle by showing how little the orangutans interact when they eat together. This doesn't define the meaning of isolation.

Distracter 2: Answer choice is the correct rhetorical function but refers to unmentioned or unrelated ideas.

This type of distracter includes the correct rhetorical structure (to define, to explain, etc.), but the object of the infinitive is not in the passage or it's from another part of the passage. (C) and (D) from the first sample question (type 1) are examples of this type of distracter.

Distracter 3: Answer choice distorts, or alters, information from the passage.

You must be careful about relying on key words alone because these words can be changed or altered slightly. Always read entire sentences carefully and make sure you understand the meaning of the whole sentence, not just one or two words. For instance, (A) from the second sample question (type 2) is an example of this type of distracter. The writer mentions the hands and feet of orangutans, but the writer does mention or imply that orangutans use them incorrectly. Orangutans use their hands and feet differently, but nothing is said about the right or best method. This choice adds unmentioned ideas and manipulates the information from the passage incorrectly. (C) from the same question is also an example of a distortion of the ideas in the passage.

Distracter 4: Answer choice refers to a plausible but unmentioned idea.

You should always check your answer with the passage because some answers might seem possible, but they are based on assumptions or inferences that are unsupported by the passage. For example, (B) in the second sample question refers to a possible reason that animals might live in the trees (the avoidance of predators), but this reason is never mentioned or suggested by any details in the passage. Therefore, never assume that something is true without checking the passage.

Practice

Now, practice the strategies you've just learned along with the general reading strategies. Read the passage below, and answer the vocabulary questions that follow. Also, you can review inference questions for the same passage.

Trains and Automobiles

Though trains saw widespread use over the course of the 19th century, the last hundred years have seen nothing but a decline in the use of railroads in the United States and a rapid growth in American car culture. While trains have begun recently to attract a little more interest from urban planners, there is no sign at all that the general public shares that interest. What accounts for this progressive loss of interest in train travel? What could have made Americans become so enamored of their cars?

Efficiency alone cannot be the answer we are looking for. Both automobiles and trains consume similar amounts of energy. The average car gets about 13 passenger-kilometers per liter of fuel, no improvement at all over trains, though if trains are forced to run with few passengers, they can actually be much less efficient than cars. Consequently, if one's objective is to conserve energy, neither mode of transportation offers any real advantage, with one exception: interurban light rail and subways are about 25 percent more fuel efficient than cars.

Both rail and automobile transportation depend on expensive infrastructure; highway construction in the United States averages several million dollars per kilometer, and can easily go much higher. Railroads are almost as expensive to build and railroad operators must also pay to maintain their locomotives and rolling stock. Therefore, there does not seem to be a particular advantage in either fuel efficiency or cost of construction and maintenance associated with either automobiles or trains.

On the other hand, trains hold very real advantages in safety. In the United States alone, more than 40,000 people die every year in car accidents, and hundreds of thousands more suffer personal and financial injury. By contrast, rail fatalities seldom number more than a few hundred per year worldwide. On a train, one need never worry whether the approaching driver is intoxicated or distracted by his cellular phone, nor does one need to worry about falling asleep at the wheel, striking a deer crossing the road, or any of the other myriad hazards that face automobile drivers on a daily basis. One would think such a safety record would attract more enthusiasm from potential passengers.

In addition to safety, with the exception of scheduling, riding a train generally offers far more peace of mind than relying on a car. Once on board the train, the passenger can read a newspaper, prepare for work, or simply relax and admire the scenery passing by. Driving, on the other hand, requires the patience to endure traffic jams and the rude person behind who drives with his horn. Then there is the maintenance, insurance, and perhaps a monthly bill the car owner has to pay.

Nevertheless, cars do offer a real advantage in versatility. It is never necessary to wait for the car—it's ready when its driver is, and is never behind schedule. Plus, there's plenty of room in the trunk for carrying groceries or skis. And a car can go all those places where no rail lines have been built. Additionally, cars can be customized and infinitely varied to suit any kind of need or taste.

However, even versatility is probably not the best answer. The truth lies in the way Americans romanticize the car. For every teenager, getting his or her driver's license is a rite of passage. Teenagers often consider themselves to be adults once they can drive. In fact, Americans have a whole body of popular culture, from dating to work to weekend vacations, built around the car—and nowhere does the train put in an appearance. Americans find the tangible, versatile car to be a marker of self-identity in a way that trains could never be.

No matter how efficient they are, how safe, or how inexpensive, trains cannot offer the thrill and empowerment provided by the automobile. No teenager dreams about cruising to pick up his or her date on the train. No businessperson wants to spend money on train tickets—he or she wants the prestige of a showy new sports car. The parent with children doesn't want to take the train either. He or she must make frequent trips to soccer practice, ballet lessons, and scout meetings and a car is the most efficient way to do so. Until cars become so expensive to purchase and operate that they are out of reach of most people, trains and other forms of transportation will always take a back seat to the automobile.

22. Why does the author mention interurban light rail in paragraph 2?

   ○   To argue that trains are superior to cars
   ○   To highlight an exception to his main point
   ○   To urge action on the part of urban planners
   ○   To highlight a modern advance in train technology

23. How does the author describe the public's declining interest in trains?

   ○   By making an analogy
   ○   By citing the results of a study
   ○   By investigating possible causes
   ○   By providing a historical narrative

24. In paragraph 4, the author describes the relative safety of trains by

   ○   Comparing riding on trains to operating a cellular phone
   ○   Describing common arguments used by train passengers
   ○   Listing automobile hazards that are not experienced on trains
   ○   Arguing that train operators are more responsible than car drivers

25. What does the author suggest about the public's declining interest in trains?

   ○   It was never shared by urban planners.
   ○   It occurred quickly after the car was invented.
   ○   It happened slowly over the course of the century.
   ○   It was due to the inefficiencies of rail transportation.

KAPLAN

26. It can be inferred from the information in paragraph 3 that railroads

- ○ Are slightly less expensive to build than highways
- ○ Were most popular in the U.S. in the past 100 years
- ○ Never had the same level of public interest as cars
- ○ Are more likely to be fuel-efficient than automobiles

Question Type 7: Coherence Questions

A coherence question asks you to insert a new sentence into the passage. You have a choice of four possible locations indicated in the passage by four symbols: [■]. When you click on one of the symbols with your mouse, the computer inserts the new sentence at that location. You don't need to drag and drop the sentence with your mouse, and you can only click on one symbol at a time. Clicking on a new symbol moves the sentence to the corresponding location.

The strategies for this question type involve many of the strategies from earlier questions in this section; referents, vocabulary, inference, and rhetorical structure are all involved in this question type. As you review the strategies for coherence questions, you may find it necessary to review the strategies for some of the related questions. First, read the following short excerpt from a passage on Esperanto. The excerpt is used in the explanation of a sample coherence question.

The Universal Language of Esperanto

[■] An artificial language, sometimes called a universal language, is an invented symbolic system intended to transcend foreign-language barriers so that people from diverse linguistic backgrounds can communicate easily among themselves. [■] An example of such a language is Esperanto, invented by a Polish ophthalmologist named Ludwig L. Zamenhof. [■] Now the best known of all artificial languages, Esperanto was first published in Russian in 1887 under the title *Mezhdunarodny yazyk*, which means "an international language." [■] Dr. Zamenhof used the pseudonym "Doktoro Esperanto" in order to disguise his identity as the author and originally called his invented language "Lingvo Internacia." However, the name "Esperanto," which translates as "hopeful," quickly caught on and eventually became the official name instead.

Esperanto was constructed rationally, with one main principle in mind: that it should above all else be easy to learn. As a result, the vocabulary, structure, spelling, and pronunciation of the language are much simpler than in most languages that have developed naturally. For example, there are only about 15,000 roots in Esperanto, derived mainly from Latin, Greek, Romance languages, and Germanic languages but these can be arranged in various combinations to produce a much larger vocabulary.

Esperanto also makes extensive use of prefixes, suffixes, and interchangeable endings to form more complex words in order to reduce the total number of words necessary to learn. The grammar of Esperanto is derived from European languages, but it has been greatly simplified and standardized with just sixteen basic rules governing the syntax and usage. *La* is the only article, all nouns end in –o, and there are no irregular verbs and no exceptions to the grammatical rules. Esperanto employs the familiar symbols of the Roman alphabet with each one standing for just one sound, greatly facilitating both spelling and pronunciation.

Strategy 1: Be familiar with basic paragraph organization and function.

Each paragraph is organized around a unique supporting point and function. The introduction may include a hook, background, and thesis statement. Each body paragraph begins with a topic sentence, which discusses one key part of the thesis, and the topic sentence is followed by detail (examples, description, explanation, etc.) that supports the topic of that paragraph. Finally, the conclusion summarizes the main idea with some general comment, recommendation, or prediction.

Depending on the paragraph where the new sentence must be inserted, pay attention to how that paragraph is organized and used in the overall passage. The following sample question includes a new sentence that must be inserted in the introduction of the excerpt above:

> Look at the four squares [■] that indicate where the following sentence would be added to the passage.
>
> Not usually meant to replace existing mother tongues, the language is instead designed to play an auxiliary role, furthering international communication.
>
> Where would the sentence best fit?

Strategy 2: Think about the overall sequence of ideas in the paragraph.

Most paragraphs begin with a general statement and then become progressively more specific, or focused. For instance, if an introduction begins with a hook, the hook is not a supporting detail for the essay; the hook introduces some idea(s) related to the main topic, but the writer then focuses the paragraph by giving background and then identifying the specific main topic in the thesis statement at the end of the introduction. Likewise, a body paragraph begins with a topic sentence, which is the most general sentence in the paragraph. After the topic sentence, the detail focuses on increasingly specific related aspects of the topic.

Since the basic progression of ideas in a paragraph is from general to specific, you must insert the new sentence in a location where it can best fit this flow of ideas. Except for the first choice above, which comes at the beginning of the introduction, each choice comes between two sentences, and you need to ensure that the new sentence can follow the sentence before and precede the sentence after. The introduction above has the following sequence of ideas: a definition of artificial languages, the current name of a specific language (Esperanto), the place and time of publication, the original name, detail related to the creator, and an English translation of the name *Esperanto*.

Strategy 3: Examine the vocabulary in the new sentence.

When you read the new sentence that you have to insert into the passage, you need to identify how general or specific the new sentence is based on the vocabulary. For instance, the new sentence from the sample question above contains the words *mother tongues, the language, auxiliary role*, and *international communication*. This vocabulary is very general since the words are not names and they are not connected to a particular time, place, or context. However, the noun *the language* includes the specific pronoun *the*, which indicates the noun is not the first mention of the idea. (The first mention of a noun usually requires a general article: a, an, some). Vocabulary and articles are cohesive devices, which are vital when you compare the new sentence to those in the passage.

Strategy 4: Compare the cohesive devices in the new sentence to be inserted and in the paragraph.

Cohesive devices are words, phrases, or expressions that a writer uses to connect ideas within a sentence, between sentences, or between paragraphs. Cohesive devices include pronouns (he, it, which, etc.), adjectives (all, this, the other, etc.), articles, conjunctions (because, if, etc.), vocabulary (repetition, synonyms, varied word form), and transitions (therefore, however, etc.). The correct location for the new sentence depends on the connections between the cohesive devices in the new sentence and those in the sentences before and after the chosen location (see also Referent Questions: Strategies 1 and 6 for more about referents and cohesive devices).

Therefore, first you must identify any useful cohesive devices in the new sentence. For example, the sentence in the sample Coherence question above includes the verb phrase *meant to replace*, the adjective/noun phrase *existing mother tongues*, and the noun *the language*. These words suggest a connection to the first sentence of the paragraph, which contains the related noun *an artificial language* and the phrases *intended to transcend* and *diverse linguistic backgrounds*. Also, the second sentence of the paragraph mentions a specific artificial language (Esperanto), so the new sentence cannot be inserted after the second sentence since the new sentence contains general, unidentified vocabulary.

Strategy 5: Identify sentences in the paragraph that cannot be separated in order to eliminate answer choices.

The cohesive devices in the passage also indicate those sentences that cannot be interrupted, or separated, by a new sentence. For example, the use of the name *Esperanto* in both the second and third sentences means that the new sentence cannot be inserted between those sentences. The new sentence doesn't discuss a specific artificial language, so the repetition of the name indicates that the third choice is incorrect. Moreover, some cohesive devices, such as transitions, indicate rhetorical structure, which cannot be interrupted.

Strategy 6: Check the rhetorical structure of the new sentence and those in the paragraph.

Rhetorical structure, such as description, definition, and explanation, refers to the use of ideas for a particular purpose. This use is an important part of the sequence of ideas. First, certain ideas logically follow others: an idea is introduced or described generally first and then it is followed by examples; a topic is identified/named before any other detail (times, location, function, origin, etc.) is mentioned; a problem is normally discussed before any solutions; instructions or the steps in a process are organized chronologically; etc. Therefore, you need to identify the rhetorical structure of the new sentence as well as the rhetorical structure of the sentences surrounding the possible choices.

Also, some of the rhetorical structures could be extended over more than one sentence. A paragraph could provide examples of a topic, and these examples are given in two or three sentences. This similarity in rhetorical structure would connect those sentences together into a series of related examples. Likewise, a problem and solution probably would not be in the same sentence.

The new sentence to be inserted in the excerpt above clarifies the purpose of an artificial language. It gives more specific detail about how all artificial languages are used. Like the general vocabulary (generic reference to all artificial languages) and cohesive devices (related vocabulary), the rhetorical structure relates the new sentence to the first sentence of the paragraph. The second sentence in the introduction names a specific language, and the subsequent sentences all discuss details related to Esperanto. A clarification about the uses of artificial languages would not fit between any of the later sentences. Therefore, you must decide between the first or second location at the beginning of the paragraph.

Strategy 7: Insert the sentence into the paragraph and check for logic and clarity.
Once you have analyzed and compared the new sentence and those in the paragraph, you can click on the ones you haven't eliminated to check each one. This gives you a chance to read the paragraph again with the new sentence inserted at a given square. You can recheck the sequence of ideas, vocabulary, cohesive devices, and rhetorical structure. For instance, the best choice for the question above is (B), the second square. The first square is wrong because it comes before the noun *language* has been mentioned, which doesn't fit the noun *the language* in the new sentence. By choosing the second square, the new sentence can follow the reference to *an artificial language* in the first sentence. Also, the new sentence continues the general discussion by giving a specific function for the language type. This fits the sequence of ideas since the next sentence begins the discussion of Esperanto.

Question Forms

You can identify a Coherence question based on the squares indicating places in the passage, and the instruction to insert, or add, a sentence to the passage. A Coherence question resembles the following form or some variation:

> Look at the four squares [■] that indicate where the following sentence would be added to the passage. [Sentence] Where would the sentence fit best?
>
> Where do you think the sentence would fit best?

Distracters

Like all reading questions, coherence questions are multiple-choice. Only one choice is correct, and the other three are incorrect answers. Distracters are incorrect answers that seem correct even though they aren't. Basically, incorrect answers for this question type break some of the recommendations in the strategies above. Certain choices might match one or more of the criteria above (sequence, vocabulary, cohesive devices, etc.) but lack others. Therefore, you shouldn't rely on just one factor when you answer Coherence questions. Coherence questions require a combination of criteria. Choosing an answer based on just one factor can lead to incorrect answers and a lower score.

Distracter 1: Answer choice follows or precedes some repeated vocabulary only.
Many students rely on key words to answer coherence questions. Although vocabulary is important for many strategies, such as cohesive devices, it can also distract you. For example, the noun *language* and the adjective *international* are key words in the new sentence that you have to insert in the paragraph. Several choices in the question might seem correct based on those key words. For example, the new sentence includes the noun the *language*, and every sentence in the introduction includes either the singular noun *language* or the plural form *languages*. This repetition should be a strong clue that you can't rely on that word alone to choose a location for the new sentence.

Moreover, the third sentence in the introduction ends with the words *international language*. Some students might incorrectly assume that this use relates to the use of the same adjective in the new sentence, which ends with the words *furthering international communication*. However, you need to pay attention to the other vocabulary of each sentence and the rhetorical structure of each one. The new sentence uses *international communication* in a discussion of the uses for all artificial languages (the noun *language* is generic in the new sentence), but the third sentence uses international language to translate the original Russian name for Esperanto. This translation relates to a discussion of a specific artificial language, not artificial languages overall.

Distracter 2: Answer choice is next to the right sentence but on the wrong side.
One of the many confusing aspects of coherence questions is that often there is a square before and after several sentences. For example, the introduction above has four consecutive possibilities. A careless or rushed student might identify the correct part of the paragraph, but choose the square on the wrong side: either before a sentence instead of after it, or after a sentence instead of before it. For example, the first square – choice (A) – comes before the right sentence. However, the new sentence should be inserted after the first sentence, not before it, due to the use of the articles (first mention = an, second mention = the).

Practice

Now, practice the strategies you've just learned along with the general reading strategies. Read the passage below, and answer the coherence questions that follow.

The Sense of Balance

One of the most important physiological senses in humans and other animals is equilibrioception, better known as the sense of balance. This is the sense that allows us to do such things as walk upright or turn in either direction without falling, and keep objects in visual focus as the body moves.

In humans, the sense of balance is maintained by a complicated set of relationships between the eyes, ears, skeletal and central nervous systems, and the brain, which processes information from all these senses. Balance problems can occur when the brain receives conflicting information from the different sense organs, or when disease affects one or more of these organs.

[■] Of all the organs that help maintain equilibrioception, perhaps the most important is the ear. To understand how the ear helps us maintain our sense of balance, it is useful to examine the ear's anatomy. [■] The outer ear—consisting of the outside, visible part of the ear (pinna), and the ear canal—primarily assists the hearing, or auditory, system. It acts as a type of preamplifier by collecting sounds from the environment and funneling them to the eardrum. The eardrum, or tympanic membrane, is a thin, flexible membrane that separates the outer from the middle ear. Its primary function is to transmit sound to the ossicles, the three small bones in the middle ear. [■] These three bones—known as the hammer (malleus), anvil (incus) and stirrup (stapes)—further transmit sound to the cochlea, which is located in the inner ear. The cochlea is a coiled, snail-shaped organ filled with a watery liquid and thousands of sensitive "hair cells," that detect different sound frequencies and transmit them to the brain, where they are interpreted.[X] Together, the outer ear, middle ear, and the cochlea, comprise the ear's hearing, or auditory, system.

[■] Within the inner ear are the sensory organs that detect orientation and movement. [■] The ear's vestibular, or balance system, includes three semicircular canals—the posterior canal, the superior canal, and the horizontal canal—and the otolith organs, which have two fluid-filled cavities called the utricle and the saccule. Movement of fluids within the semicircular canals trigger hair cells that signal the brain about the direction and speed of rotation of the head, for example wether we are nodding, or looking left or right. [■] The otolith organs detect our linear orientation, for example, our speed and relationship to gravity when we are walking in a straight line. The brain interprets the information it receives from the ear's vestibular system, and combines it with the information it receives from other sense organs. For example, the eyes send visual orientation about the body's position in relation to its surroundings. [■] The brain compares this information with information from the vestibular and skeletal systems (joints, skin, muscles), to maintain its overall orientation and balance.

When balance is impaired, an individual has trouble maintaining his or her orientation to the environment. [■] Balance impairment may occur temporarily when the brain tries to process conflicting information from the various sense organs—for example the motion sickness an individual experiences from trying to read while riding in a car—or it may become chronic as a result of diseases that affect the organs of the vestibular system, or injury to the part of the brain that processes balance-related information. A certain amount of deterioration in equilibrioception is also natural, due to aging. [■] As people age, the hair cells in the cochlea begin to die, resulting in hearing loss. [■] Likewise, the hair cells in the vestibular organs also die off with time. It is thought that one of the reasons many elderly people fall and break their hips and other bones, is due to the impairment to the balance system caused by the deterioration of these hair cells. [■] Regardless, many chronic balance disorders may be treated with medication, physical therapy, or surgery. A person experiencing ongoing dizziness or loss of balance should always be evaluated by a physician.

27. In paragraph 3, look at the four squares [■] that indicate where the following sentence would be added to the passage.

> The ear is generally divided into three main sections: the outer ear, the middle ear, and the inner ear.

Where would the sentence best fit?

28. In paragraph 4, look at the four squares [■] that indicate where the following sentence would be added to the passage.

> These organs are entirely separate from the hearing organs.

Where would the sentence best fit?

29. In paragraph 5, look at the four squares [■] that indicate where the following sentence would be added to the passage.

> Symptoms of balance impairment may include dizziness, nausea, a sense of falling, visual blurring, or a number of other indicators.

Where would the sentence best fit?

Question Type 8: Paraphrasing Questions

A paraphrase is a restatement of ideas in new words and new sentence structure with the same meaning. You've already seen paraphrasing in many question types already since correct answers often paraphrase some or all of the relevant information from the passage. This is particularly important in detail questions and NOT/EXCEPT questions as well as inference and rhetorical structure questions. However, paraphrasing is also tested in its own question type. A paraphrasing question refers to a highlighted sentence in the passage, and usually the question also specifies the paragraph by number. Like all reading questions, paraphrasing questions are multiple-choice questions, and only one choice is correct.

To answer paraphrasing questions correctly, you need to be aware of the related methods of paraphrasing and the kind of language variations to look for. First, read the excerpt below, which includes a sample paraphrasing question used in the explanation of the strategies. Then, review the strategies for identifying the correct paraphrase and eliminating distracters.

The Eye of the Storm

Did you know that both hurricanes and tornadoes have eyes? As a meteorological term, *eye* describes a circular region at the center of a severe wind storm. The eye of a hurricane and the eye of a tornado have one thing in common: the air circulating around them moves in a counterclockwise direction in the northern hemisphere; in the southern hemisphere, both storms spin clockwise. However, the eye of a hurricane and the eye of a tornado differ in various ways based on size and speed.

The eye of a hurricane is calm, compared with the rest of the storm. In the eye of a hurricane, you may experience light, variable winds. This calmness contrasts with the surrounding destructive winds that create a swirling spiral with speeds up to 200 miles per hour. These violent winds are actually thunderstorms that form a ring around the eye. Also, hurricanes are much bigger than tornadoes, and therefore, their eye is much larger. It can be several miles across, even as much as 50 or more miles in diameter.

The eye of a tornado, on the other hand, is far from calm. A tornado's eye is composed of strong ascending and descending air currents that can lift up and destroy almost everything in their paths. Smaller than the hurricane's eye, the eye of a tornado may be only a few feet or yards in diameter. Because this area is so small and irregular, it is difficult to observe. The only real accounts we have of what the eye of a tornado is like come from people who found themselves in a tornado's path. Survivors who have experienced the center of a tornado report a brief silence in the eye, and a strange, blue glow. Looking up in the eye of a tornado, one might see a hollow column with a slick, opaque surface, resembling the inside of a pipe.

Strategy 1: Be familiar with the methods of paraphrasing.

The following chart reviews the various ways of paraphrasing a sentence. Although each method is discussed individually below, most paraphrases involve a combination of the following methods:

Paraphrasing Method	Examples from the Excerpt (paraphrasing in *italics*)
Synonym A synonym is a word, phrase, or expression with the same meaning as another word, phrase, or expression.	As a meteorological term, "eye" describes a circular region at the center of a severe wind storm. As a meteorological *concept*, "eye" *refers to a round area in the middle* of an *extreme* wind *disturbance*.
Voice Voice refers to the use of active verbs (active voice) or passive verbs (passive voice).	In the eye of a hurricane, you may experience light, variable winds. In the eye of a hurricane, light, variable winds *may be experienced*.
Variation of Word Form and Structure Word form refers to a word's part of speech, or type: noun, adjective, adverb, verb, infinitive, etc. Usually, when the word form is changed, the sentence structure must also be changed accordingly.	Because this area is so small and irregular, it is difficult to observe. *People have difficulty* observing this small and irregular area.
Phrase/Clause Structure This is a specific variation of word form and structure. A clause requires at least one subject and one verb; a phrase does not contain either a subject or verb. Paraphrases often alternate between them.	However, the eye of a hurricane and the eye of a tornado differ in various ways based on size and speed. However, the eye of a hurricane and the eye of a tornado differ in various ways based on *how big they are* and *how fast they go*.

Strategy 2: Focus on the words that add the most meaning and purpose in the original sentence.

A paraphrasing question does not repeat the original sentence. For example, look at the following sample question about the highlighted sentence in the excerpt above:

> Which of the following best expresses the essential information in the highlighted sentence? *Incorrect* answer choices change the meaning in important ways or leave out essential information.
>
> ○ The ascending air currents in the tornado's eye cause more damage than the descending currents.
>
> ○ Wind is dangerous and can destroy almost anything.
>
> ○ The tornado's eye can be destroyed by violent air currents that move up and down.
>
> ○ The eye of a tornado has powerful rising and falling winds that are capable of causing a great deal of damage.

Once you identify the question type based on the key words (essential information, highlighted sentence) and form, return to the passage and find the highlighted sentence. Reread the sentence closely and carefully. Pay attention to the key words, or content words, in the sentence; the content words are those words that give the essential meaning in the sentence: nouns, adjectives, adverbs, verbs, infinitives, gerunds, adverb conjunctions (because, since, after, etc.), and transitions.

Also, eliminate the less essential function words; these words perform a grammatical function in the sentence, but they add little or no meaning to the sentence. Function words include articles, many prepositions, relative pronouns, auxiliary verbs (*be* in the passive, *have* in the perfect tense), and coordinating conjunctions in parallel structure (and, but, yet). You could return to some of these words later when you finalize your choice, but you want to focus on the most meaningful words first. In the highlighted sentence above (It is composed of strong ascending and descending air currents that vacuum up and destroy everything in their paths), the essential words are *composed, strong, ascending, descending, air currents, vacuum up, destroy, everything.*

Strategy 3: Identify the relationships among the ideas and the rhetorical structure of the sentence.

It's important that you identify clearly any actors, objects, actions, or states. You need to be certain about what or who is causing any action or state in the sentence (actors) as well as what or who is receiving any action (objects). Since voice can vary without affecting meaning, the active or passive voice of the verbs is an important part of this. Finally, note any specific details, such as descriptions, comparisons, reasons, places, times, and sequences. These ideas are often rearranged and distorted in distracters, or incorrect answers, so you need to identify them correctly when you analyze the original. For example, in the highlighted sentence above, air is moving up and down and its movement is both forming the eye and causing a lot of destruction. Moreover, the rhetorical structure of the sentence involves the description of the eye of the tornado and its effects. The correct paraphrase can use different words and structure, but it must still have the same rhetorical structure, the same details, and the same relationship among them.

Strategy 4: Create a simplified sentence in your own words on scrap paper.
Since the original sentence is not repeated in the question, you want to avoid going back and forth between the answer choices and the passage. Although this is possible since the reading passage is visible with every question, you will waste time if you do it repeatedly. Later, once you've made a choice, you can jump back to the passage to make a final check, but right away you should avoid going back and forth and rereading the original and the choices too often.

Therefore, record the essential words from the highlighted sentence on a scrap piece of note paper that is provided to you at the beginning of the test. Depending on the length of the original sentence and the difficulty of the vocabulary, you could start by copying just the key words in fragmented note form. However, it's best if you can briefly paraphrase the sentence in your own words right away.

Don't worry about grammar or style since this sentence is not marked and won't count for or against your final score. All you need is an accurate record of the basic main idea of the sentence, so you can reference it once you return to the question. For example, a paraphrase of the highlighted sentence above could look like the any of the following:

> Composition, eye, tornado, wind, up, down, destruction (key words in fragments only)
>
> The composition of the eye is violent wind going up and down, and many things are destroyed.
>
> Great gusts of wind move up and down and form the eye of the tornado, destroying a lot.
>
> Very strong wind forms the tornado's eye and causes great destruction.

Strategy 5: Select the answer choice that best matches your paraphrased record of the original.
The correct choice doesn't have to match your paraphrase exactly since there are always several ways of restating something. Instead, look for similarities based on the methods outlined in Strategy 1: repetition, synonyms, variations of word form, and phrases and clauses with the same meaning. Also, it is very important that the actors, objects, actions, and states are the same. As you will learn below, the distracters for this question type rearrange and add to the details in the sentence. Based on the sample paraphrases in Strategy 5, the best choice is (D).

Strategy 6: Quickly verify your answer choice with original in the passage.
Your own paraphrase is useful for identifying the best choice(s) efficiently, but it shouldn't be the final guide to the best answer. Once you have eliminated as many choices as possible (based on actors, objects, rhetorical structure, etc.), you can compare your final choice with the passage. You should verify that you haven't omitted or misunderstood anything from the original. For example, (D) remains the best choice for the sample question above.

Question Forms

A paraphrasing question can be recognized based on the reference to a highlighted sentence in the passage, and the use of key words, such as *essential information*. You can identify this question type based on the following forms, or some variation:

> Which of the following best expresses the essential information in the highlighted sentence? *Incorrect* answer choices change the meaning in important ways or leave out essential information.

Distracters

A paraphrasing question is a multiple-choice question with only one correct answer. The other three choices are incorrect answers, or distracters. A distracter is an incorrect answer that is not obviously incorrect; it may seem correct due to various factors, but it is still incorrect. You need to be aware of the types of distracters for Paraphrasing questions in order to avoid them.

Like the distracters for many question types, one of the common patterns in distracters for paraphrasing questions is the repetition of one or more ideas from the original. All correct answers for multiple-choice questions restate some or all of the information from the passage, but this repetition is more obviously wrong in this question type because it involves paraphrasing. At the same time, some ideas are not easily paraphrased, such as dates and names, so repetition alone is not proof that a choice is incorrect but it is a strong clue. Therefore, you should always be wary or suspicious of answer choices that repeat too many ideas from the passage. The following strategies explain the various distracters and how to avoid them. Although each distracter is discussed separately, many distracters use a combination of methods.

Distracter 1: Answer choice rearranges ideas from the original.
You need to pay careful attention to the specific relationship among the ideas (actor and object, cause and effect, etc.) in any answer choice because this distracter uses the same basic ideas but puts them in an incorrect order or relationship. This distortion of the original sentence could involve switching the actor and object, or changing a cause into an effect, among many other modifications. Like so many distracters, this distracter often repeats ideas from the passage. Choice (C) is an example of this type of distracter. The answer choice uses a correct idea from the original (the air currents destroy things) and correctly paraphrases the movement of the air currents (ascending and descending = that move up and down), but the sentence uses a passive verb and incorrectly states that the currents destroy the eye. In the original, the air currents both form the eye and cause destruction.

Distracter 2: Answer choice adds unmentioned ideas.
You must identify the key content words in the original sentence, and make sure that you don't choose an answer that includes ideas that are not in the original. This type of distracter can also involve the kind of rearrangement described in Distracter 1. (A) is an example of

this kind of distracter; it repeats some words from the original (ascending, descending, air currents), and the choice incorrectly adds the comparative adjective *more*, which is not in the original. Also, the incorrect comparison of the air currents represents a rearrangement of the ideas similar to that in Distracter 1 above.

Distracter 3: Answer choice omits key information.

Some distracters can correctly paraphrase part of the original but leave out important meaning. This is another reason that you need to be certain about the key ideas in the original sentence. Because this distracter leaves out key information, the vocabulary in the answer choice might seem too general, too uncertain, or too unclear. Answer choice (B) is an example of this type of distracter. The choice refers to the unspecified noun *wind*, which is far too general. The wind in the original is identified as part of a tornado and described as moving in two different directions (up and down). None of that information is included in (B).

Practice

Now, practice the strategies you've just learned along with the general reading strategies. Read the passage below, and answer the paraphrasing questions that follow. Also, you can review coherence questions for the same passage.

Semiotics in American Pop Culture

The visual images, signs, and symbols, of pop culture are important for you to understand if you want to understand Americans. It is not enough to be able to speak and understand the language when you come to the U.S.A; you also need to be able to interpret the messages contained within the signs and symbols. These signs and symbols are shorthand representations of the culture's abstract ideas and concepts. But how does one interpret these signs and symbols of a culture? Is there a road map?

The study and interpretation of signs and symbols is the province of a field known as semiotics. Semiotics is a special branch of communication studies, which specifically combines the techniques of sociolinguistic analysis with anthropological analysis. ■ When a person interprets a sign or symbol, that person engages in an act known as semiotic decoding. ■ An example of semiotic decoding is reading. ■ The words and letters are symbols that must be interpreted. To a semiologist, anything can be taken as a sign to be decoded and analyzed for meaning. ■

Semiotics, as a field of study, was formally established in the late nineteenth and early twentieth centuries mainly through the writings and teachings of two men: Charles Sanders Pierce, an American philosopher and physicist, and Ferdinand de Sausurre, a Swiss linguist. Roland Barthes, a Frenchman, furthered the semiotic field in the 1950s by using semiotic techniques to analyze pop culture. To apply the theories of semiotics to pop culture is actually a lot of fun for anyone to do.

One area of interest in semiotics involves decoding the signs of cultural identity. A typical American university dormitory room or an American college student's bedroom at home provides a good picture of the identity or profile of that person. What decorates the walls? Is it a picture of James Dean? Che Guevara? Madonna? It is a shot of Michael Jordan? What is on the shelves? A teddy bear collection? Or football paraphernalia? Is there a CD system in the room? What music titles are there? ■ Is there a predominance of sixties retrograde collections like the Beatles, Simon and Garfunkel, Motown and the Doors? ■ Or seventies disco music or crossover country rock? ■ Or eighties rap and new wave? ■ Is it easy listening music, salsa, blues, jazz or soul? Is there a computer in the room? What type is it? A Macintosh® or a Dell®? All of these commercial symbols carry with them images and styles that suggest you are one type of person or another.

It is possible to conduct your own semiotic analysis. Every decoration choice a person makes in a room tells us something about that person's identity. Remember, a cultural sign gets its meaning from the system in which it appears. Its significance does not lie in its usefulness but rather in its symbolism—in the image that it projects, and that image always has a social significance. To decode what is in a person's room, you have to ask what that person is trying to say with these objects—and what that person wants other people to think about him or her. Try doing this with one of your friends.

Americans often refer to different decades in our history when certain fads or fashions were in vogue. The decade or time frame in which a given style was popular provides the key to the system for explaining it. ■ Evian® water for instance, only came into fashion in the health-conscious, body-building eighties. ■ American cultural trends change with every decade. ■ Americans speak of the sixties, the seventies, the eighties. ■ You will find classic rock stations on the radio that only play music from a certain decade They carry a certain image by doing so. Fashions and styles are dated by the decade in which they gained prominence. Sometimes we look at someone and say "he's so seventies!" or "she's so eighties!" Decade to decade shifts in styles in America have been occurring for much of American history, but clearly reached a crescendo in the twentieth century—the age of mass marketing…

30. Which of the following best expresses the essential information in the highlighted sentence in paragraph 1? Incorrect answer choices change the meaning in important ways or leave out essential information.

   ○ Interpreting the meaning of American signs and symbols is a process similar to learning to speak a language.

   ○ It is as important for visitors to understand American signs and symbols as it is for them to speak and understand the language.

   ○ People visiting the United States may not be able to interpret the signs and symbols as easily as they speak and understand the language.

   ○ A person arriving in the United States must not confuse the culture's signs and symbols with the meanings these symbols have in his or her own culture.

31. Which of the following best expresses the essential information in the highlighted sentence in paragraph 3? Incorrect answer choices change the meaning in important ways or leave out essential information.

   ○ People are primarily attracted to a product's symbolic meaning.

   ○ Commercial symbols are interpreted differently by different people.

   ○ Some types of commercial symbols convey more meaning than others.

   ○ The commercial products a person buys provide information about that person.

32. Which of the following best expresses the essential information in the highlighted sentence in paragraph 5? Incorrect answer choices change the meaning in important ways or leave out essential information.

    ○ Styles are easiest to decode in the decade during which they are popular.

    ○ The system that surrounds popular styles changes at least once every decade.

    ○ The popular styles of a decade provide a key for understanding that period of time.

    ○ To understand a given style, one must understand the period of time when it was popular.

33. In paragraph 1, look at the four squares [■] that indicate where the following sentence would be added to the passage.

    > When you read, you look at graphic symbols and place meanings on them.

    Where would the sentence best fit?

34. In paragraph 3, look at the four squares [■] that indicate where the following sentence would be added to the passage.

    > An American's music collection can tell more about that person than anything else.

    Where would the sentence best fit?

35. In paragraph 5, look at the four squares [■] that indicate where the following sentence would be added to the passage.

    > Another French bottled water, Perrier®, is more associated with the preppie cocktail parties and mixers of the seventies.

    Where would the sentence best fit?

Question Type 9: Drag-n-Drop Table Completion Questions

A table completion question asks you to categorize the details from the passage. It provides a mixed list of choices on the left side of the table and two distinct categories on the right side. The list of choices contains the key details from the passage as well as two distracters, or incorrect choices. Each category is based on a type, or kind, from the passage, and the name of one type appears at the top of each category. You must use the drag-n-drop function of the mouse to move the correct choices into the appropriate category.

The mouse's drag-n-drop function allows you to move an item on the computer screen. With your mouse, you must first place the arrow over a key point from the list on the left side of the table. You can move the item by clicking once with the left tab of the mouse and holding the tab down while the arrow is directly over the item. Then while pressing down on the mouse, you can move, or drag, the key point to the appropriate category on the right side of the table. Once you have chosen a category, you can place, or drop, the idea in the category by releasing the left tab of the mouse.

Table completion questions have an uneven number of choices, usually seven or nine; two of those choices are distracters, or incorrect answers, which should not be placed in either category. The remaining choices (usually five or seven) must be correctly placed in the appropriate category. Moreover, each category contains a specific number of empty spaces, so you can easily recognize the number of correct choices for each category. Depending on the number of choices, this question is worth 3 or 4 points, and the question indicates the value in the instructions.

In order to receive full points for this question, you must identify the correct details from a list of paraphrases, and connect the details to one of two supporting points from the passage. The strategies for this question involve a good initial reading of the passage, an outline of the passage, and scanning for details. First, read the following excerpt of a passage, which is used for a sample table completion question. Then review the strategies that explain how to answer the sample question.

Jazz and Bebop

For a jazz musician living in New York City in the early 1940s, the most interesting place to spend the hours between midnight and dawn was a Harlem nightclub called Minton's. After finishing their jobs at other clubs, young musicians like Charlie Parker, Dizzy Gillespie, and Thelonious Monk would gather at Minton's and play jam sessions—informal performances featuring lengthy group and solo improvisations. The all-night sessions resulted in the birth of modern jazz. Working together, these African-American artists forged a new sound, known as bebop.

Before bebop, swing had been America's popular form of jazz. Led by such virtuoso instrumentalists as Benny Goodman, Count Basie, and the Dorsey Brothers, swing bands were primarily dance bands. They were large, with 12 to 16 musicians on average, and arrangers were usually key to their success. Swing bands concentrated on tight and precise ensemble playing. As the swing style developed, musicians began to incorporate more technically and harmonically advanced approaches to the music, which influenced the younger musicians that came to form the bebop sound.

Unlike swing, bebop was not dance music. It was often blindingly quick, incorporating difficult, irregular rhythms and discordant sounds that jazz audiences had never heard before. Bebop was based on a 12-note scale, opening up new harmonic opportunities. The musicians who pioneered bebop shared two common elements—a vision of the new music's possibilities and astonishing improvisational skill, the ability to play or compose a musical line on the spur of the moment.

Like many revolutions, unfortunately, the bebop movement encountered heavy resistance. Opposition came from older jazz musicians, and from a general public alienated by the music's complexity. Furthermore, due to the government ban on recording that was in effect during the early years of World War II, the creative ferment that produced bebop remains largely undocumented today.

Strategy 1: Anticipate this question type based on the organization of details in the passage.

A table completion question is based on the fact that multiple details can be related to different people, places, times, types, and so on. Therefore, you should recognize whenever a passage contains two broad ideas that each involve distinct details, for example: the reading passage is organized according separate types; the passage mentions two people who do several different things; and multiple events occur in two places. For example, the excerpt above refers to two types of jazz, swing, and bebop. This should be a clue that you may have to categorize details according to these types later on.

Of course, you can never be certain which questions will come with the passage, but the characteristics mentioned above are good clues that a Table Completion question could come. Look at the following sample Table Completion question:

Directions: Select the appropriate phrases from the answer choices and match them to the style of music to which they relate. TWO of the answer choices will NOT be used. This question is worth 4 points.

Answer Choices	Swing
Grew out of classical music traditions	•
Much of the music was not recorded	•
Primarily a form of dance music	•
Concentrated on tight ensemble playing	
Innovative harmonies, based on a twelve-note scale	**Bebop**
Opposed by the U.S. government	•
New, fast, challenging rhythms	•
Not well received by the general public	•
Usually consisted of more than a dozen musicians	•

Strategy 2: Record any organization of detail in your outline.
An outline is very useful for many question types, but especially the table completion questions (and summary questions discussed later). Since you have to locate and verify more choices in this question than you do in most multiple-choice questions, an outline can help you find information more quickly. Your outline can function like a map, guiding you to the right locations in the passage. For example, look at the following outline for the excerpt above. The outline clearly shows how the passage is organized by musical style: swing and bebop.

> #1 – jazz, NYC, Minton's, jam, informal
> #2 – swing = dance, 12-16 players, arrangers, precise, ↑ technical
> #3 – bebop = not dance, irregular, discordant, 12 note, more improve,
> #4 – bop not popular, too complex, gov ban WWII = no records

Strategy 3: Identify key words in the categories.
The name or title of each category is an important clue to the location of the relevant detail in the passage. Each category in the question could be just a word or a longer expression of several words. Identify the key idea in each category. For example, each category in the sample table above is just one word (Swing and Bebop), but other categories could be longer. Based on the categories and the outline, you should look primarily in the second and third *categories* for the relevant information, but some ideas could be found elsewhere.

Strategy 4: Be prepared to recognize restatements, not quotes.
As in all question types, the correct answers are paraphrased from the passage. Of course, not all ideas are easily restated, but paraphrasing is the common pattern in most Reading (and Listening) question types. Moreover, as you can review below, many distracters repeat, or copy, information directly from the passage.

Strategy 5: Use your memory and outline to place as many details as possible.
Try to categorize a few details right away based on your first reading and your outline. If you read the passage thoroughly the first time, instead of skimming it too quickly, you may remember enough to place a few choices correctly right away. You can always verify your categorization later, but you should try to limit the amount of scanning and rereading you have to do. For example, the sample outline above contains a few clues that identify the proper place for several details:

> Primarily a form of dance music = Swing
> Innovative harmonies, based on a twelve-note scale = Bebop
> Usually consisted of more than a dozen musicians = Swing
> Not well received the general public = Bebop

Strategy 6: Identify key ideas in the remaining choices, and scan for them in the passage.
Choose a unique or identifying idea from each answer choice that you can't categorize right away. Using your outline as a guide, scan the passage to find the relevant information. Read the passage carefully; don't skim it. The reading passages often contain complex information

that is very mixed. You may need to read one or more sentences a few times to ensure that you understand the information clearly. For example, the outline did not immediately identify the location for the following details:

Grew out of classical music traditions

Much of the music was not recorded

Concentrated on tight ensemble playing

Opposed by the U.S. government

New, fast, challenging rhythms

The key ideas in the choices are *classical*, *much not recorded*, *tight ensemble*, *opposed/U.S. government*, and *new, fast, challenging*.

Strategy 7: Don't place all choices; skip the right number of incorrect answers based on the instructions.

The instructions at the top of the table always indicate the number of choices that you should not choose. This information is emphasized with capital letters (TWO of the answer choices will NOT be used). You can eliminate the choices that are not mentioned in the passage or the choices that distort stated information from the passage (see also Distracters below). The correct answers for the sample table are as follows:

Swing
• Primarily a form of dance music
• Concentrated on tight ensemble playing
• Usually consisted of more than a dozen musicians

Bebop
• Much of the music was not recorded
• Innovative harmonies, based on a 12-note scale
• New, fast, challenging rhythms
• Not well received by the general public

Question Forms

You can recognize a table completion question based on the organization of the table (list of choices on the left and two or more categories on the right) and based on the need to categorize details. Review the following form for table completion questions:

Select the appropriate phrases from the answer choices and match them to the person/type/ purpose. TWO of the answer choices will NOT be used.

Drag your answer choices to the spaces where they belong. To review the passage, click on **View Text.**

Answer Choices	Category 1
Choice 1	
Choice 2	
Choice 3	
	Category 2
Choice 4	
Choice 5	
Choice 6	
Choice 7	

Distracters

Out of the total number of choices, a table completion question has two distracters, or incorrect answers, which should not be placed in either category. Since table completion questions involve details, the distracters resemble those for Detail questions, among other question types. Review the following explanations of the distracters for this question type.

Distracter 1: Answer choice includes unmentioned ideas.
It's important to make sure that all ideas in an answer choice relate to some information in the passage. You shouldn't rely on just one word when choosing your answers. Some of an answer choice could be mentioned in the passage, but if certain ideas have no relation, then the answer choice is a distracter. The first choice in the list (Grew out of classical music traditions) refers to classical music, which is unmentioned.

Distracter 2: Answer choice rearranges, or distorts, ideas from the passage.

Some answer choices can contain ideas from the passage, but they are in an incorrect order or relationship. Often, the answer choice changes the relationship of actor and object, or the choice uses a detail, such as time or location, from another unrelated idea. Also, the choice may contain vocabulary that is repeated throughout the passage, so students might incorrectly assume that it must be true if it is mentioned so often. The sixth choice in the list for the sample table above is an example of this type of distracter (Opposed by the U.S. government). According to the passage, the public opposed bebop due to its complexity, and the government banned new recording to save material during World War II. The distracter rearranges these details and incorrectly states that the government opposed the new music.

Practice

Now, read the passage below and practice your reading strategies. The passage is followed by a table completion question as well as coherence and paraphrasing questions.

Two Economic Giants

The last two decades have witnessed an important shift in the global distribution of economic power. ■ Since the middle of the 20th century, the United States has been the paramount economic force in the Western Hemisphere, with a peerless industrial base and strong technology and service sectors. ■ In the last 20 years, and especially in the last 10, China has emerged as a powerhouse, eclipsing Japan in the Pacific as Asia's largest economy and rapidly overtaking the global lead of the United States. ■ As a consequence, many economic parallels can now be added to existing geographic similarities between China and the United States. ■ Despite these similarities, there remain a number of significant economic and demographic disparities between these two economic giants.

Both China and the United States are very large nations. At 9.5 million square kilometers, China is only slightly smaller than the United States, which occupies 9.6 million km sq. Both countries occupy a similar range of latitude, with an attendant diversity of climatic conditions, and both have long coastlines and good ports. Like the United States, China is gifted with diverse natural resources, including metals and timber, and significant domestic energy sources, especially fossil fuels and hydropower.

Physical similarities are now joined by economic congruities. China's economic base is coming to rely on manufacturing and other industry. The country is quickly catching up with the United States in this area, with approximately US$2.7 trillion gross domestic product (GDP) in 2006, as compared to the American $13 trillion GDP. Like the U.S. government, China's government depends heavily on deficit spending: the U.S. public debt is currently about 60 percent of the American GDP, and China's has reached just over 20 percent of the Chinese GDP. However, annual inflation of consumer prices is below 3 percent in both countries.

■ China's economy, though rapidly industrializing, cannot be considered a free market economy based on the U.S. model, since major segments of the Chinese economy are government-run enterprises sheltered from competition. ■ Dependent on the mood of foreign consumers and investors, China's export-driven economy needs to develop a large and stable domestic market before it can achieve the diversity and resiliency of the U.S. economy. Not all of the differences between the two economies place China at a disadvantage, however. ■ For example, China's 10.7 percent annual growth in 2006 was triple that of the United States; the country enjoys a positive balance of payments, at over

$200 billion in 2006, in contrast to the massive $582 billion U.S. trade deficit in the same year; and China's capital investment rate far exceeds that of the United States. ■

The huge and increasingly well-educated Chinese workforce (which at 750 million is five times the size of America's) can be given much of the credit for China's rapid growth. Foreign investment finds the abundance of inexpensive labor an important advantage over the expensive U.S. labor market. Nevertheless, at least 10 percent of the urban Chinese population is unemployed, probably more in rural regions, whereas the U.S. unemployment rate hovers around 4.5 percent. A further consequence of China's large population is that, distributed among so many people, its $2.7 trillion gross domestic product is only around $2,000 per capita, compared to just over $42,000 per capita in the United States.

Persistent American fears of being overshadowed economically by China are not entirely unfounded; the Asian nation is indeed growing very rapidly. However, a developing energy shortage promises to slow that growth, and as Chinese labor costs rise, as they inevitably must, the country will become less attractive to foreign investment. By that time, however, China may well have acquired a healthy component of domestic capital, and its increasingly affluent population will in turn become a market for American exports. The next few years are likely to see the economic connection between China and the United States develop into a prosperous partnership.

36. Directions: The author of the passage mentions similarities and disparities between the United States and China. Select the appropriate phrases from the answer choices and match them to the category to which they relate. TWO of the answer choices will NOT be used. This question is worth 4 points.

Drag your answer choices to the spaces where they belong. To review the passage, click on **View Text.**

Answer Choices	True for BOTH China and the United States
✓ More than 9 million square kilometers in size	•
✓ Heavy reliance on deficit spending by the government	•
	•
Positive balance of payments in trade	
Largest amount of trade is with immediate neighbors	
✓ Consumption of significant amounts of fossil fuels	**True for China ONLY**
Per capita gross domestic product below $10,000	•
National gross domestic product below $5 trillion	•
Major segments of the economy face little competition	•
Low annual consumer inflation figures	•

Practice: Review

37. Which of the following best expresses the essential information in the highlighted sentence in paragraph 3? Incorrect answer choices change the meaning in important ways or leave out essential information.

   ○ China's export dependency has weakened its ability to develop a strong domestic market.

   ○ China needs to attract more foreign investors before it can achieve the U.S.'s economic strength.

   ○ Whether China's economy continues to grow will depend a great deal on its ability to increase exports.

   ○ China's economy is too reliant on outside economic forces, and it can't enjoy the strengths of the U.S. economy until it improves domestically.

38. Which of the following best expresses the essential information in the highlighted sentence in paragraph 4? Incorrect answer choices change the meaning in important ways or leave out essential information.

   ○ China's rapid growth provides a real challenge to the U.S.'s economic supremacy.

   ○ Fear of China has been growing in the United States, due to China's rapid economic growth.

   ○ While China's economy is growing quickly, it will be a long time before it overshadows the United States.

   ○ U.S. economists fear that China's economy is growing just as the economy of the United States has begun slowing down.

39. In paragraph 1, look at the four squares [■] that indicate where the following sentence would be added to the passage.

   > Yet the United States is by no means the world's only economic dynamo.

   Where would the sentence best fit?

40. In paragraph 3, look at the four squares [■] that indicate where the following sentence would be added to the passage.

   > Despite these similarities, a number of important differences must be noted, both in the two countries' economies and their demographic bases.

   Where would the sentence best fit?

Question Type 10: Drag-n-Drop Summary Questions

A drag-n-drop summary question asks you to choose three statements from a list of six possible choices. The question provides a restatement of the passage's main idea, or thesis, and the three correct choices are restatements of the key supporting points from the passage. Together, the correct answers form a summary of the passage and support the thesis statement provided with the question. A summary question is worth 2 points, which is indicated in the instructions of the question.

As the name suggests, this question type requires you to use the drag-n-drop function of your mouse. Like the table completion question above, a summary question requires you to use your mouse to choose a statement and move, or drag, it into an empty space. However, unlike the table completion question, this question type has only one list of empty spaces, not two categories.

In order to answer a summary question correctly, you need to understand the author's main idea and recognize the statements that best support it. Summary questions require three correct choices, and they include three incorrect choices, or distracters. These distracters can be too specific or unmentioned in the passage. The following strategies explain how to identify the correct answers and avoid distracters. First, read the excerpt below, which is used for a sample Summary question in the strategy explanation.

Fermentation

Fermentation is the key to winemaking, and is a process that occurs naturally within grapes as long as the skin of the grapes has been broken. This is because a grape's skin is covered by a type of yeast that consumes the fructose and glucose sugars found inside the fruit. This reaction creates alcohol, and careful management of the process can vary its concentration in the wine.

First, a winemaker's job is not as simple as just crushing the grapes because alcohol is not the only by-product of fermentation. Hydrogen sulfide, a compound known for its rotten-egg like odor, may also be created. Many other by-products, which may or may not be desirable, can also surface. Therefore, the winemaker has to guide the fermentation process and control its by-products to achieve the desired taste.

One way a winemaker can help the fermentation process is by regulating temperature. Fermentation creates heat, but yeast cannot tolerate much heat. When temperatures reach more than 98°F (37°C), the yeast becomes dormant, and fermentation ceases. This is why wine is usually stored in cool, underground cellars, or in large vats which can be hosed down with cold water. Therefore, a winemaker's first task is to keep the wine at a low temperature until it reaches the desired stage.

Adding sugar is another method of changing the outcome of a wine. A winemaker may put sugar or sweet unfermented grape juice in his wine, not to sweeten it but to raise its alcohol content. The yeast eats the extra sugar, and converts it into more alcohol. However, if the winemaker wants a stronger and sweeter wine, he may add even more sugar. If the alcohol level rises past a certain point, the yeast die—essentially poisoned by the very alcohol they are creating—and will no longer convert sugar to alcohol; the remaining sugar will then add sweetness, but not alcoholic strength.

Strategy 1: Record paragraph topics in your outline.

As you know, you should create a general outline of the passage when you first read it. Since your outline should only be a broad record, record mostly key supporting points in your outline, not too many details. Most often, the key supporting points are paragraph topics, so focus on these when you make your outline. Your notes should serve as a reminder, not a replacement for the original passage, so make sure to read the passage while you make notes.

Organize your outline to ensure that you can recognize the different paragraph topics when you review it later (see # in sample below), and group any details under the appropriate paragraph topic. Your notes can guide your scanning when you return to the passage, and if your memory is good, they may also allow you to make some choices right away.

Look at the following sample outline for the passage above. Notice how efficient notes rely on good paraphrasing and summary skills. It isn't always necessary to record exact words, and only key ideas, not all, are necessary:

> #1 – ferm
> key, natural, broken grapes/yeast eat sugar → alcohol
>
> #2 – more than crush
> control other by-prod/Hydro-sulf
>
> #3 – temp
> ferm makes heat, >98F = block yeast/ferm, cool
>
> #4 – add sugar
> yeast + sugar = more alc, ↑ kills yeast

Strategy 2: Use the thesis in the question as a guide.

A summary question includes a restatement of the thesis after the instructions. After you read the instructions carefully, use the thesis as a clue. It may not only remind you of the passage that you just read, it can also suggest the best choices. Remember that the best choices are key supporting points of the thesis, not just details from the passage. As the instructions always mention, one or more distracters could be a minor detail, so one or more answers could be wrong even though they are mentioned in the passage.

Look at the following sample summary question:

> An introductory sentence for a brief summary of the passage is provided below. Complete the summary by selecting THREE answer choices that express the most important ideas in the passage. Some answer choices do not belong in the summary because they express ideas that are not presented in the passages, or are minor ideas in the passage. *This question is worth 2 points.*

Drag your answer choices to the spaces where they belong. To review the passage, click on **View Text.**

Winemakers supervise fermentation to produce wine.

- _____
- _____
- _____

Answer Choices	
• Wine is created naturally. • Not all the results of fermentation are beneficial. • The yeast must be protected throughout the process.	• The enormous variety of wines in the world results from different fermentation techniques. • Grapes contain the basic ingredients for fermentation. • One by-product of fermentation has a bad odor.

The thesis is one clue to the best choices. Therefore, identify any key words in the thesis. For example, the thesis above restates the main idea of the passage, which is that wine requires careful management of the fermentation process. This main idea is reflected in the verb *supervise,* which is a restatement of the phrase *careful management of the process* in the original passage above. This should remind you of the emphasis placed on monitoring, controlling, and limiting various factors. This was a key theme in each paragraph of the passage, so your key points should reflect this importance of supervision. Supervision also implies that some things must be avoided, controlled, or stopped. The key word and its related ideas should help you choose your ideas.

Strategy 3: Be prepared for paraphrased ideas, not quotes.
Each correct answer choice in summary questions is a restatement of information from the passage. As you check your outline, think about the ideas, not the specific vocabulary, and try to use your understanding of the whole passage. As you'll see later, many distracters repeat vocabulary from the passage.

Strategy 4: Use your memory, outline, and the thesis to select as many choices as possible.
If you've read the passage well and recorded the key points in a well-organized outline, you should be able to identify at least one correct choice right away. The sample outline above indicates that fermentation is a natural process, and that the yeast and sugar in the grapes

create alcohol (natural, broken grapes/yeast eat sugar → alcohol). This is an example of a key point that is paraphrased in the list. Moreover, a general understanding of the passage should remind you that fermentation is a natural process used to create wine, but winemaking itself is not natural. You can understand this based on the fact that winemaking requires so much supervision of the fermentation process. Therefore, fermentation is natural, but winemaking is a product of human intervention.

Strategy 5: Eliminate overly specific choices based on vocabulary.

Since you know that one or more choices could be too specific, examine any choices that use specific vocabulary, such as names, titles, and dates. For example, the last choice in the right column refers to the odor of a particular by-product (one by-product of fermentation has a bad odor), which is a very specific reference in this context. Of course, your judgment of how specific the vocabulary is also depends on the main idea and other choices. You can compare any choice with the others to determine its level of specificity. The last choice mentioned earlier seems very specific when compared with the second choice on the left side (not all the results of fermentation are beneficial). Since you must select the best choices for a summary, the more general choice is probably the best, but you can always check any choice by scanning the passage.

Strategy 6: Identify key ideas in the remaining choices, and scan the passage.

Depending on how many choices you aren't sure about, you can either remember the key ideas, or note them down quickly on your scarp paper. As the instructions to the question indicate, you must press the *View Text* button to see the passage again. With your outline and key ideas, scan the passage for the relevant information related to the remaining questions. You might scan for the following information from the choices in the sample question: *results of fermentation*, *yeast must be protected*, and *variety of wines*.

Strategy 7: Be prepared to combine key points from different paragraphs.

Students sometimes focus too much on paragraph topics when they answer this type of question. As a result, they incorrectly assume that each correct answer must refer to only one paragraph. This is not true, however. In fact, because the answer choices represent a summary of the passage, one or more correct answer choices could combine information from more than one part of the passage. These separate ideas must have some similarity or connection, but you have to recognize that relationship even though the ideas are mentioned separately. Combining several ideas into one is a large part of summarizing. For example, the sample notes above indicate that yeast can be harmed by heat and also by alcohol level. This repetition of the dangers to yeast is a strong clue about one correct answer in the sample Summary question.

The correct choices for the sample question are the following:

> - Grapes contain the basic ingredients for fermentation.
> - Not all the results of fermentation are beneficial.
> - The yeast must be protected throughout the process.

Question Forms

You can recognize a Summary question based on the table, the instruction to choose three out of six choices, and the paraphrased thesis statement below the instructions. Identify a Summary question based on the following form:

An introductory sentence for a brief summary of the passage is provided below. Complete the summary by selecting THREE answer choices that express the most important ideas in the passage. Some answer choices do not belong in the summary because they express ideas that are not presented in the passage, or are minor ideas in the passage. This question is worth 2 points.

Drag your answer choices to the spaces where they belong. To review the passage, click on **View Text.**

(A restatement of the passage's main ideas appears here)

- _____
- _____
- _____

Answer Choices	
Choice 1	Choice 4
Choice 2	Choice 5
Choice 3	Choice 6

Distracters

A summary question has six choices; three are correct, and three are incorrect answers, or distracters. In all question types, a distracter can seem correct for a variety of reasons. In a summary question, some distracters are mentioned in the passage while others are not. Review the following strategies for identifying and avoiding distracters in Summary questions.

Distracter 1: Answer choice is a minor detail in the passage.

This is arguably the most difficult distracter for this question type because it can be found in the passage. It's important to remember that you must choose major supporting points, which are the ideas that provide the most support for the thesis. The last choice on the right side is an example of this kind of distracter (One by-product of fermentation has a bad odor). This choice represents an example of the undesirable results of fermentation, but it supports the key point that a winemaker must remove certain by-products of fermentation. This key point is paraphrased in the correct choice *Not all the results of fermentation are beneficial.*

Distracter 2: Answer choice distorts information from the passage.

Summary questions can include answers that manipulate ideas from the passage in a similar way to the distracters in Detail questions, among others. Be careful not to rely on just one key word or phrase when you choose your answers. Pay close attention to the specific relationship of the ideas. The first answer choice on the left side is an example of this type (Wine is created naturally). The natural quality of fermentation, not winemaking, is mentioned in the passage.

Distracter 3: Answer choice is unmentioned.

Although this distracter might seem obvious, it can sometimes be tricky or confusing. Often, this answer seems plausible, or likely, in a broader context. For example, the first choice on the right side demonstrates this quality: The enormous variety of wines in the world results from different fermentation techniques. In that incorrect choice, the ideas seem logical and possible. Although the truth might be different (wine variety also involves the type of grapes used and their location), the answer is wrong because the ideas are off-topic. The different types of wine are never mentioned or explained in the passage.

Practice

The Interpretation of Dreams

Sigmund Freud published his groundbreaking book, *The Interpretation of Dreams*, in 1899, after years of studying the brain as a neurologist. His book marked the real beginnings of research into the human mind, and led to the development of modern psychoanalysis.

Freud's basic insight that our minds preserve memories and emotions in ways that are not always consciously available to us profoundly and permanently transformed how humans viewed themselves. One of Freud's most important discoveries was that the repressed thoughts and emotions buried below the level of conscious awareness emerged in dreams, and that dreams could be used to uncover lost feelings.

Before Freud, dreams were dealt with in religious or metaphysical ways. In the Mesopotamian epic, *Gilgamesh*, written in 2700 B.C., dreams heralded the arrival of gods. In Babylon, Assyria, and Egypt, dreams were considered prophetic; rulers were known to heed their dreams before making military and other major decisions. The transition from a religious conception of dreams to an attempt at scientific explanation began in Ancient Greece. Pythagoras believed nightmares were caused by spoiled food, and Aristotle considered dreams to be manifestations of the soul, both internal to the individual, as opposed to supernatural interpretations. The European Renaissance initiated scientific inquiry into all manner of things, but it was centuries before scientific principles were applied to the interpretation of dreams. From the 16th to the 19th centuries, the brain was studied as an anatomical object, and by Freud's time neurologists had already begun to discover correlations between areas of the brain and specific neurological activities. But it remained for Freud to open up the mind, as opposed to the brain, to modern scientific investigation.

Freud's revelation that dreams might contain useful information came in 1895, when he dreamed about a patient of his whose treatment had not yielded its expected results. By carefully making conscious associations with the imagery in his dream, Freud was able to interpret the dream as representing an attempt by his mind to protect itself from the disappointment generated by his inability to heal his patient. From this experience, Freud concluded that parts of the human mind worked outside of our conscious awareness, and that dreams had meaning. Freud believed that, at the core, dreams represented a disguised fulfillment of suppressed or repressed wishes. This definition of Freud's emphasizes two key points: first, that dreams are disguised, and second, that the wishes represented in dreams are repressed. The logical conclusion Freud arrived at was that, by analyzing dreams, the meaning of the dream could be lifted into conscious awareness, and the dreamer liberated from self-censoring repression. Freud concluded that dreams were "the royal road to the unconscious," and suggested that they existed to keep our mental troubles from waking us up at night. Freud further proposed that the interpretation of dreams might be used as a means to understand, and ultimately treat, mental maladies. In his effort to understand dreams, Freud developed psychoanalysis, laying the foundation for all modern psychotherapies.

A century later, few psychoanalysts believe dreams primarily exist to protect sleep, but the central idea, that dreams are messages from a part of our mind outside our conscious control, is stronger than ever. And scientists are finding that dreams provide many insights into the mind's inner workings. In the last few years, brain-imaging techniques have shown what parts of the brain are active during dreaming.

Scientists are discovering that, when people dream, the parts of the brain involved in recognizing objects and processing images, and the areas involved in processing emotions and emotional memories, are as active as they are when the mind is awake, which explains why dreams can feel so realistic and powerful. And both experience and research show that logical thought is almost completely absent from dreams, which helps to explain their irrational nature.

Freud's book marked the real beginnings of scientific research into the mind, and the development of truer understanding of mental health problems. Without Freud, neurology might have remained limited to mapping neural connections, and psychology might never have developed insights that every day save millions of people from the misery of mental illness.

Practice: Drag-n-Drop Summary

41. An introductory sentence for a brief summary of the passage is provided below. Complete the summary by selecting THREE answer choices that express the most important ideas in the passage. Some answer choices do not belong in the summary because they express ideas that are not presented in the passage, or are minor ideas in the passage. This question is worth 2 points.

> Drag your answer choices to the spaces where they belong. To review the passage, click on **View Text**.

Sigmund Freud's contributions to our understanding of mental processes, especially regarding the role of dreams, cannot be understated.

- _____
- _____
- _____

Answer Choices

• Freud's revelation that dreams were important began with his conscious analysis of his own dreams.	• Freud's book, The Interpretation of Dreams, was hotly debated within the field of neurology.
• The history of human dream analysis shows a progression from religious to scientific interpretations.	• Freud's discovery that the mind preserves memories and emotions outside of conscious awareness formed the basis of modern psychoanalysis.
• The irrational nature of dreams is due to the fact that the part of the brain that controls logic is inactive during sleep.	• The scientific investigation of the the mind, as opposed to the brain, begins with Freud.

Practice: Review

42. Look at the word **prophetic** in paragraph 2. In stating that dreams were considered prophetic, the author means that they were considered

- ○ holy
- ○ intuitive
- ○ predictive
- ○ inspirational

43. Look at the word **maladies** in paragraph 3. The word maladies in the passage is closest in meaning to

    ○ cures
    ○ illnesses
    ○ behaviors
    ○ explanations

44. Look at the word **itself** in paragraph 3. The word itself in the passage refers to Freud's

    ○ mind
    ○ dream
    ○ disappointment
    ○ imagery

45. Look at the word **which** in paragraph 5. The word which in the passage refers to

    ○ the awake mind
    ○ the fact that parts of the brain remain active during dreams
    ○ the realistic feeling of dreams
    ○ the processing of images and emotions

IMPROVE-YOUR-SCORE STRATEGIES

There are many activities and habits that you can practice in order to improve your reading skills. The following strategies review some of the most useful things you can do.

Strategy 1: Read often.
This strategy is first because it is the most important. Like many skills, including all language skills, reading improves with regular practice. Moreover, you can create a series of related benefits: The more you read, the more you learn. The more you know, the better you can read.

Strategy 2: Read more than just TOEFL practice exercises.
It is common for TOEFL students to do many practice tests and TOEFL exercises. Unfortunately, this isn't the best way to improve general language ability. Exercises that are similar to those on the TOEFL, such as multiple-choice questions, are tests. Doing tests or practice tests is necessary to practice certain test-taking skills and strategies, but tests shouldn't be the only things you read (and listen to). Broaden your exposure to English by reading magazines (serious magazines), newspapers, and current affairs websites.

Strategy 3: Read about similar or related topics.

Learning new ideas is empowering and motivating. However, many TOEFL students study a wide variety of topics, which limits how much they can eventually learn and remember later. Variety is necessary to keep you interested, but you should focus your reading on a set of ideas, issues, or developments that you can follow.

Repetition is a powerful tool for learning. If you read many unrelated passages that never review or expand on previous reading, then you ultimately forget a lot. However, if you read related articles or passages that discuss ideas you've recently read about, then you can reinforce your earlier reading and build on it. This helps you learn more.

Strategy 4: Read books.

Books are created for just the kind of related, connected reading that can help you build your vocabulary and expand your knowledge. Read nonfiction if possible since the TOEFL is a non-fiction exam. Although creative artists can be the topic of a reading passage (and lecture), you are not expected to read or listen to fiction or poetry on the exam.

Fiction and poetry are excellent ways to entertain yourself while reading, and it's important to have fun and enjoy your reading practice. However, your main focus should be on non-fiction. At the same time, don't attempt to read too much, which can be discouraging and frustrating. Instead, focus on short discussions of a topic that interests you.

Strategy 5: Read topics that interest you.

Too many students assume that TOEFL preparation must be boring because it is academic and advanced. They might not bother to look for articles and books that are personally interesting or relevant. Although your reading material must be informative and factual, it does not have to be boring to you personally. Read topics that interest you.

Strategy 6: Improve your background knowledge.

The TOEFL iBT is not a knowledge-based exam, which means that it isn't designed to test specific knowledge about any particular field or area. The test is designed to be accessible to as many educated people as possible, and the test writers try not to give any preference to any one field or area.

This does not change the fact, however, that you must still have an understanding of fundamental concepts in world history, science, business, and the arts. The TOEFL is written to test your ability to study in a college or university classroom, so you must possess the background knowledge necessary to study at that level. For instance, basic atomic structure (an atom with a nucleus containing neutrons and protons, and electrons circling around it) is a fundamental concept in chemistry, physics, and biology. Moreover, the Periodic Table (a list of elements ordered by increasing atomic weight) is another basic piece of background knowledge.

Having some background knowledge in an area simply means having some familiarity with the terminology and concepts. You'll need to analyze your own abilities in certain academic areas. There's no need to become an expert, but you should be familiar with the key concepts, famous people, and major discoveries in a field.

Strategy 7: Use an English language dictionary, but don't overuse it.
There is no way to avoid the fact that reading depends on vocabulary. Without a good vocabulary, reading will always be difficult. As you read, you'll need to build your vocabulary. An English-English dictionary is a vital part of that learning process.

First, it's important that you not use a dual language dictionary, or a dictionary in both English and your native language. Although these certainly have their uses, they provide translation, not explanation in English. Since the TOEFL iBT is not a translation test, don't learn the meaning of new words just in your own language. As you've seen in the strategies and practice earlier, the vocabulary questions and paraphrasing all involve English to English understanding. Therefore, you need to learn how to explain and discuss ideas in English.

Second, don't overuse the dictionary. Don't go straight to the dictionary immediately every time you find an unfamiliar word. Try to learn the word in context of the text around it.

Strategy 8: Record new vocabulary in a detailed journal.
Students often record the meaning or definition of new words in a simple list. This is good, but not good enough. English words often have specific relationships with other words or expressions. When students record only a definition, they often can't use the new word later because they don't know enough about the right context or necessary structure.

Some words change meaning slightly or greatly depending on the subject, object, or context. The verb *fight*, for instance, could mean a physical confrontation for a boxer or soldier (*He fought hard but his opponent was too good*), or an emotional, psychological effort for a cancer patient (*He fought hard, but the disease spread too quickly and he died*). You need to notice the variation in meaning between different contexts.

Moreover, some words are combined with specific words. The adjective *interested* is always paired with the preposition *in* (*I am interested in space travel*). This regular pairing is called *collocation*, and it's an important feature of many new words. Make sure to record these details by writing down a sample sentence along with your definition. Although this might seem like too much work, you will benefit immensely when you review your list later.

ANSWER KEY

On the actual TOEFL, the multiple-choice answers are not letters (A), (B), (C), (D), as you may have seen on other exams. Beside each choice is an empty circle; you select an answer by clicking on the circle with your mouse and darkening the circle. For that reason, the answer choices in this book have no letters in the practice.

For the purposes of identification here, however, the choices in the practice are identified in the answer key as letters. For each multiple-choice question in the practice:

$\bigcirc$ = (A)

$\bigcirc$ = (B)

$\bigcirc$ = (C)

$\bigcirc$ = (D)

Each question number in the answer key below is followed immediately by the correct answer, and then a brief explanation.

For questions that involve a table or list, the table is recreated with the correct additions.

1. (D)

The best answer is (D); according to paragraph 2, homeschooling provides "...plenty of room for spontaneous discussion..." (A) is unmentioned; homeschooling is not associated with the development of parenting skills in the children. (B) is also unmentioned; self-direction in the workplace is not a benefit that is mentioned. (C) is a distracter. According to paragraph 2: "...learning experiences that classroom logistics make difficult..."

2. (C)

(C) is correct; paragraph 1 calls public schools "crowded," and in paragraph 3 children "find distractions." (A) uses stated detail incorrectly from paragraph 4: "...less likely to become involved with gangs or drugs." (B) also misuses stated information from paragraph 6: "...87th percentile on standardized exams." (D) resembles but distorts detail in paragraph 3: "...most lessons are aimed at the middle level of ability."

3. (C)

(C) is correct based on paragraph 3: "...can readily make adjustments...as the child requires." (A) is unmentioned: paragraph 5 mentions that parents need patience, knowledge of the material, and energy, not training. (B) is also unmentioned since paragraph 3 mentions the benefits of moving at an appropriate level, not at a higher level. (D) is also unmentioned.

4. (D)

(D) is correct because paragraph 2 states that larger states have more electoral votes, but there is no mention that it influences national policy. (A) correctly restates ideas from paragraph

4: "...creating a gulf between themselves and a significant fraction of the electorate." (B) also accurately paraphrases paragraph 4: "...many members of political minorities don't bother to vote at all." (C) is also based on paragraph 4: "...they avoid campaigning in small states."

5. (B)

(B) is best since there is no mention of the population's size at the time of the nation's founding. (A) since the passage states: "Each state functioned a lot like an independent country." (C) correctly reflects the following information: "Communication and transportation systems between disparate parts of the country were extremely poor." (D) also correctly relates to the passage: "Communication and transportation systems between disparate parts of the country were extremely poor."

6. (A)

(A) is correct; although paragraph 4 states that "...powerful party organizations make national political campaigns easy to conduct," there is no mention that people are loyal to their parties. (B) correctly restates paragraph 4: "Americans have developed a very strong sense of national identity..." (C) also accurately reflects paragraph 4: "Americans...demand to play a direct role in the selection of their leaders." (D) is a good restatement of paragraph 2: "...the unfortunate consequence of blinding most citizens to its workings."

7. (C)

The best choice is (C). The "winner-take all" aspect of the system is mentioned in paragraph 4, which best supports the main idea: that the system should be abolished. (A) describes how the modern electoral system functions, but does not directly support the abolishment of the system. Likewise, (B) only describes why the Electoral College was created; it does not directly support the main idea. (D) is unmentioned; there is no mention of people who would "cling to the Electoral College."

8. (D)

(D) is the correct choice based on paragraph 3: "So the Electoral College was provided as a solution." (A) misuses ideas from paragraph 4: "...powerful party organizations..." (B) is a modern, not historical, attitude. (C) is a distracter based on the hook in paragraph 1: "Neither a parliamentary system like that of the United Kingdom or Japan, nor a system of direct popular vote as in France or South Korea..."

9. (B)

The correct answer is (B). The pronoun is plural and it must refer backwards to a preceding noun. Based on the context, the stones are held in place by the keystone, and the sentence begins with the noun *stone*. (A) comes after the pronoun, and *others* cannot replace a noun in a later clause or sentence. (C) is plural, but the pronoun is modified by *all*, which can only describe three or more nouns, not just two. The correct adjective for two is *both*. Also, the sentence doesn't begin with the noun *pile*. (D) is wrong because the sentences do not discuss multiple groups of people or actors. Also, the Romans are building the arch, so they can't be part of its foundation.

10 (A)

(A) is correct, for the parallel structure (as they have been throughout) indicates that the pronoun *they* receives the same action as a previous subject. *Wooden bridges* is the only subject receiving an action in the sentence. (B) is part of the time phrase for the parallel action, so it can't also be the subject. Similarly, (C) is part of a time phrase for the first verb *were built,* so it can't also be the subject of the parallel action later in the same sentence. (D) doesn't fit the immediate context; the parallel verb *have been* is passive and parallel with the verb *were built,* so *people* cannot be the subject.

11. (B)

(B) is correct because the pronoun them is the object of the infinitive *to strengthen*. Based on the context, the bridges need to be fixed or improved. (A) is part of the time phrase (during the first decades) for the action of the sentence, and time periods cannot become stronger or better physically. (C) might seem possible since failures are negative and the infinitive *to strengthen* is positive, but in this context the infinitive suggests adding power to the failure, which isn't desirable. Failures are corrected and avoided, not strengthened. To strengthen a failure suggests that the failure will become even more negative and harmful. (D) is modified by the infinitive phrase *to strengthen them*: the infinitive describes the kind of effort that was made (effort = to strengthen them), so the efforts cannot be strengthened since the two parts of speech refer to the same action.

12. (D)

(D) is correct. Since the pronoun *it* is described as the replacement of other materials, the antecedent must be a material. (A) might seem possible because a design can replace other designs, but the rest of the sentence doesn't support this choice; the change involved material, not specific designs. (B) also seems correct because both *revolution* and *it* are subjects. However, a pronoun and antecedent don't need the same function. Also, the revolution itself is described later in the sentence as the introduction of steel, so revolution is too general, and the revolution is not a material. In this context, (C) is incorrect because the time period cannot replace anything.

13. (C)

(C) is best. *Kinetic* means "pertaining to motion, caused by motion, or characterized by motion." This is closest in meaning to *dynamic,* which means "vigorously active or energetic." (A) is wrong since kinetic has nothing to do with weight, and weight is not implied or mentioned in the sentence. (B) is the opposite of "characterized by motion." (D) is a distracter based on *random,* which means characterized by unpredictability.

14. (D)

The best answer is (D) since one of the definitions of specific is "with individual qualities that allow a distinction to be made." Something that is exclusive is limited to only one thing. Osmosis is defined as a distinct type of diffusion, meaning that other types of diffusion are excluded. (A) is wrong because *usually* involves custom or habit, neither of which is mentioned. (B) is a reference to certainty, but there is no discussion of how well anything is known or understood. (C) is wrong in this context; accuracy would involve precision, measurement and experiment, but at this point in the passage, the writer is explaining general concepts.

15. (A)

The best choice is (A) because *withstand* means to be 'strong enough to resist something and remain unchanged." It is closest in meaning to *endure,* which is to "tolerate something without giving up." Choice (B) since *oppose* is argumentative or combative, which is impossible given the subject *cells.* (C) is a distracter: *protect* might seem related to invasion and security, but the turgor pressure is the protection and it is mentioned after the vocabulary. Also, *protect* doesn't work with the object invasion since it means that the invasion is protected, not the cell. (D) is out of context; the cells and water are not in a competition or race.

16. (C)

The correct answer is choice (C) since the constant motion explains why the food coloring and water mix completely. (A) is wrong; because the pronoun is connected to molecular motion, it can't explain the color, which is the result of light reflection and absorption. (B) doesn't fit the structure; the pronoun can't refer to the constant motion because the motion is part of the adverb clause that gives the reason for this. The ideas in choice (D) are too far away.

17. (B)

(B) is the correct choice. Which is a relative pronoun that modifies the entire independent clause describing the tendency of molecules to move from high to low concentration. The tendency explains why the coloring diffuses throughout the water. (A) is too far away; the rate of diffusion is mentioned in an earlier sentence, and *which* must modify something within its own sentence.

(C) is in the wrong clause. The movement in both directions is part of the adverb clause that precedes the independent clause in that sentence. *Which* must be next to the word, phrase or clause that it modifies/replaces. *Which* can't jump across the independent clause and modify the adverb clause. (D) is a distracter because it is incomplete. The less concentrated molecular field is only half of the reason that the coloring diffuses in the water.

18. (B)

(B) is correct. *Biology* is an excellent choice because it immediately precedes the pronoun. Biology is an academic field or area of study, so it can be modified by where. (A) is grammatically impossible; *water* is the subject of the dependent clause, so grammatically *where* cannot refer to it. (C) is also structurally impossible. Osmosis is the subject of the independent clause, so it's on the wrong side of the verb is. *Where* cannot modify any word or phrase on that side of the verb. Finally, (D) is also incorrect based on the grammar of relative pronouns; *solvent* is part of the dependent clause that begins with *where,* so it can't be the antecedent for *where.*

19. (C)

The correct choice is (C) based on the fact that communities that took part in the games developed environmental programs afterwards. (A) is a distracter based on paragraph 2: the toxic waste dump was reclaimed, not discovered. (B) takes the inference too far. The author stresses positive environmental outcomes from Sydney, though there is some criticism of Games in general. (D) is too broad.

20. (C)

(C) is correct based on the key word even in paragraph 4: "Even Southern cities…now have hockey teams." *Even* shows surprise. (A) distorts ideas from paragraph 4: "American successes in speed skating and hockey," "50 percent increase in the number of ice arenas…in the last 15 years." (B) adds incorrect ideas (always) to information from paragraph 4: "figure skating drama" (D) also distorts ideas from paragraph 4: the "growth in interest" in "Tampa Bay, Florida, and Atlanta, Georgia."

21. (A)

(A) is correct based on paragraph 2: The "impact" is "complex." Tourism in general increases but in some areas it may fall. (B) is too broad—only 2 examples are given. (C) takes the inference too far based on paragraph 2: "Long-term effects are generally positive." (D) adds unmentioned ideas to information from paragraph 2: "the Games tend to improve the international image of the host city." No mention of host cities' satisfaction or otherwise.

22. (B)

The correct answer is (B) since light rail is mentioned as an exception to the energy parity between cars and trains. (A) is wrong; the paragraph mainly discusses the comparable fuel efficiency of cars and trains, not the superiority of one over the other. (C) is incorrect; the paragraph does not urge action of any kind. (D) is off-topic; modern advances are not the focus of the passage.

23. (C)

The answer is (C). Paragraph 1 asks "What accounts for this…loss of interest in train travel?" Subsequent paragraphs explore possible reasons. (A) refers to the wrong rhetorical structure; some comparisons are made in the final paragraph, but the public's declining interest is not expressed through an analogy. (B) is unmentioned. (D) is incorrect since the passage is not structured chronologically.

24. (C)

The correct choice is (C), which is the main rhetorical structure used to support the argument that trains are safe. (A) is a distracter based on the following: "On a train, one need never worry whether the approaching driver is…distracted by his cellular phone." (B) refers to the wrong type of passage; no arguments are mentioned. (D) is unmentioned. The safety of trains is highlighted; train operators are never mentioned.

25. (C)

The answer is (C) based on paragraph 1: "...the last hundred years have seen nothing but a decline in the use of railroads..." (A) is a distracter based on paragraph 1: "...trains have begun recently to attract a little more interest from urban planners..." (B) is also a distracter based on paragraph 1: "A rapid growth in car culture" is mentioned, but the public's declining interest in trains is described as "progressive." (D) is based on paragraph 2: "Efficiency alone cannot be the answer we are looking for."

26. (A)

The answer is (A) based on paragraph 3: "Railroads are almost as expensive to build..." (B) is wrong based on paragraph 1: Trains saw "widespread use" over the course of the 19th century, but have declined over the last 100 years. (C) is a distracter based on paragraph 1: Widespread use implies that people were pretty interested in them. (D) is a distracter based on paragraph 2: "...if trains are forced to run with few passengers, they can actually be much less efficient than cars."

27. (B)

(B), or the second square, is the correct answer. Because the inserted sentence gives extra information about the ear, it should come after the ear has been introduced, but before the sentences about the eardrum or the cochlea, both of which are at a narrower level of detail. (A) comes before the ear has been mentioned as a topic. This is too soon to mention its basic parts. The third choice—(C)—comes after a new detail (function), and the square comes between the antecedent bones (preceding sentence) and referent these (following). The fourth square, (D), could seem right since the following sentence mentions two sections, but the inserted sentence is introductory/general, not summarizing and it doesn't mention the cochlea.

28. (B)

(B), the second square, is the best answer. The inserted sentence gives extra information about the sensory organs that detect orientation and movement. It is not a topic sentence for the paragraph, nor does it logically follow the discussion of the semicircular canals or the example of the eyes, both of which are more specific than the inserted sentence. (A) is wrong; the first square has no preceding sentence. Therefore, there is no antecedent for *these organs*. (C) is incorrect since the third square follows the discussion of one the canals and fluid, and it precedes the mention of the otolith organs. The inserted sentence would interrupt this flow of ideas. The fourth square interrupts an antecedent-referent connection; the following sentence includes the referent *this*, so (D) is wrong.

29. (A)

Choice (A), or the first square, is correct. The inserted sentence gives further information about balance impairment. It does not logically introduce the topic, but because of its lack of any of the ideas mentioned later, it does not logically belong after the discussion of aging, dying hair cells, or why elderly people fall. (B) is wrong because the inserted sentence doesn't mention aging, and it would interrupt the flow of ideas between the sentence before the second square and sentence after it. (C) is incorrect; the sentence after the third square includes also, which connects that action to the preceding sentence. The sentence after the fourth square includes the transition regardless, which rhetorically links the sentence before and after the fourth square, so (D) is incorrect.

30. (B)

The correct answer is (B). The main point of the sentence is the importance of being able to interpret American signs and symbols. (A) is wrong since the original sentence mentions "speaking and understanding" the language, not learning how to speak it. (C) is incorrect because the original sentence does not describe the ease with which a visitor may learn to interpret signs and symbols, only the importance of learning to do so. (D) is wrong; the original sentence does not make any comparisons between American signs and symbols and the signs and symbols contained in a foreign person's culture.

31. (D)

The best answer is (D); According to the sentence, a person's "type" is reflected in the commercial symbols a person has about them. (A) is off-topic; the main point of the sentence is how commercial symbols reflect the people who own them, not how people are attracted to products. (B) includes unmentioned ideas; the sentence has nothing to do with interpreting signs and symbols. (D) is too general/vague; all commercial symbols convey meaning.

32. (D)

Choice (D) is correct. The main point is that the timeframe provides the system for decoding the semiotic messages communicated by popular styles. The sentence does not suggest that a decade's styles can be more easily decoded during the decade they are popular, so (A) is wrong. (B) is incorrect because the main point of the sentence is not how often systems change. (C) rearranges ideas; the focus of the sentence is the opposite: it is the decade that is used to understand the style.

33. (C)

The third square, or (C), is correct. The sentence provides more information about reading, so it logically follows the sentence that first mentions reading as an example of semiotic decoding. It does not logically fit anywhere else in the paragraph. The first square, (A), could be distracting because of sociolinguistic analysis, but this is not the same as reading. So the inserted sentence would seem off-topic. (B), or the second square, might seem correct because of the verb interprets, but this is also not identical to reading. The fourth square, (D), comes too late after the discussion of reading. The preceding sentence broadens the discussion to anything, which would make the inserted sentence off-topic.

34. (A)

(A), or the first square, is correct. The first square is the best choice because this sentence introduces the topic of an American's music collection. It logically precedes all the sentences that provide examples of different types of music. The second square, choice (B), comes between two examples of musical artists/styles. The inserted sentence is too general in comparison to the list of examples. Likewise, the third square—choice (C)—interrupts a series of examples. Like the previous two other incorrect choices, the fourth square (D) follows an example of a musical style (rap), which is too specific.

35. (B)

(B), the second square, is the best answer. The second square follows the reference to a French brand of bottled water. The word another in the inserted sentence indicates that a different bottled water was previously mentioned. The best place for this sentence is after the sentence that introduces Evian. (A), the first choice, is too early since it doesn't mention any bottled water. Due to the phrase another French bottled water, the inserted sentence must follow a sentence containing that noun or a synonym. (C), the third choice, follows the mention of American cultural trends, which is too general and relates too well to the detail in the following sentence. The fourth square, (D), adds detail to the previous sentence. The inserted sentence would interrupt that flow of ideas, and it would be off-topic.

36. See tables below

    Answer

True for BOTH China and the United States
More than 9 million square kilometers in size
Heavy reliance on deficit spending by the government
Low annual consumer inflation figures

True for China ONLY
National gross domestic product below $5 trillion
Positive balance of payments in trade
Per capita gross domestic product below $10,000
Major segments of the economy face little competition

Explanation

True for BOTH China and the United States
paragraph 2: China has 9.5 million square kilometers, the U.S. has 9.6.
paragraph 3: "Like the U.S. government, China's government depends heavily on deficit spending."
paragraph 3: "...annual inflation of consumer inflation is below 3 percent in both countries."

True for China ONLY
paragraph 4: "...enjoys a positive balance of payments...in contrast to the massive...U.S. trade deficit."
paragraph 5: China's per capita GDP is $2,000
paragraph 3: China's national GDP is $2.7 trillion; U.S. GDP is $13 trillion
paragraph 4: Government control of major segments of China's economy shelter them from competition

Distracters

paragraph 2: Possession, not consumption, of fossil fuels is mentioned.

Trade with immediate neighbors is not mentioned.

37. (D)

The correct answer is (D). The comparison to the U.S. is that the U.S. has a more diverse and resilient domestic economy. (A) is wrong since the highlighted sentence does not describe a cause/effect relationship between China's export dependency and the weakness of its domestic economy. (B) is incorrect because the sentence states the fact of China's dependence upon foreign consumers and investors, not that foreign investors hold the answer to China's economic future. Also, the sentence does not state that China needs to increase its exports, so (D) is wrong.

38. (A)

The best choice is (A). The sentence states that American fears are realistic, due to China's rapid growth. (B) is wrong, for American fears are described as persistent (meaning constant, or lasting), not growing. (C) is incorrect; the sentence does not suggest a time period for China's economy to overshadow that of the United States. (D) refers to unmentioned ideas; China's rapid growth is mentioned, but the slowing of U.S. economic growth is not.

39. (B)

The second square, or choice (B), is the best answer. The conjunction yet signals a contrast to the preceding thought. The only place in the paragraph where it logically fits is after the description of the U.S.'s economic strength, and before the discussion of China's emergence. The first square—(A)—is wrong; the first square comes before the United States is mentioned, so the yet in the inserted sentence has nothing to contrast with. (C), or the third square, follows the first mention of China, so the inserted sentence is off-topic. The fourth square (D) interrupts the connection between referent these differences in the following sentence and the antecedents parallels/similarities in the preceding one.

40. (A)

(A), or the first square, is correct. The previous paragraph discusses similarities in the two nations' economies. This sentence introduces the topic of differences. The rest of the paragraph provides examples of these differences. The only place this sentence logically fits is at the start of the paragraph. The second square, or (B), is wrong; the second square follows the first mention of details related to differences, so the inserted sentence is too general. Also, the choice is too far from the previous paragraph, so the phrase despite these similarities is also too far. Likewise, the third square (C) follows the mention of beneficial differences, which is too specific. Although some students may want to choose it because of for example, the example relates to positive differences, which is more specific than the inserted sentence. Finally, the fourth square, (D), follows a long series of detailed sentences.

41.

• Minor point. Freud's self-analysis in paragraph 3 is an interesting idea that reflects his intellect and focus, but it isn't central to the main idea.	• Unmentioned. Although it is possible to assume that the new ideas created debate, there is no mention or suggestion of it in the passage.
• Correct. Freud's contribution to dream analysis played a major role in the transition from religious to scientific interpretations, so this is a major part of the main idea. (Paragraphs 1, 2, 5)	• Correct. Freud's discovery about the preservation of memories and emotions is central to his ideas. (Paragraphs 1, 3, 5)
• Minor point. Although it is mentioned in paragraph 4, the inactivity of a part of the brain during sleep is not a major point.	• Correct. Applying the scientific method to studying mental, as opposed to merely physical phenomena, is one of Freud's major contributions. This is one of the main ideas of the passage. (Paragraphs 1, 2, 3, 5)

42. (C)

The correct answer is (C). A prediction is a statement about future events or states, so this choice fits the earlier verb heralded, and the context of the sentence. If rulers paid attention to the dreams before acting, then the dreams must have had some relevance to the future. (A) is a distracter because of the earlier mention of gods. Certainly, in the epic story, the dreams are holy since they are directly related to awareness of the gods. But *holy* is only partially relevant and misses the key idea of prediction in the verb *heralded*. Also, the vocabulary word appears in a sentence that involves military and other ideas, not only religious ones. (B) is wrong because intuition is the perception of truth without rational or logical thought, but people only intuit truth when they are awake, and the truth is not primarily about the future. (D) is a distracter since inspiration is motivation, which might seem relevant to the decisions and actions in the sentence. However, there is no suggestion that the dreams made the rulers act; the dreams only affected how they acted.

43. (B)

The best choice is (B); A malady is a physical or psychological disorder or disease, and illness fits the verb treat and the related vocabulary in the conclusion. (A) is a distracter. Cures are solutions to diseases, and the use of *treat* in the sentence could confuse a student. However, the vocabulary word is the object of *treat,* which means that the vocabulary word requires treatment, so it must be negative, undesirable, problematic, etc. Also, Freud's work is connected to mental health problems and mental illness in the conclusion, which are later clues about the uses of his work. (C) doesn't fit the adjective *mental.* One cannot behave mentally. Behavior is physical action that is directed by the mind, but it doesn't occur in the mind. (Choice (D) is a distracter. The use of understand in the sentence might suggest explanation, but in this context, people were not trying to understand the explanations; they were trying to understand the maladies.

44. (A)

The right answer is (A). *Itself* is the object of the infinitive *to protect,* so the antecedent must not only be able to offer protection but also able to be self-protecting. The mind is capable of this. Also, the previous phrase *by the mind,* which modifies *attempt,* identifies the actor making the attempt. Choice (B) fits none of the above structure and vocabulary clues. Also, dreams offer no protection. (B) is wrong since *disappointment* is the object of *from* and the phrase modifies the infinitive that includes the pronoun, so this noun isn't doing anything; it is describing the action by identifying the threat or problem. (D) refers to mental pictures, which offer no protection. Also, the noun is very far from the pronoun, which isn't proof but a strong clue that it is wrong, especially since it is not a part of the noun and phrase *attempt by the mind.*

45. (B)

The correct choice is (B). The pronoun *which* refers to the scientific discovery: that, when people dream, the parts of the brain involved in recognizing objects and processing images, and the areas involved in processing emotions and emotional memories, are as active as they are when the mind is awake. The pronoun refers to the whole noun clause, not a word or phrase. The fact explains the realism and power of dreams. (A) is wrong; this is a reference to the adverb clause when the mind is awake. Although the pronoun refers back to a clause, this adverb clause is only part of a complete idea or fact. One can recognize that based on the incomplete meaning (what happens at that time?) and the incomplete structure (what is the main clause?). (C) is only part of the fact, so it is incomplete. (D) is also only part of the fact, so it is incomplete.

Chapter 2: **Listening**

The Listening section of the TOEFL is the second part of the exam. This section includes both conversations and lectures: two or three conversations with five questions per conversation, and between four and six lectures with six questions per lecture. Depending on the number of passages, this section lasts 60 or 90 minutes. At the start of the section, an instruction appears, telling you to put on your headphones. Once you are wearing your headphones, you are also shown how to adjust the volume by clicking on a volume icon with your mouse. You can adjust the volume at any time during the exam. Once you begin the first passage, the time remaining is indicated by a digital timer at the top of the screen, and you have the option of showing or hiding the time. Also, you can change your answer as often as you like, but you must finalize your answer before you can move on to the next question. Once you finalize your answer, you cannot return to the question later. When the Listening section is over, it is followed by a mandatory 10-minute break.

There are several things you need to do to perform well on the Listening section of the TOEFL. You must:

- Familiarize yourself with the question types
- Have the right skills to listen intelligently
- Take good notes
- Avoid incorrect answers

This section gives you strategies for listening, note-taking, and answering each type of question that appears on the exam. Once you review the strategies, you can practice each question type for both a conversation and a lecture.

GENERAL LISTENING STRATEGIES: CONVERSATIONS AND LECTURES

The following strategies refer to skills that are always important anytime you are listening during the TOEFL.

Strategy 1: Listen actively.

To listen actively means you should think about what you are hearing. One of your major challenges as a TOEFL student is that you must do many things while listening. You must recognize more than just names, places, and times. While you listen and take notes, you must also anticipate, learn, connect ideas, organize, generalize, infer, assume, and conclude.

To get the right information from a conversation or lecture, it helps if you ask yourself the right questions before you begin listening. The strategies in this section will teach you what questions you should have in mind.

For example, while listening, ask yourself the following:

- What are they talking about?
- How do they feel about it?
- What are the problems or key points?
- Why did he say that?
- What are they going to talk about next?

Strategy 2: Anticipate key points based on the main idea.

Anticipating means thinking about what might or will come next, and it is a part of active listening. Anticipation makes listening easier. Once you hear the main topic, you can expect certain details. For example, a student who wants to talk about his term paper might have problems with the topic, organization, due date, length, bibliography, or a partner, etc. Similarly, a professor who gives a lecture on sharks might discuss their appearance, abilities, evolution, migration, reproduction, diet, or reasons for studying them, and so on. Knowing the possibilities makes it easier to hear what the speaker says.

Strategy 3: Try to infer meaning whenever possible.

To infer means to understand unstated ideas, and inference is a part of active listening. You must infer often during the TOEFL because meaning, organization, attitude, connections among ideas, and purpose are not always stated explicitly. Instead, they are implied. To imply is to communicate unstated meaning. Speakers imply meaning and listeners infer it. Look at the following conversation:

Man: Have you tried the steak in the cafeteria?

Woman: Actually, I'm a vegetarian.

The woman doesn't say directly that she didn't eat the steak, but she implies it by saying she's a vegetarian, who is a person that eats only vegetables (and possibly fish). Based on that fact, the man and you can infer that she probably didn't try the steak. A correct inference is always based on a stated idea.

Inference is an important skill on the TOEFL; it is involved in most question types. You may have to infer an opinion, attitude, purpose, or connection among statements.

Strategy 4: Write notes in fragments, abbreviations, initials, symbols, and acronyms.
While listening, you must write notes quickly and efficiently. Because you have so little time, ignore spelling, verb tenses, and proper grammar. For example, mathematical symbols, shapes, and arrows are useful for representing many ideas. Look at the following chart (see also the example for Strategy 5):

Symbol	Possible Representation/Meaning/Use
+	and, also, as well, moreover, furthermore, with, together, etc.
=	be, become, look, sound, taste, smell, feel, verb + like
≠	be + not, not like, dissimilar, not alike
$x < y$	x less/lower/weaker than y, x not as…as y
$x > y$	x more/bigger/greater/stronger than y, y not as…as x
→	become, change/transform into affect, result in, cause, create, make move/relocate to, immigrate/emigrate to
↑	increase, rise, jump, spike, go higher/up, raise, lift, improve, be/become popular
↓	decrease, fall, go lower/down, hurt/ruin, be/become unpopular
×	not accept, reject, not like/enjoy, disapprove of, stop, destroy, ruin
?	question, not know, unsure, uncertain, must find out/learn/study, debate

To take good notes, you must develop your own system of abbreviation and symbols.

Strategy 5: Use your own words instead of copying the speaker exactly.
Many students waste time writing the exact words a speaker says, but this isn't always necessary. Understanding is a part of active listening, and if you understand an idea, you can record it in shorter or simpler words. Often, this requires you to be able to understand academic vocabulary and paraphrase it in your notes. Look at the following sample:

Original Passage	Notes
John Hancock was the first person to add his signature to the Declaration of Independence.	JH = first sign DI
The U.S. population exploded around the turn of the 20th century.	US pop ↑ 1900

Strategy 6: Don't try to record everything in notes.

Notes can't replace memory; notes only help you remember, so don't try to write a transcript of the conversation or lecture. The most important things to do are to listen, think, and understand.

Strategy 7: After the main idea, focus on key supporting points, not minor ones.

The most important ideas are those that say the most about what the speakers are discussing. While listening to a history lecture about a famous battle, for example, focus on the main actors (attackers, defenders, civilians), actions (attacking, defending, surrendering), and sequence. A general's marital status (single or married), on the other hand, probably doesn't have a major connection to the battle.

Strategy 8: Connect details to the main idea and supporting points.

The TOEFL tests how much you understand, not just how much you remember. If the professor refers to a woman, what is her importance? What did she do? What happened to her? Is she compared to anyone? Why is she mentioned? Be sure to remember or record her importance in your notes.

Likewise, if a student complains that he has football practice in two days, what does this relate to? Is this a good thing, or a problem? If it is an obstacle, what does it stop him from doing? If the detail doesn't relate to anything, then it is a minor detail.

Strategy 9: Pay attention to cohesive devices.

Cohesive devices can help you recognize organization, anticipate ideas, make inferences, and recognize connections among the main idea, supporting points, and details. As you've already learned, a speaker (or writer) uses cohesive devices to make connections among the various ideas in a sentence and a paragraph. Cohesive devices include pronouns, adjectives, articles, transitions, synonyms, repetition, and variations of word form. Look at the following paragraph:

> Newton's law of inertia is perhaps the easiest law to understand, but it is often misunderstood. Many people assume that the law states that a body at rest will remain at rest until another force acts on it. This is true. If you place an apple on a perfectly flat table, the apple will not move on its own. However, if the table is not perfectly flat, the apple will start to roll because gravity pulls it down. But, this is only part of the law. Actually, the law states that…

The professor here explains a scientific principle of mechanics developed by Isaac Newton. He uses cohesive devices throughout, and you can follow the explanation more easily if you are aware of them. The pronoun *it* in Sentence 1 refers back to "Newton's law of inertia." In Sentence 2, *the law* uses the specific article to refer to the same law while *it* refers to "a body at rest." Sentence 3 uses the pronoun *this* to refer to the law's statement that a body at rest will remain at rest until another force acts on it. The adverb *actually* in the last sentence is a transitional device that signals a clarification or correction.

Strategy 10: Be prepared for the unique elements of spoken English.
Although the speakers in the Listening section use academic vocabulary and speak about academic topics, the lectures and conversations in the TOEFL have elements that are not found in academic writing. In other words, the speakers in the Listening section do not sound like they are reading an essay. Instead, their speech is more natural and resembles everyday spoken English, which can include some or all of the following elements: interruptions, confusion and clarification, self-correction, and fragments.

Interruptions
An interruption occurs when a listener in a conversation or lecture says something before the speaker has finished making a point or observation. Any speaker (a student, professor, or school official) could be interrupted, and usually, an interruption comes in the middle of a speaker's sentence. An interruption is always done as politely as possible in TOEFL passages. However, interruptions are relatively rare; most speakers can finish their sentences since frequent interruptions are always impolite. Nevertheless, a speaker could be interrupted by a question or comment, and the speaker will have to resolve the interruption before returning to the original topic. You will have to remember what was said before the interruption in order to understand fully what is said after. Some common vocabulary for interruptions includes:

- Excuse me but/I'm sorry but
- I don't want to be rude, but
- Hold on/Hang on
- Can I interrupt/Can I just say
- Let me interject here
- I just want to ask/say

Confusion and Clarification
Any person (a student, professor, or school official) could be confused or uncertain about some information or concept. The confused person could ask for an explanation or someone else could identify an error in the person's statement. Regardless of how the confusion is identified, it is followed by some clarification, which is an explanation that removes the confusion or misunderstanding. Moreover, a person might interrupt a speaker to ask for clarification, but misunderstandings don't always involve interruptions. You can recognize

this problem based on the following vocabulary and related word forms:

- Confuse/confusing/confused/confusion
- Misunderstand/misunderstood/misunderstanding
- Not + know/get/understand/comprehend/follow
- Uncertain/vague/unclear/foggy
- Incorrect/wrong/mistaken

Self-Correction

Any person, but especially a lecturer, could misspeak (speak incorrectly), often by using the wrong word. The speaker always recognizes his or her own error and immediately corrects the mistake. Essentially, when a person misspeaks, the person interrupts himself or herself and then restates the idea(s) correctly. Self-correction could involve some of the following key words or expressions:

- No/Actually/Hang on/Hold on
- That's not exactly right
- That's not really true
- Let me rephrase that
- Let me start again/start over

Fragments

Speakers in both conversations and lectures don't always speak in complete sentences. Sometimes, a speaker uses a fragment, which is an incomplete sentence, such as a phrase (*in my class, to buy a new book, sitting on the lawn,* etc.), a clause (*because I'm a freshman,* etc.), or even just one word. Fragments are more common in conversations, but they can also occur in lectures with student comments. The fragment could be a question or an answer, so pay attention to the intonation of the speaker (a question = rising; a statement = flat or falling). Moreover, context is extremely important for understanding fragments since the fragment is based on an assumed connection to the previous statement or question. The importance of context is one reason that active listening is so important. For example, look at the following excerpt:

Man:	Wow, you've got a nice tan. Where've you been?
Woman:	Florida.
Man:	On a vacation?
Woman:	Actually, to do research. I'm in marine biology.

The woman identifies where she visited with only one word (Florida). Also, the man inquires about her purpose with just a prepositional phrase (on a vacation), and the woman answers him with an infinitive phrase (to do research). Conversations and lectures usually contain only a few fragments, but you must be prepared for them because you can get lost quickly if you aren't paying attention.

Strategy 11: Avoid answer choices that repeat too many words exactly from the lecture or conversation.

Most of the Listening questions are multiple-choice questions (others require you to reorder a list or complete a table). The multiple-choice questions are followed by four possible answer choices, and you must choose one or more correct answers. Though some vocabulary such as names might be repeated, the correct answer usually paraphrases ideas from a conversation or lecture.

To paraphrase means to restate ideas in new words and structure with the same meaning. Paraphrasing is done for many types of questions (Main Idea, Detail, Purpose, etc). Incorrect answers, or distracters, usually repeat some words exactly from the passage in order to make the choice seem correct. You must be aware of this trick and stay away from it; it is common for most question types.

GENERAL LISTENING STRATEGIES: CONVERSATIONS

These next strategies refer to skills that are particularly important when you listen to a conversation during the TOEFL.

Strategy 1: Pay attention to organization.
The basic organization of a conversation is the turn. A *turn* occurs when one person speaks and the other responds. A conversation usually has many turns. Look at the following turn:

Student:	Good morning, Professor. I hope I'm not disturbing you.
Professor:	Of course not, Steve. Come in.

Strategy 2: Be prepared for delayed answers in conversations.
One turn might be interrupted by other questions or statements, so a turn might last several statements. For example, a request or question might require one or more other turns before it is answered. Look at the following example:

Student:	Hello. I'm looking for census data. Which aisle is it in?
Librarian:	What year do you want?
Student:	Oh, don't worry; Just point me in the right direction and I'll find it myself.
Librarian:	We keep the records for all years before 1970 in digital format. Everything else is in books on this floor.
Student:	Well, I need data for the 1990s.
Librarian:	Then look in the stacks down Aisle 6.
Student:	Thanks.

KAPLAN

The student's first question is not answered right away. The librarian first asks about which year the student wants. Therefore, the first turn doesn't end until another one is completed. (Also, notice how the student assumes that the librarian wants to get the information for him, which is why he says, "Don't worry...I'll find it myself." You must infer the purpose of that statement.

Another example of possible inference is the detail "digital format," which implies that the data is stored on computers.)

Strategy 3: Create one column for each speaker in your notes to avoid confusion when you read them.
It's important to take organized notes. If you can't connect the ideas in your notes to the correct speaker, then your notes could be confusing rather than helpful. In order to help you stay organized, create one column for each speaker in the conversation. Although names are useful, don't worry if you miss the names of the speakers in the conversation since you won't be asked to name any of the speakers. Names are more important for distinguishing between the speakers and other people who could be discussed in the conversation. Depending on the type of conversation, identify each speaker as *Professor* and *Student, Man* and *Woman,* or *Student 1* and *Student 2.* Regardless of the method you use, make sure you can connect the ideas in your notes to the correct speaker.

Strategy 4: Focus on the student's need or desire.
A student in the TOEFL conversations always has a reason to see a professor or school official. Many possible reasons exist, such as to request information, a delay or extension, a change, a clarification of a misunderstanding, or advice about a problem. The student is usually very direct about the reason for the conversation, using a recognizable expression such as *I want to discuss* or *I need to know*, and so on. (See also Main Idea Questions: Strategy 4)

However, you should be also prepared to infer the student's reason since the student might not be completely direct or explicit about his or her reason. Of course, a student can't be completely vague or confusing since that would be rude. Nevertheless, the student might be indirect and the professor or official could infer the student's need or desire based on one or more details, so you should be prepared to make the same inference. For example, look at the following exchange between a student and a professor:

Professor:	How's your work on your project going, Tom? The deadline is only a few days away.
Student:	Well, that's what I want to talk to you about, professor. Is that date set in stone? I'm hoping there's some flexibility; something's happened to my experiment.

In the excerpt above, the student's key words clearly imply that he wants an extension (*set in stone* is an idiom meaning definite or unchanging; *flexibility* is a noun which here refers to the possibility of change), but he doesn't explicitly refer to changing, moving, or extending the deadline. The professor could then infer the student's request and answer accordingly. This kind of indirect speech is a reason that active listening is so important.

Strategy 5: Listen for the official or teacher's suggestions or comments.
Think about how the other person reacts to the student's request. As always, be prepared to infer the other speaker's suggestions, directions, or comments to the student. The other speaker might recommend another person, offer to do something, or the student might be required to act. Listen for more than one possibility, and any conditions that go with these options. Remember that a reaction or comment could extend over several turns, so don't rely on brief or concise answers to requests or questions. The reactions may take several turns because of complicating factors, which could come from the student or the other speaker.

Strategy 6: Listen for complicating factors.
Usually, a solution isn't easy or immediate (otherwise the conversation would be very short!). The student or teacher will have extra problems or responsibilities, which will make a solution more difficult. Complicating factors could include the rules of the school, schedules and deadlines, the demands of another class or professor, the needs of another student, or family responsibilities, among many other possibilities. You may also have to infer the effect of a complicating factor on the student's need or desire.

Strategy 7: Use the questions of each speaker as clues to anticipate information.
As you've already learned, anticipating information means being aware of what the speakers might say next. Although the number of questions in each conversation can vary, both speakers in a conversation will have some questions for each other, which is natural when two people don't know everything about each other. Questions include clues to the information in the other speaker's answer since the answer must at least be about the same topic as the question. These clues can come from the vocabulary in the question, especially the verb, as well as from the pronoun used: what (noun thing), who/whom (person), whose (possession), which (choice), when (time), where (place), why (reason), and how (method/ manner).

Strategy 8: Pay attention to any conclusions or final decisions.
Unlike a lecture, a conversation normally does not end with key words like *in conclusion* or *to wrap up*. Although such words are possible, speakers rarely will signal the end of the conversation in the same way that a professor would end a lecture. Since the TOEFL conversations in the Listening section involve a student's need or desire, the key words that signal the conclusion to the conversation relate to the student's final decision or plan regarding his or her request. These words could include verbs like *decide, intend, plan,* or *choose* (as well as any related word forms, such as *decision, choices,* etc.). However, the key words will also depend on the topic; recognizing any decision or choice depends on your knowledge of what the student is asking or looking for. For example, if a student wants an extension for an assignment, you can recognize the conclusion based on vocabulary that indicates a resolution to that problem, such as the verbs *take, postpone* or *extend,* and a date or amount of time.

GENERAL LISTENING STRATEGIES: LECTURES

The next strategies refer to skills that are especially important when you listen to a lecture.

Strategy 1: Use the typical organization of a lecture to help you follow the lecture and connect ideas.

A lecture is organized into paragraphs. Look at the following:

First paragraph = Introduction

The opening statements, or hook, may mention the main topic or ideas related to it. The last sentence is the thesis statement, which will mention the main topic and the supporting points, usually with parallel structure. Also, the thesis may state or imply the speaker's attitude and opinion about the main topic, so it is important to recognize it. However, there may be no thesis statement.

Middle paragraphs = Body paragraphs

Each body paragraph begins with a topic sentence, which gives one key supporting point for the thesis. The supporting points are important ideas, so you should record them in your notes. As long as you understand the main topic, you can identify the topic sentences by key words (*First, Second, Next, Another, One more*) and by the fact that each topic sentence will contain ideas that are elements or parts of the main topic. This identification requires understanding and is another part of active listening.

Another way to recognize supporting points is by contrast with the detail. The topic sentences contain general information, and they are followed by detail (examples, facts, description, definition, explanation). Once a lecturer mentions general ideas after giving some explanations or examples, then the speaker has just begun another topic, or supporting point.

Last paragraph = Conclusion

The conclusion may or may not begin with clue words (*Finally, To sum up, To wrap up, Generally*). And the conclusion may or may not paraphrase the thesis. It will give the lecturer's conclusions, observations, opinions, and/or predictions related to the main topic. These are not always stated directly, however, so you should always be prepared to infer. Also, the speaker might suggest or mention the topic for the next lecture.

Strategy 2: For a lecture with comments, use turns as clues to key points.

An academic discussion follows the same basic lecture format (introduction, body paragraphs, and conclusion), but it has multiple speakers taking turns. The professor always begins and is periodically interrupted by students. The questions and answers are clues to topic shifts and new supporting points.

Strategy 3: Be prepared to stop taking notes during dense parts of a lecture or discussion.
Most lectures and discussions include parts that involve too many details (actors, actions, parts, steps, etc) to be recorded accurately in notes. During these dense parts, rely on memory and focus on understanding the concepts.

LISTENING STRATEGIES: QUESTION TYPES

The Listening section of the TOEFL has eight types of questions. Most question types are multiple-choice, but some involve a list or table. Most multiple-choice questions require only one correct answer, but others might require two or more. If more than one correct answer is required, the question will clearly indicate that.

Read the question carefully to understand the type of question and the number of correct answers required.

Now, you can review each type of question, learn strategies to recognize and record important information for that question, and learn how to avoid incorrect answers, or distracters. For each question type, you can then practice listening, note-taking, and answering sample questions for both a conversation and a lecture.

Main Idea Questions

A main idea question asks about the overall focus of a conversation or lecture. The main idea includes the topic of the conversation or lecture and what the speakers say about it.

Use the following strategies to help prepare you for this task.

Strategy 1: Be familiar with some of the possible topics and the related vocabulary.
The main idea of a conversation is the student's need or problem and how it will be addressed. A conversation always occurs either between a student and a professor during office hours, or between a student and school official for a specific school service. Therefore, you can anticipate at least some of the possible topics by thinking about the common reasons to see a professor and by thinking about the various services available at a college o university. Look at the following partial list:

Conversation in a Professor's Office	Conversation for a School Service
Homework, project, essay is late	Bursar: fees, tuition, pay, postpone, cancel
Topic is too vague, narrow	Loans, bursaries, scholarships: apply, qualify, get
Problems with group work, partners	Academic credits: transfer, requirements
Need an extension, postponement	Registrar: records, major, minor, drop class
Don't have sources, bibliography, citation	Library: find, borrow, return, late fees/fine
Missed lecture, class notes, test	Labs: fees, procedures, equipment, safety
Need make up assignment, make up exam	Sports: facilities, teams, gym, try out, join
Drop, join, observe a class	Dormitories: co-ed, location, change, rules
Mark/grade is too low, unfair, vague	Parking: location, theft, fees, schedule
Plagiarism	Transportation: bus, train, schedule, fares
	Security: lights, walk-safe, guards
	Appointments, directions

The main idea of a lecture is the topic of the lecture and how the lecturer discusses it. The topic will be from an academic field, and you should be familiar with the names of academic fields taught at American colleges and universities. The following are some, but not all, of the possibilities:

- History (American and World)
- Anthropology
- Archeology
- Paleontology
- Math
- Chemistry
- Physics
- Biology
- Sociology
- Psychology
- English Literature
- Fine Arts (painting, sculpture, etc)

Any topic from these fields could be used in a variety of different lectures. A lecture might be about the Statue of Liberty, but a speaker could argue for its removal, explain its construction, discuss its symbolic significance, give a history of its voyage from France, compare it to the Eiffel Tower, or describe its restoration. Each of these lectures would have a different main idea about the same thing, the Statue of Liberty.

Strategy 2: Use the opening remarks, or hook, as a clue, but don't confuse the hook and the main topic.

A lecturer might or might not use a hook. A hook is a way for a speaker to get the audience's attention and to get the audience to begin thinking about the main topic. It might refer to the main topic directly, or it might relate to the main topic through some common idea. If a professor is giving a lecture on Isaac Newton and his laws of motion, she might begin with one of the following:

A quote or paraphrase

Albert Einstein once said, "Imagination is more important than knowledge." Nobody better demonstrates this than Isaac Newton, who essentially invented modern physics when he…

A fact (historical, scientific, statistical, etc)

Today, we know that the moon orbits the earth due to the same force that pulls a dropped pencil to the floor. The man who discovered this was named Isaac Newton, and he….

A rhetorical question

Have you ever wondered where rainbows come from? Isaac Newton wondered about such things all his life. His observations became laws of classical mechanics, such as….

Art (books, painting, popular culture, etc)

I'm sure everyone has seen Hollywood movies, such as *Back to the Future*, and is familiar with the stereotypical scientist. This awkward but brilliant person usually has untidy, white hair and huge eyes. Isaac Newton fit this description perfectly. His study of mechanics became…

A personal anecdote

I was sitting under a tree this summer, when an apple fell to the ground in front of me. I thought of Isaac Newton, who…

Strategy 3: Listen for a thesis statement after the hook as a clue to the main idea and supporting points of a lecture.
A thesis may or may not follow the hook. If it does, it may or may not involve parallel structure. If it has parallel structure, this is an excellent clue to the topics of the body.

Look at the following introduction for a lecture on Isaac Newton:

Albert Einstein once said, "Imagination is more important than knowledge." Nobody better demonstrates this than Isaac Newton, who essentially invented modern physics by developing the first proven laws of physics. Through his own observations and experiments, Newton wrote three laws of motion: the law of inertia, acceleration, and reaction.

The parallel nouns (inertia, acceleration, and reaction) will be subtopics, or the topics of body paragraphs later in the lecture. You can use this parallel structure as a clue to the repeated and related vocabulary as you listen.

Strategy 4: Use a student's opening request or question as a clue to the main idea of a conversation.

Conversations don't have hooks or thesis statements, but after exchanging polite greetings with someone, the student will use recognizable language to introduce his or her reason for coming. Look at the following list of possible expressions, phrases, and key words:

I'd like to discuss/I want to discuss/I want to talk about

I was wondering if/I was hoping that

I hope I'm not bothering you, but

Is it possible if/Would it be possible if/Is there a possibility that

Would it be a problem if/Is it a problem if/Is it too much trouble if

Can I talk to you about/Can we discuss

Can you help me with

I need/want/hope for/am looking for

I can't/am unable to/have trouble with/have difficulty with/don't understand

It is impossible/difficult/confusing/a problem for me

Strategy 5: Anticipate key points and details based on the introduction.

Recognizing possible ideas makes it easier to hear what is actually said. So think about what you hear at the beginning of a lecture or conversation. Once you know the topic, the supporting points shouldn't be a total surprise. If a student has a problem with a roommate, you may hear comments about the roommate's habits, beliefs, traits, or actions in the body of the conversation.

Likewise, if a professor is discussing Newton's laws, you might hear about early but incorrect theories, experiments, instruments, a sequence of procedures, data/results, proof/evidence, publications and books, practical uses/applications, current scholarly opinion, etc.

Strategy 6: Note the type of lecture or academic discussion.

In other words, determine if the passage is primarily an argument, a description, an explanation, a comparison, step-by-step instructions, a historical narrative, etc. As you learned in General Listening Strategies, the main idea is more than just the subject of the lecture or conversation. You must also recognize which ideas related to the subject are emphasized, how those ideas are connected and organized, and what comments or opinions are expressed about them. Ask yourself, *What is the speaker focusing on? How does the speaker feel about the topic? How is the passage organized?* This information is also necessary for purpose questions, rhetorical structure (organization) questions, and attitude questions.

Strategy 7: Follow repeated and related vocabulary to understand the main idea.
Some vocabulary will be repeated, but you can't rely on repetition alone. You must be able to recognize related vocabulary. During a conversation about changing roommates, you might hear the word roommate often, but you will have to catch other vocabulary related to the roommate (*noisy, unhelpful, plays his music too loud*), suggestions (*speak to him, change your schedule*), and solutions (*fill out a form, see the dorm manager*). The need to connect this vocabulary is one reason that active listening is so important.

Strategy 8: Don't confuse the supporting points/details with the main idea.
Remember that the body of a conversation or lecture presents parts of the main idea, so don't focus on just one detail. A student who complains about his roommate might mention his roommate's taste in music, but this is only one reason for wanting a change, not the main topic.

Likewise, a lecture on Newton's laws of physics might include an experiment to test the effect of gravity, but the experiment isn't the main topic; it supports one part of it, which is the law of gravity.

Strategy 9: Listen for the main topic to be repeated in the conclusion, but don't rely on the conclusion to paraphrase the thesis or summarize the whole conversation or lecture.
In the conclusion, speakers give final opinions, make suggestions, predictions, or decisions about the main topic, so you may hear the main topic repeated. A school official might decide to talk to a student's roommate about the complaints. Likewise, a professor might conclude that Newton's laws of physics are the greatest individual scientific achievement.

Main Idea Question Forms

A main idea question can be identified by the following question forms and related vocabulary:

- What is the talk mainly about?
- What are the speakers mainly discussing?
- What aspect/part/type/element of…does the professor mainly discuss?
- What does the student ask for/need/want from…?

Distracters

Distracters are incorrect answers. main idea questions have four answer choices; one is correct and the other three are distracters. The distracters for main idea questions are the following:

Distracter 1: The answer choice is too specific.
Usually, this distracter refers to one or more supporting points from the conversation or lecture, but not all of them. It would refer to the main idea of the Newton lecture as Newton's two laws of physics, or it could refer to Newton's observation of the moon's orbit. The correct main idea must include all the key points.

This distracter also repeats the exact same words as those in the conversation or lecture. Whenever possible, the correct answer paraphrases ideas from the conversation or lecture.

Distracter 2: The answer choice is too general or vague.

This distracter refers to a topic that resembles the correct one but is more general than the correct one. Sometimes, it can be recognized by the fact that the topic would be impossible to discuss in a few minutes. This distracter could refer to the Newton lecture as the physical laws of the universe. The topic of physical laws is correct but is more general than Newton's laws.

Distracter 3: The answer choice refers to the wrong kind of passage.

It is important to understand and recognize the different types of passages because this distracter could refer to one of many different types. For example, this distracter could refer to an explanation of Newton's laws as an argument against those laws.

Distracter 4: The answer choice rearranges ideas from the conversation or lecture.

This distracter includes ideas from the listening, but they have been reorganized. Possibilities include rearranging subjects, objects, descriptions, comparisons, causes and effects, times, and locations, among others. It could refer to Einstein's four laws of physics rather than Newton's laws.

Distracter 5: The answer choice is the opposite of the correct one.

This distracter might add or remove *no/not*, or use antonyms for ideas expressed in the passage. Instead of the student's request to change his roommate, this distracter might refer to the student's desire to keep his roommate.

Practice

Now, practice main idea questions for a conversation and then a lecture. Using the strategies you've learned, take notes in the space provided and answer the multiple-choice question(s) after each passage.

Notes – Conversation

_____

_____

_____

_____

_____

Play CD 1, Track 1 for the following conversation.

1. Which of the following statements best describes the student's situation?

   ○ His stress over his term paper has caused him to feel ill.

   ○ His professor is unwilling to extend the deadline on his term paper.

   ○ He needs more time to write his term paper because he has been ill.

   ○ He needs to take time off from school so he can take care of his health.

Notes – Lecture

_____

_____

_____

_____

Play CD 1, Track 2 for the following lecture.

2. Which aspect of the cougar study do the speakers mainly discuss?

   ○ How to divide the work that needs to be done

   ○ A change that has been made to the research plan

   ○ The best ways to conduct the cougar study

   ○ Which aspect of cougar behavior should be the study's main focus

3. What is the main reason the speakers give for studying cougars?

   ○ Because of their solitary nature, cougars have been more difficult to study than other wild cats.

   ○ A lack of knowledge about cougars makes it difficult to determine the best cougar management strategies.

   ○ The future of wildlife conservation depends on how humans interact with large predators, such as cougars.

   ○ Recent advances in research methods have made it easier to study cougars without influencing their natural behavior.

4. What aspect of the cougar management debate do the speakers mainly address?

○ The effects it has on cougar management techniques

○ The strongly opposing views of people involved in the debate

○ The impact it has on the ability of researchers to study cougars

○ The similarity between the cougar debate and other conservation issues

Purpose Questions

A purpose question asks about the speakers' purpose, or what the speaker hopes to accomplish in the conversation or lecture. It is more common for conversations, but it may be asked for a lecture. A speaker's purpose includes the main idea and the type of passage it is. Therefore, this question type is another way of asking about the main idea.

Use the following strategies to help prepare you for this task.

Strategy 1: Understand the main idea (see also main idea questions above).

For conversations, this is basically what the student wants or needs from the professor or school official. Since it is more polite to be clear and direct with people in authority, a student in a conversation is usually explicit about his or her purpose, and often states it early. For lectures, the purpose is directly related to the topic, opinion, attitude, and key points of the speakers. Lecturers may also be direct about their goals for a class, but not always.

Strategy 2: Recognize the kind of passage.

This relates more to lectures because lectures can be clearly understood as arguments, comparisons, explanations, etc. Often the purpose is simply an infinitive form of the passage type. For example, if a lecturer gives an argument, then her purpose is "to argue."

Strategy 3: Place the lecture in a broader context.

A lecturer's purpose may also relate to the course curriculum or the semester. Therefore, you should listen for clues to how the lecture fits into the class schedule and the course as a whole. For example, a professor may give a lecture that reviews material for a test. This directly relates to what the professor wants to accomplish, which could be "to prepare students for an upcoming exam."

Strategy 4: Be prepared to identify the purpose as an infinitive phrase.

Purpose is expressed as "infinitive + main topic/main idea." For example, if a student wants a new roommate, the purpose of the conversation could be "to change his roommate" or "to replace his roommate." Likewise, if a lecturer gives an explanation of Newton's laws of motion, his purpose could be "to explain a scientist's theories about mechanical laws."

Purpose Question Forms

You can recognize a purpose question by the question form and the related vocabulary, especially the interrogative pronoun "why."

Why does the student go to see his/her professor?

Why is the professor giving the lecture?

Why are the speakers discussing/talking about X?

Distracters

Distracters are incorrect answers. Purpose questions have four answer choices; one is correct and the other three are distracters. Each distracter may be one of the following:

Distracter 1: The infinitive refers to the wrong action or state.

This is why you must correctly identify the student's need or the type of lecture. A student who complains and asks for a new roommate wants to change/to alter his roommate. However, he doesn't want to disqualify or to expel his roommate. Likewise, if a professor explains Newton's laws of physics, the professor's purpose could be "to explain a scientist's three fundamental principles of mechanics." However, it is not "to compare three laws of nature" or "to argue that nature works according to laws."

Distracter 2: The purpose is too specific (see also Main Idea: Distracter 1)

Since the purpose is most often expressed as an infinitive phrase (infinitive + main topic), a distracter could use the correct infinitive but a main topic that is too specific. Like one type of main idea distracter, this type of purpose distracter includes parts or elements of the conversation or lecture topic, but not the entire topic. For example, if a student complains about his roommate (loud music, bad manners, and lack of cleanliness) and requests a new one, this distracter could refer to the purpose of the conversation as "to complain about his roommate's taste in music." After a lecture in which a professor explains Newton's three laws of motion (inertia, acceleration, reaction), this type of distracter could describe the purpose as "to illustrate why a body at rest will remain at rest until acted upon," which relates only to the first law of inertia.

Distracter 3: The purpose is too general (see also Main Idea: Distracter 2)

Much like Strategy 2 above, this distracter uses a correct infinitive and an incorrect object. However, this object is broader than the actual topic of the conversation or lecture. The purpose of the conversation about a new roommate could be described as "to complain about the tenants of the dormitory." Since the noun *tenants* is plural, this choice would be too general if the student complains about only one person, his roommate. Likewise, in a lecture about Newton's laws of motion, this distracter would refer to the lecture's purpose as "to explain the natural laws of the universe." Although motion, or mechanics, is a part of the universe, there is a lot about the universe that is not included in Newton's classical mechanics, such as atomic energy. Therefore, a reference to the entire universe would be much too broad.

Distracter 4: The purpose rearranges ideas (see also Main Idea: Distracter 4)

Like the similar distracter in other question types, such as detail questions and rhetorical function among others, this distracter type repeats familiar ideas (names, actions, states, etc.) from the passage and rearranges their relationship with each other. For example, for the student's conversation with his professor about changing his roommate, this distracter might refer to the student's purpose as "to avoid getting kicked out by his roommate" even though the student wants to get rid of the roommate, not the other way around.

Distracter 5: The purpose reverses a positive or negative, or uses antonyms (see also Main Idea: Distracter 5)

You should always read each answer choice carefully and pay close attention to the vocabulary used. If you skim the choice too quickly and carelessly, you might miss the addition or omission of a negative, such as *no* or *not,* which is a small difference that can change an answer immensely. For instance, this distracter type could state that the student's purpose in complaining about his noisy, rude roommate was "to explain his reasons for not changing his roommate." Likewise, this kind of distracter could use an antonym for *change* or *replace* and say that the purpose behind the student's conversation was "to continue living with his roommate."

Practice

Now, practice purpose questions for a conversation and a lecture. Using the strategies you've learned, take notes in the space provided and answer the multiple-choice question for each passage.

Notes – Conversation

_____

_____

_____

_____

_____

Play CD 1, Track 3 for the following conversation.

5. Why does the student approach the Financial Aid officer?

○ To ask for help in applying for an in-school scholarship

○ To explain why she is having trouble covering tuition costs

○ To find financial assistance that doesn't come from the school

○ To request help in using the Internet

Notes – Lecture

_____

_____

_____

_____

Play CD 1, Track 4 for the following lecture.

6. What is the main purpose of this lecture?

○ To showcase the major inventions of Ansel Adams

○ To explain the main reasons for Adams's modern appeal

○ To describe Adams's role in furthering and popularizing photography

○ To show the influence of the natural world upon Adams's photographs

Detail Questions

Detail questions ask about the people, places, things, definitions, descriptions, and reasons mentioned in the conversation or lecture. However, the TOEFL is not primarily a memory test. Although memory helps immensely, you must do more than simply remember a person's name or a location. You must understand how the detail supports what the speaker is saying.

Strategy 1: Be familiar with the various types of detail and the related vocabulary.
The chart below lists the major details and some, but not all, of the relevant vocabulary.

KAPLAN

Type of Detail	Clues/Key Words
Description: Physical traits (material, parts, size, shape, color, design, arrangement, orientation), taste, smell, sound, touch Personality Ability Quality Origin, nationality, source Category, etc	Adjectives: big, round, blue, pungent, loud, coarse, tilted, African, competent, etc Adverbs: quickly, hesitantly, etc Verbs: be, become, have, possess, compose, comprise, make up, sound, look, taste, feel, smell, etc. Nouns: an introvert, a bully, a fool, etc.
Definitions: A definition is a particular kind of description. A definition describes what a noun is. It is unique to that noun, and can't be used for any other noun. For example, "Claustrophobia is a fear of enclosed spaces." (Only claustrophobia can be described as a fear of closed spaces.)	Verbs: be, mean, refer to, be referred to as, define, be defined as, be described as, be called, be named, involve, include, relate to, etc Nouns: definition
Numbers: Date, data, measurement, amount, quantity, range, approximation, rate, ratio, percentage, statistic, equation, minimum, maximum, average.	Numbers: 1, 2, etc Ranges: dozens, hundreds, thousands, etc; from X to Y, between X and Y, around X. Ratios: X out of Y, X over Y, X per Y. Verbs: be, measure, count, amount to, etc Adjectives: few, little, many, most, close to, almost, near, approximate, etc Adverbs: nearly, approximately, etc
Comparisons: Similarities Simile, metaphor (see also Rhetorical Structure)	Verbs: be, look, sound + like, etc Prep/adjectives: like, alike, similar to, same as, equal to, comparative adjective + than Conjunctions: as, as + adj/adv + as
Contrasts: Differences	Unlike, not alike, dissimilar to, different than, unequal to, not as…as

Locations: Place, position, direction, area, range, zone, town, village, city, country, continent, etc.	Verbs: be, inhabit, be located, be situated, be found, sit, lie, reside, live, etc Prepositions: in, at, on, by, beside, next to, above, below, around, across, etc Conjunctions: where
Times: Beginning, duration, ending, order/ sequence, frequency	Adjectives: early/earlier, late/later, etc. Verbs: be, begin/start, continue/last for, go on, finish, end, etc. Nouns: o'clock, seconds, minutes, hours, days, weeks, Tuesday, March, etc Prepositions: in, at, on, for, during, before, after, since, etc Conjunctions: before, after, since, while Adverbs: first, second, next, then, finally, last, rarely, seldom, occasionally, often, etc
Examples: Specific nouns, actions, or states that belong to general types, kinds, classes, etc. A Mercedes is a type of car. A car is an example of a vehicle. A vehicle could be a an example of a purchase, possession, etc.	Prepositions: such as, for example, for instance, in particular, particularly, namely, specifically Adj + nouns: One type/kind/example, another type…
Explanation: Reasons, causes and effects, intentions, desires, methods, results	Prepositions: because of, due to, thanks to, as a result of, after, before, with Conjunctions: because, after, before, since, if, when, as. Infinitives: in order to + verb, to + verb Adverbs: therefore, as a result, then Verbs: cause, lead to, result in, create, make, want, hope, plan, etc.

Strategy 2: Identify the main idea, and anticipate the subpoints, or supporting points (see also General Listening Strategies).

As you know, you should always listen for the main idea, which is the subject of the conversation or lecture and the particular points the speakers make about it. At the same time, anticipation is an important part of active listening. Based on the opening remarks and the introduction, you should always try to anticipate how the speaker might or will support the conversation or lecture. These strategies are especially important for detail questions. A clear understanding of the main idea and supporting points will help you identify the most important details. There are a lot of details in a listening passage, and students can easily get distracted if they waste too much time trying to record everything. Remember that you must focus on active listening and comprehension. As long as you have a clear understanding of the main idea and purpose of the passage, you will be better able to recognize which ideas you should record and which ones you should commit to memory.

Strategy 3: Focus on the most important details, not minor ones.

The important details are those that are most necessary to understand the main topic. For example, in a lecture about Newton's laws of motion, the subpoints or supporting points would be the laws, and the most important details would be those details that explain the laws, such as the experiments, the results of an experiment, observations of nature, and so on. Newton's date of birth, social life, and marital status would be minor details because they don't support the discussion of the laws.

Strategy 4: Use the turns in a conversation or academic discussion to identify important points and details (see also General Listening Strategies).

Questions are excellent clues because the pronouns are directly related to the answer. For example, if a student asks, "Whom do I see about this?" you know the answer will mention a person because of "whom."

Strategy 5: Relate a detail to the main idea (see also General Listening Strategies).

You must understand how a detail supports the main idea. For example, if a professor mentions a school official, will this person help the student? How? Why is the person mentioned?

Strategy 6: Listen for key words.

A word is important if it supports the main idea or topic. Therefore, key words depend on the main idea. They can be any part of speech (noun, verb, adjective, adverb, infinitive, preposition, or gerund). For example, if a student wants to change roommates, the key words will relate to problems with the roommate (noisy, doesn't clean, inconsiderate), possible solutions (talk to him, study in the library), and final decisions or plans (fill out a form, meet with an advisor).

Distracters

Distracters are incorrect answers. Detail questions are followed by four choices; one or two are correct, and the others are distracters. Read the question carefully to know whether you should choose one or two choices. Distracters may be one of the following:

Distracter 1: The answer choice repeats exact vocabulary, phrases, and clauses from the passage.

The correct answer paraphrases information from the conversation or lecture. However, remember that certain information, such as names, is difficult to paraphrase.

Distracter 2: The answer choice rearranges details.
Some choices refer to detail from the passage, but alter the subject, object, place, time, order, comparison, positive/negative nature. This kind of distracter, for example, will describe the student as noisy even though the roommate is noisy.

Distracter 3: The answer choice doesn't match the main idea.
Impossible or unlikely details are easier to avoid when you clearly understand the main idea. For example, if a student wants a new roommate and is having problems with the roommate, then it is more logical that noise is caused by the roommate, not the student who is complaining.

Detail Question Forms

You can recognize detail questions by the following question forms and related vocabulary, especially the interrogative pronouns.

According to the professor, what/who/whom/which/why/where/when/how…?

What/who/whom/which/why/where/when/how does the professor…?

According to the student, what/who/whom/which/why/where/when/how…?

What/who/whom/which/why/where/when/how does the student…?

Practice

Now, practice Detail questions for a conversation and a lecture. Using the strategies you've learned, take notes in the space provided and answer the multiple-choice questions for each passage.

Notes – Conversation

_____

_____

_____

_____

_____

Play CD 1, Track 5 for the following conversation

7. What does the staff person give the student?

○ A document to complete

○ A receipt for his payment

○ A contract for his insurance plan

○ A Web site address to visit for more information

8. According to the staff person, which of the following two forms of health care are covered by the Vital Savings Plan?

○ Dental

○ Fitness

○ Acupuncture

○ Prescription drugs

Notes – Lecture

_____

_____

_____

_____

Play CD 1, Track 6 for the following lecture.

9. According to the professor, one of Robert Moses' core beliefs about cities was that they

○ had ceased to be intimate or pedestrian-friendly

○ were built around cars rather than people

○ had many problems stemming from overpopulation

○ were unhealthy to live in because of automobile traffic

10. What are two common complaints about Robert Moses mentioned by the professor?

Click on 2 answers.

    ⊖ He built unlivable apartment buildings for public housing.

    ⊖ He encouraged the development of suburbs across the country.

    ⊖ He damaged New York City's public transportation system.

    ⊖ He put highways in the middle of residential areas.

Speaker's Attitude Questions

Attitude questions ask about a speaker's emotional state, level of certainty, confidence, or agreement/disagreement. Because attitude depends on intonation, emphasis, and word choice, they may replay excerpts from the conversation or lecture. However, not all attitude questions replay parts of the passage, so active listening and good notes are as important as always.

Use the following strategies to help prepare you for this task.

Strategy 1: Be prepared to infer a speaker's attitude.
It is highly unlikely that a speaker will express his attitude explicitly in a statement like "I am unhappy about…" or "I have a different opinion than you." Instead, you have to understand attitude through inference.

Strategy 2: Use context as a clue to inference.
Context refers to the situation of the speakers. It includes the identity of the speakers, and the time, place, reason, main topic, and details of the conversation or lecture. Any inference must logically fit the context. If a student visits a professor to complain about a noisy, lazy roommate, she is annoyed or frustrated, not happy or satisfied.

Likewise, if two students argue about what a painting means, you can infer that they have different interpretations and see the painting differently.

Strategy 3: Listen for key words of attitude and connect them to infer a speaker's feelings or opinion.
A person's choice of words gives clues about feeling and opinion. If a professor gives a lecture on Isaac Newton and uses words such as brilliant, amazing, genius, and revolutionary, he has a positive attitude toward Isaac Newton.

Key words may be any part of speech, and some, but not all, include the following:

Adjectives/Adverbs

Possibility: possible/possibly, maybe, potential/potentially, etc.

Assumption: likely, probably, etc.

Certainty: clear/clearly, certain/certainly, definite/definitely, absolute/absolutely, etc.

Positive: beautiful, smart, helpful, kind, etc.

Negative: ugly, ignorant, hurtful, mean, etc.

Modals

Possibility: may, might, could

Assumption: should, must

Certainty: will

Verbs

Possibility: guess, estimate, approximate, think

Assumption: assume

Certainty: know (for sure), be positive that

Positive: like, enjoy, agree with, support, appreciate, etc.

Negative: dislike, disagree with, object to, oppose, fear, etc.

Strategy 4: Listen for intonation and emphasis.
Rising intonation usually indicates positive feeling, interest, acceptance, surprise, or a question. Falling intonation suggests negativity, sarcasm, disinterest, rejection, or disapproval.

Strategy 5: Be familiar with possible answer choices and their subtle differences in meaning and degree.
Like many questions on the TOEFL, attitude questions test your vocabulary. Adjectives in English can be very specific; so many answer choices may be close in meaning but not exactly the same. Consult a dictionary or thesaurus about the following, identify synonyms and differences of degree, and create a list of others:

Happy:	pleased, glad, enthusiastic, thrilled, joyful, jubilant, ecstatic
Thankful:	appreciative, relieved
Neutral:	objective, apathetic, uninvolved, uncaring, cold, dismissive
Sad:	unhappy, despondent, depressed, bleak, hopeless
Disappointed:	let down, devastated

Anxious:	concerned, worried
Uncomfortable:	awkward, stiff
Inappropriate:	rude, impolite, disrespectful, insulting, demeaning
Surprised:	shocked, staggered
Afraid:	fearful, terrified, petrified
Annoyed:	bothered, frustrated
Upset:	angry, furious, incensed

Strategy 6: Pay attention to changes in attitude.
Sometimes, a speaker's attitude changes during a conversation or lecture. The attitude could improve or worsen. A student might be frustrated at first, but at the end of the conversation, the same student could be satisfied. Or, a student could hold one opinion at the beginning of an academic discussion, and then change that opinion by the end. The possibility of change is one reason that active listening and key words are so important.

Attitude Question Forms

You can recognize attitude questions by the following question forms and related vocabulary:

What is the student's attitude toward…?

What is the student's opinion of…?

What is the professor's attitude toward…?

What is the professor's opinion of…?

What does the student imply about…?

What does the professor imply about…?

Listen again to part of the conversation/lecture. Then answer the question

[An excerpt is heard]

What does the student/professor mean when s/he says this:

[A specific part of the excerpt is repeated]

Distracters

Distracters are incorrect answers. Attitude questions have four answer choices; one is correct and the other three are distracters. The distracters for attitude questions are the following:

Distracter 1: Answer choice is too extreme.
Many adjectives are close in meaning but can be weaker (happy) or stronger (ecstatic). This distracter refers to an emotion that is too strong. Although strong emotions are possible on the test, passages don't include excessive or threatening attitudes, such as fury.

Distracter 2: Answer choice is out of context.

This distracter might be easy or hard to recognize. The distracter is easy to see if the attitude is obviously inappropriate. For example, if a student complains about his noisy, rude roommate, the student is obviously not happy or comfortable. However, many adjectives are close in meaning but each may refer to a specific situation. For example, the adjectives "satisfied" and "relieved" are both positive, but the former describes a person who gets what he or she wants and the latter describes a person who no longer has to worry about something.

Practice

Now, practice attitude questions for a conversation and a lecture. Using the strategies you've learned, take notes in the spaces provided and answer the multiple-choice questions for each passage.

Notes – Conversation

_____

_____

_____

_____

Play CD 1, Track 7 for the following conversation. Question 12 for this conversation is Track 8 on the same CD. For that question, you must play the corresponding track.

11. What is the professor's opinion of the student's initial research suggestion?

　　○ It is an original and exciting topic.

　　○ It will probably take too long to research.

　　○ It might lead to other interesting discussions.

　　○ It may be more appropriate for a later project.

12. Listen again to part of the conversation. Then answer the question.

Track 8
What can be inferred about the student when he says this?

　　○ He is concerned that other students will use his idea.

　　○ He finds the wolf spider's behavior unusual and exciting.

　　○ He disagrees with the professor's opinion about his project.

　　○ He is frustrated with his inability to find information about the wolf spider.

Notes – Lecture

_____

_____

_____

_____

Play CD 1, Track 9 for the following lecture. Questions 14 to 15 for this lecture are Tracks 10 to 11 on the same CD. For each question, you must play the corresponding track.

13. What is the professor's opinion of the theory of plate tectonics?

   ○ It can neither be proven nor disproved.

   ○ It is only one of several competing theories.

   ○ It is not accurately understood by most people.

   ○ It adequately explains the earth's geological changes.

14. Listen to the conversation. Then answer the question.

Track 10
What can be inferred about the professor when he says this?

   ○ He finds it surprising that Wegener's theory was not proposed before the 1900s.

   ○ He believes Wegener's contribution to mapmaking to be his major contribution.

   ○ He does not think Wegener's theory has been greatly advanced since it was proposed.

   ○ He believes Wegener did not receive enough credit for his accomplishment.

15. Listen again to part of the conversation. Then answer the question.

Track 11
What does the man mean?

   ○ He is unsure if his answer is correct.

   ○ He believes the woman's answer is incomplete.

   ○ He thinks the theory is difficult to understand.

   ○ He disagrees with the description the woman provides.

Rhetorical Function Questions

Rhetorical function refers to a statement that indicates a speaker's intention, reaction, or attempt to influence the listener. Rhetorical function questions test your ability to recognize such ideas as a speaker's plan, apology, compliment, criticism, complaint, or joke, among many others. As always, you must do more than hear what is said. You must also understand why it is said.

Rhetorical function questions depend on exactly how something is said. Therefore, these questions often, but not always, replay excerpts from the conversation or lecture. Although it's possible for you to make a note about a speaker's apology, for example, you don't always have to rely on your notes because the apology could be replayed as an excerpt. In order to prepare for such questions, it is best to listen actively and then relate a statement to what was said before and what is said after.

Use the following strategies to help prepare you for this task.

Strategy 1: Be prepared to infer.
Speakers don't always specify why they say something, so don't rely on key words or structures.

Strategy 2: Pay attention to the context.
Think about what purpose is most logical based on the surrounding statements, actions, states, times, places, key points, and vocabulary. For example, if a student goes to see the librarian and the student wants to avoid paying fines, the student might suggest things about his situation, finances, or work load as a way to gain sympathy.

Strategy 3: Recognize the attitudes of the speakers.
Think about what kind of feelings the speakers seem to be showing through intonation and vocabulary (also see attitude questions). If a professor describes a concept as extremely complex and sounds doubtful about the class's preparation, she might decide to delay the discussion of it until later.

Strategy 4: Distinguish between rhetorical and interrogative questions.
There are two general types of questions: interrogative and rhetorical. An *interrogative* question is a question used to gather information. When someone asks an interrogative question, he wants an answer. Rhetoric is the art of speaking or writing to persuade, and a *rhetorical* question is one designed to influence the listener. There is no answer expected when a rhetorical question is asked.

Context and tone are usually required to distinguish the difference between an interrogative and rhetorical question. "How old are you?" can be either rhetorical or interrogative given the circumstances. Two students getting to know each other often want to find out each other's age. Under these circumstances, the student asking "How old are you?" is asking an interrogative question. Most likely, she would stress the word *old* when asking the question: "How OLD are you?"

Now let's say that a college student is misbehaving in class. Because of that student's childish behavior, the professor might ask the same question, but with an emphasis on different words: "HOW old ARE you?" This gives the question a tone of sarcasm. The professor doesn't really want to know the student's age; he is trying to discourage the student's bad behavior by embarrassing him and implying that he is acting like a child.

Other purposes of rhetorical questions are the following:

To express an opinion: "How can they charge so much?" could be a reaction to a very high price.

To express a feeling: "Why me?" could be a response to the third bad thing that has happened in one day.

To make a request or give a command: "Would you stop talking?" could be a teacher's response to students in class.

Strategy 5: Pay attention to stress and intonation.
Although stress is a common and natural part of English, speakers often stress certain words to convey specific meaning. If a teacher is giving a demonstration of a chemical experiment, she might say the following (stress added):

Professor: Now, once you add the powder, the TUBE will become VERY HOT.

The professor is not just describing the reaction; she is also warning the students to be careful. Because the statement is serious, the teacher would use a falling, or low, tone (see also Strategy 4 above).

Question Forms

A rhetorical function question can be recognized by the following forms and related vocabulary:

Why does the student say/mention/refer to X?
Why does the professor say/mention/refer to X?

Listen again to part of the conversation/lecture…
[An excerpt is heard]
What does the student/professor mean when she/he says this:
[A specific part of the excerpt is repeated]

Distracters

Distracters are incorrect answers. Rhetorical function questions have four answer choices; one is correct and the other three are distracters.

Distracter 1: The answer choice repeats vocabulary, but it is impossible or highly unlikely.

This type of distracter repeats a word, phrase, or whole expression from the passage, but the answer choice is illogical based on the context. For example, look at the following excerpt from a conversation between a student and professor:

Professor:	Hello, John. What can I do for you?
Student:	Actually, Professor, I'd hate to sound inconsiderate, but can I run back to my dorm for a second? I just realized that my notes are still on my desk.
Professor:	(falling intonation) As you can see, John, there is a line of students outside my office waiting to see me during these office hours.

A rhetorical function question could ask, "Why does the professor refer to the students waiting to speak with him?" As always, the context (time, place, identity, tone, surrounding statements) is extremely important for understanding the rhetorical function of the professor's statement. The statement follows the student's request and the falling intonation indicates a negative attitude. If you ignore or miss the context, you may choose an incorrect answer, such as "to recommend that he borrow notes from a student in line." This choice is an example of Distracter 1: it repeats the nouns *notes* and *line* from the conversation, but the choice is not logical. It is very unlikely that strangers would be carrying information that the student could use to discuss his own problem, so there's no reason for the professor to make the suggestion. Instead, the professor's statement is best understood as a negative reaction or response to the student's request. A correct answer could be that the professor wants "to insist that the student not waste time" or "to show disapproval of the student's request."

Distracter 2: The answer choice is a possible rhetorical function in a different context.

This distracter type is more complicated because it refers to a possible understanding of a statement or question in a different situation. However, the choice is still wrong based on the specific context of the conversation or lecture that you hear in the test. For example, in the excerpt above, the professor's reference to the line could indicate pride or surprise in different contexts, so an example of Distracter 2 could describe the function of the professor's statement as "to boast about his popularity" or "to show surprise at the number of students."

However, a boast would be off-topic or irrelevant following the student's request to leave quickly and return with notes. Also, a boast would probably come from someone else, not a professor, since it seems inappropriate for a professor to brag about his popularity (although it isn't impossible). Moreover, the same statement could indicate surprise, but in the excerpt above, there are no clues that the line up is unexpected, and an expression of surprise also would seem off-topic given the student's request. Finally, in order to be correct, the distracters would also require a slight change in intonation. The falling intonation indicated in the transcript suggests displeasure, and an expression of pride or surprise would require a more positive, rising, or fluctuating intonation.

Practice

Now, practice Rhetorical Function questions for a conversation and a lecture. Also, for the lecture, you can review main idea and detail questions. Using the strategies you've learned, take notes in the spaces provided and answer the multiple-choice questions.

Notes – Conversation

_____

_____

_____

_____

Play CD 1, Track 12 for the following conversation. Question 17 for this conversation is Track 13 on the same CD. For that question, you must play the corresponding track.

16. Why does the professor mention how long he has been teaching the course?

○ To emphasize the uniqueness of the student

○ To establish himself as an authority on the subject

○ To signal his unwillingness to change the assignment

○ To express his preference for public school education

17. Listen to the conversation. Then answer the question.

Track 13
Why does the professor ask this?

○ He wishes to hear the student's suggestion.

○ He would like the student to reconsider his opinion.

○ He wants to know how much the student already knows.

○ He is interested in having a conversation about homeschooling.

Notes – Lecture

_____

_____

_____

_____

Play CD 1, Track 14 for the following lecture. Questions 18 to 19 for this lecture are Tracks 15 to 16 on the same CD. For each question, you must play the corresponding track.

18. Listen again to part of the conversation. Then answer the question.

Track 15
Why does the student say this?

○ To clarify his understanding of what the professor said

○ To indicate that he does know the author's full name

○ To demonstrate his knowledge of history to the professor

○ To signal that he does not see the relevance of the professor's point

19. Listen again to part of the conversation. Then answer the question.

Track 16
What does the student mean to say?

○ She is correcting a statement the professor made.

○ She is disputing the accuracy of the events in _Beowulf_.

○ She is demonstrating her knowledge of the subject matter.

○ She is suggesting a different way to look at the events in _Beowulf_.

20. Which aspect of _Beowulf_ does the professor mainly address?

○ Its plot details

○ Its central themes

○ Its historical accuracy

○ Its influence on the English language

21. What is the professor's main point about *Beowulf*?

   ○ Its status as a foundation of English literature has not been historically appreciated.

   ○ It presents moral choices that are universal to people across cultures and historical periods.

   ○ It is foundational because it represents the beginnings of a uniquely English worldview.

   ○ It reveals a continuity between modern English literature and pre-Christian Germanic traditions.

22. According to the lecture, what two value systems does the author of the *Beowulf* try to reconcile? *Choose 2 answers.*

   ○ Christian forgiveness

   ○ Belief in the supernatural

   ○ Allegiance to a strong king

   ○ Anglo-Saxon warrior honor

Rhetorical Structure Questions

Rhetorical structure questions test your understanding of how the statements or ideas in a conversation or lecture relate to each other. There are two main types of rhetorical structure questions. The first type is general and asks about the overall organization of a passage. The second type is more specific and asks about how a particular statement or idea supports a conversation or lecture.

Question Type 1: Organization of a Passage

This first type of rhetorical question asks about how the speaker organizes the whole passage. It's more common for lectures, but it may be asked about a conversation as well. Also, because this type refers to the entire passage, it doesn't replay excerpts from the conversation or lecture.

Use the following strategies to help prepare you for this task.

Strategy 1: Be familiar with the possible methods of organizing a conversation or lecture.
These methods are some, but not all, of the possible ways to organize a conversation or lecture:

A general category (family, class, species) and specific types

Definition and examples

Cause(s) and effect(s)

Problem and solution(s)

Smallest to the biggest

Nearest to the farthest

Argument and refutation(s)

Hypothesis/experiment and results

Historical narrative/chronology of events

Chronological sequence of steps (the first step to the last)

Cheapest to the most expensive

Most likely to the least likely

Strategy 2: Be prepared to infer.
Though it is possible, a speaker will probably not state the organization directly. Therefore, don't rely on an explicit, obvious statement such as, "Let's organize our discussion by type" or "I want to begin with the best solution and then continue until the worst." You will probably have to recognize how the passage is organized on your own.

Strategy 3: Remember the basic outline of conversations and lectures as you take notes (see also General Listening Strategies).
People in conversations take turns, so the organization will not be revealed all at once. It will appear as each speaker takes his or her turns. A lecture, on the other hand, is organized in paragraphs, so the organization will be recognizable by the topic of each body paragraph, which is a supporting point for the main idea.

Strategy 4: Listen for key words.
A speaker might use vocabulary that gives you a clue about the organization. For example, a professor giving a lecture on Newton's laws of physics might discuss each law as follows:

Body paragraph 1

Newton first thought that…He tested this theory by…His observations showed…

Body paragraph 2

Next, Newton had always believed…He conducted an experiment that…Finally, he saw how…

Body paragraph 3

Finally, Isaac Newton theorized that…However, when he studied…, he learned that…

In the above quotes, notice how Newton first believes something, then tests that belief, and finally learns if it is true or not.

Strategy 5: Take notes on key points, not every detail (see also General Listening Strategies).

The supporting points, not minor details, are clues to the organization. For example, the key points from the Newton lecture could be the experiments, observations, and proof of his theories. However, details about his personal life (marriage, hobbies, etc) would be minor since they would reveal little or nothing about the key points.

Strategy 6: Look for paraphrased ideas in the answer choices, not repeated vocabulary.

The correct answer choice will not repeat vocabulary exactly. For example, it could refer to the organization above as "hypothesis followed by observation and conclusions." Although the vocabulary relates to that used in the passage, it isn't exactly the same.

Rhetorical Structure Question Forms

You can recognize organization questions by the following question forms and related vocabulary:

> In what order does the professor tell/talk about/mention/discuss X?
> In what order does the student tell/talk about/mention/discuss X?

> How does the professor talk about/explain/discuss X?
> How does the student talk about/explain/discuss X?

> How does the professor organize X?
> How does the student organize X?

Distracters

Distracters are incorrect answers. Rhetorical structure questions have four answer choices; one is correct and the other three are distracters. The following paragraphs describe distracters for organization.

Distracter 1: The answer choice reverses the correct organization.

This type of distracter can be especially confusing because it uses the correct ideas, but puts them in the opposite order. If a professor first offers the solution that costs the most money and then offers solutions that cost less and less money, this distracter would describe the organization as "from the least to the most expensive."

Distracter 2: The answer choice refers to an unrelated idea that might seem connected but is out of context.

If the professor organizes his suggestions from the most to the least expensive, the conversation is organized according to cost. Cost involves money, and this type of distracter would use an idea related to money, such as fee, salary, and debt. Therefore, it might say that the conversation is organized "from the highest to the lowest salary."

Distracter 3: The answer choice uses minor details.
This type of distracter will incorrectly base the organization on minor details. If a student complains about his roommate by mentioning characteristics and giving examples of his behavior but the student also mentions that his roommate is on the football team, this distracter would say that the conversation is organized "by hobby."

Practice

Now, practice rhetorical structure – organization questions and review main idea, purpose and rhetorical purpose questions for a conversation and a lecture. Using the strategies you've learned, take notes in the spaces provided and answer the multiple-choice questions for each passage.

Notes – Conversation

_____

_____

_____

_____

Play CD 1, Track 17 for the following conversation. Questions 24 to 25 for this conversation are Tracks 18 to 19 on the same CD. For each question, you must play the corresponding track.

23. How does the advisor organize the information that she presents to the student?

   ○ By connecting the student's actions with his intentions

   ○ By contrasting the student's resume with other resumes

   ○ By providing a set of guidelines the student should follow

   ○ By highlighting the relevance of resume content to specific jobs

24. Listen to a part of the conversation. Then answer the question.

Track 18
Why does the advisor say this?

   ○ To suggest a job the student might wish to apply for

   ○ To convince the student to remove a job from his resume

   ○ To recommend a better organization for the resume

   ○ To recommend he include his experience as an ice cream server

25. Listen to part of the conversation. Then answer the question.

Track 19

Why does the student mention his graduation date?

- ○ To highlight an important job qualification
- ○ To explain why his resume has so few jobs
- ○ To emphasize the limited amount of time he has
- ○ To clarify why he has never visited Career Services

26. What are the speakers mainly discussing?

- ○ How to improve the student's resume
- ○ Jobs that the student would like to apply for
- ○ Reasons why the student has not been hired for a job
- ○ Experiences the student would like to highlight on his resume

27. Why does the student go to see the advisor?

- ○ To fulfill a requirement for graduation
- ○ To respond to the advisor's invitation
- ○ To discuss his possible career options
- ○ To elicit help writing a resume for a job

Notes – Lecture

_____

_____

_____

_____

_____

Play CD 1, Track 20 for the following lecture. Questions 31 to 32 for this lecture are Tracks 21 to 22 on the same CD. For each question, you must play the corresponding track.

28. How does the professor clarify the points she makes about using simplified texts?
    ○ By showing different methods for using simplified texts in the classroom

    ○ By providing examples of texts that are not significantly changed by being simplified

    ○ By discussing recent scientific research that supports the benefits of reading simplified texts

    ○ By comparing the psychological benefits of reading texts in their simplified and original forms

29. What is the professor's main point?

    ○ Intensive and extensive reading of literature improves all aspects of language learning.

    ○ The people who suggest that simplifying texts devalue them have not carefully examined the benefits.

    ○ Modifying classic literature is more beneficial and less problematic than some people think.

    ○ The reading of great literature in its original form can help to teach EFL students about important elements of English culture and history.

30. What is the main purpose of this lecture?

    ○ To support the creation of more simplified educational texts

    ○ To advocate that teachers use simplified texts in their classrooms

    ○ To urge EFL students to read simplified texts as much as possible

    ○ To familiarize teachers with both sides of the simplified texts debate

31. Listen again to part of the lecture. Then answer the question.

Track 21
Why does the professor say this?

    ○ To introduce the central point of her argument

    ○ To establish the benefits of reading classic texts

    ○ To present a comparison between two points of view

    ○ To bring up an interesting departure from her main point

32. Listen again to a part of the lecture. Then answer the question.

Track 22
What does the professor suggest?

○ She does not believe these terms are well defined.

○ She believes the distinction is worth exploring further.

○ She would like someone in the class to propose an answer.

○ She wants all the students to answer the question for themselves.

Rhetorical Structure Question Type 2: Organization of Details

The second type of rhetorical structure question is similar to the first but more specific. It asks about a particular statement or idea and asks you how the idea helps you understand the conversation or lecture. This question can ask you about introductions/hooks, examples, explanations, descriptions, digressions, and conclusions. As you listen, you must recognize what these supporting points do in the conversation or lecture.

Unlike the first type of rhetorical structure question, this second type may replay an excerpt from the conversation or lecture. However, it may not replay a part of the conversation or lecture, so you must listen actively and take detailed notes to be prepared for this question.

Use the following strategies to help prepare you for this task.

Strategy 1: Be familiar with hooks.
You must be able to recognize the various possible hooks for lectures: an anecdote, a quote, a fact, etc (see also main idea questions: Strategy 2).

Strategy 2: Be familiar with the types of detail and their related vocabulary.
You must be able to recognize descriptions, definitions, examples, comparisons, contrasts, locations, times, and explanations (see also detail questions: Strategy 1).

Strategy 3: Listen actively and connect ideas.
As you've already learned, you must recognize how details support the main idea, and how they relate to each other. If a speaker compares X and Y, what concept is she trying to explain or clarify? If a speaker describes a person, does that description give you a reason for what the person did or experienced? As always, think about what you hear.

Strategy 4: Distinguish literal and metaphorical language.
Many words in English have both literal and metaphorical meanings. To avoid confusion, you must be able to recognize when a word or phrase is being used literally or metaphorically. *Literally* means "truly" or "actually." Literal language is very specific and doesn't refer to another meaning. If someone mixes two chemicals and the new mixture explodes, the explosion is an instantaneous, violent production of energy that destroys the chemicals, the instruments and possibly harms people. One can say that the experiment literally exploded.

However, the verb *explode* also has other more metaphorical meanings. Metaphors express ideas that aren't actually true. If a student gets angry and explodes in a chemistry class, that person yells or screams; the explosion is a sudden, emotional outburst, not an instantaneous production of energy that destroys the person physically. Therefore, one could not say that the student literally exploded in class. Also, explode could also mean "increase suddenly," so a speaker could refer to a population explosion. Again, you could not say that the population literally exploded.

You must use common sense, logic, and the context of the conversation or lecture to recognize literal and metaphorical language.

Strategy 5: Recognize similes and metaphors, but don't take them too far.

A *simile* is a type of comparison using "like/not like" or "as...as." A professor in a sociology class might say, "Drug addiction is like a cancer in society." The simile is meant to imply something about drug addiction through association. The professor is expressing something about the harmful effects of drug addiction in society: it spreads to many parts, ruins people's health, kills people, and ultimately destroys societies. This is similar to cancer, which is a disease that spreads to many parts of the body, ruins people's health, and can kill them.

The simile does not mean that drug addiction and cancer are exactly the same. A simile is not meant to express exact similarity, only partial or metaphorical similarity. So if you assume that drug addicts have cancer, you are taking the simile too far.

A metaphor is similar to a simile, except it does not use "like" or "as...as." A metaphor is an implied comparison, and of course it is metaphorical, not literal. If a student says, "My roommate is a lion on the football field," the student is not literally saying that his roommate becomes an animal. Instead, the student implies that his roommate is like a lion in some way; the roommate is strong, powerful, difficult to stop, etc. Because metaphors do not use clue words, you must use your imagination and infer their meaning.

Strategy 6: Recognize idioms but don't take them literally.

Idioms are common examples of metaphorical language. An idiom is a set, or fixed, expression, which means that the meaning and choice of words don't change even though the tense can. Also, an idiom's meaning is not the same as the literal meaning of the words in it. For example, "It's raining cats and dogs" is a common idiom that always means, "It is raining hard." The meaning does not involve actual cats and dogs. You should become as familiar as possible with idioms that relate to campus life and academic fields.

Strategy 7: Distinguish between the real and the hypothetical.

Speakers often describe situations involving people, actions, or states in order to support their point. You must be able to recognize if the actions or states are real (actually occurred), or if they are hypothetical, or imaginary.

The Real

Speakers in both conversations and lectures often give anecdotes. An anecdote is a brief description of a real event or activity that occurred at some point in the past. It is not a complete story, but a short scene or part of a story that is used to support a speaker's idea. An anecdote stands out from the rest of a conversation or lecture because the anecdote usually occurred before the conversation or at a different time than the other actions or states discussed in a lecture. Look at the following example from the middle of a conversation between a school official and a student complaining about his roommate:

Student: I've already spoken to him several times and tried to reason with him, but I don't think I'm getting through.

Official: He's not taking things seriously?

Student: I just don't think he cares. Last weekend, I went home to visit my parents. When I returned, he had obviously had a fun weekend. The dishes were piled to the ceiling and every glass had been used and left out. The place smelled like a dump because he hadn't even taken out the garbage.

Official: I see…

The student's description of his return to the apartment is a description of an event that actually happened to the student before the conversation with the official. In the context of the conversation, it is a real event.

The Unreal, Imaginary, or Hypothetical

Hypothetical means imaginary, and when someone speaks hypothetically, the person describes an imaginary situation in the past, present, or future. Look at the following example from a lecture on Newton's laws of physics:

Newton's law of inertia is perhaps the easiest law to understand, but it is often misunderstood. Many people assume that the law states that a body at rest will remain at rest until another force acts on it. This is true. If you place an apple on a perfectly flat table, the apple will not move on its own. However, if the table is not perfectly flat, the apple will start to roll because gravity pulls it down. But, this is only part of the law. Actually, the law states that a body will tend to stay in the same state, rest or motion, unless another force acts on it. In other words, a ball like the one on the table will remain stationary until something, gravity or a child's hand, moves it. But, it is also true that a rolling apple will roll forever unless an opposing force acts on it. The reason an apple cannot roll forever is a combination of gravity and friction. Friction creates an opposing force that slows down and eventually stops the apple. But, imagine the apple floating through outer space. In space, there is no friction, so the apple could theoretically float forever.

The professor refers to an apple on a table, an apple rolling on Earth, a child's hand, and an apple in outer space. Each of these examples is hypothetical, or imaginary. It is important to recognize that the professor is not referring to something he is doing in class or something he wants the students to do. Moreover, he does not have an apple or table in class while speaking. The students are only expected to imagine what he says in order to understand his topic, the law of inertia.

Strategy 8: Use key words to recognize the unreal or hypothetical.
Though imaginary situations often have no clue words, when people speak hypothetically, they can use some of the following vocabulary:

> Let's imagine that/if…
> Imagine if/that…
> Let's suppose that…
> Suppose if/that…
> Picture…
> Let's say that…
> So,…
> Hypothetically speaking,…

Strategy 9: Be aware of the possible rhetorical actions and their meanings.
A statement may perform any of following actions: introduce the main topic, give background, change subtopics, give an example/exemplify, explain, illustrate/demonstrate an idea, compare, contrast, elaborate/develop/expand, describe, define, clarify, refute, digress from the topic, or conclude the conversation or lecture.

To Introduce

An introduction often begins with a hook, which may be a rhetorical question, quote, fact, an anecdote, background, or a brief description (see also main idea questions: Strategy 2). In the following example, a movie is mentioned to introduce the main topic, time travel.

> Have any of you seen the Hollywood movie *Back to the Future*? Well, in the film, a scientist uses a sports car to travel back in time. Although that method has never been seriously proposed, time travel has fascinated scientists for centuries.

To Give/Provide Background

Background refers to any essential information that a listener might be unfamiliar with. Although the background is not the focus or main topic, the listener may still need to know it in order to understand the speaker's attitude, purpose, or main idea. It is always given at the beginning of a conversation or lecture. In the following example, England is discussed as background to the ideals adopted in the United States.

When the Puritans first left England and sailed to the colonies, they left behind political and religious persecution. The King of England was also the head of the Church of England, and he refused to allow any religions, Protestant or Catholic, to deny his authority and the supremacy of his church. The Puritans were considered rebels for wanting to lead, organize and spread their own religious affairs, and many were imprisoned or worse. It is no wonder that freedom of religion became one of the founding principles of the country that grew out of those colonies, the United States of America. Along with religious freedom, the newly formed country also protected...

To Change Topics

A speaker starts or changes topics by using a rhetorical question or certain key words, such as the following:

First, at first, one + noun

Next, the next, another, one more, some other

Second, third, fourth, etc

Moreover, furthermore, also, on the other hand

To Describe

A description gives you detail about appearance, texture, smell, sound, taste, etc (see detail questions: Strategy 1). There are many ways of describing something. A comparison, contrast, simile and metaphor are all methods of description.

To Define

A definition is a description that is unique to one idea, and cannot describe any other (see also detail questions: Strategy 1). For example, the following is a definition of DNA:

Deoxyribose nucleic acid, or DNA, is the genetic code used to manufacture proteins in a cell.

It is important to distinguish a definition from a description because they are not the same. Look at these two statements:

NASA's Space Shuttle is a manned, two-winged aircraft that is carried into outer space on rockets and then returns to Earth on its own power.

NASA's Space Shuttle is a spectacular achievement of technology and innovation.

The first statement can only refer to the Space Shuttle, so it is a definition. However, the second statement could also describe the Hubble Space telescope, the International Space Station or the tunnel between England and France, so it is only a description.

To Give an Example/To Exemplify

Examples are more specific than the previous idea, which is often a general class, family, species, or type. A vehicle is a general noun, and a car is a specific type, or example, of a vehicle. A bus, motorcycle, and truck are also specific types of vehicles. Examples are best identified by the contrast of general and specific and by key words or phrases (see also detail questions: Strategy 1).

To Explain

An explanation gives reasons, causes, methods, functions, steps, or results (see also detail questions: Strategy 1). When speakers explain, they show why or how something exists or happens. Although explanations often involve certain key words (*due to, because, if, therefore, the reason why,* etc), explanations don't require them. A speaker can mention a fact, and follow it with a reason without ever using a key word. This is a major reason why you must always listen actively. Also, many kinds of statements can explain, such as descriptions, comparisons, contrasts, and examples.

Student:	Whom should I see about this?
Professor:	You'll have to see the Dean of Arts, Professor Miller. His office is down the hall from mine, and his office hours are posted outside his office.
Student:	I'll check them right away, but maybe I'll try to catch him in the cafeteria. I see him there all the time.
Professor:	Stick to his office hours; he's a very busy man without a lot of patience.

The professor describes Prof. Miller as busy and impatient in order to explain why the student should see him only during office hours.

To Illustrate/To Demonstrate

Specifically, to *illustrate* means to show an idea in pictures since an illustration is a picture. If a professor wants to give a student directions and the professor draws a map with arrows, the professor is illustrating his instructions. However, generally "to illustrate" means to help someone to understand an idea with examples.

Similarly, to *demonstrate* specifically means to perform or to show an idea with the body, experiments, or instruments since a demonstration is a performance or presentation. If a chemistry professor adds one liquid to another liquid in front of the class, the professor is demonstrating the reaction. However, generally "to demonstrate" also means to help someone to understand an idea with examples.

To Elaborate/To Develop/To Expand

An elaboration is a continuation of the same idea in great detail. If a speaker continues to give a lot of detail (examples, description, facts, data, etc), especially after the speaker has already given some, then the speaker is elaborating.

To develop and *to expand* are synonyms for "to elaborate" since they also mean "to give more detail" about something.

To Clarify

A clarification is a statement that removes or ends confusion. Usually, a speaker clarifies an idea that someone misunderstands. Therefore, a clarification often follows a question, request, or incorrect comment. However, a speaker can also anticipate confusion and clarify an idea before anyone gets confused. Many types of statements can clarify an idea; a description, definition, comparison, contrast, example, and explanation can all clarify.

Student:	Professor, I'm concerned about my bibliography for my research paper. There have been no books published on my topic, only some magazine and newspaper articles. It is very recent and contemporary.
Professor:	Well, that sounds fine.
Student:	But, the assignment says that the bibliography must include at least five academic sources.
Professor:	By academic, I don't mean just books. As long as the articles are serious and informed, they'll be fine.

The student had incorrectly assumed that he must use books, but the professor clarifies the requirements by explaining that "academic" has a broader meaning than just books.

To Refute

To refute means "to argue against" or "to prove wrong." A refutation is an opposing or countering argument. A speaker refutes another person's argument or opinion by explaining what is wrong with it. An argument can be refuted based on incorrect information, missing information, an incorrect assumption, a misinterpretation of the facts, and poor logic. You can recognize a refutation by the language of argument (argue, believe, think, etc), the language of contrast (however, but, on the other hand) and the fact that the speaker is explaining the error(s) in another's opinion.

Professor:	Gun lobbyists argue that a citizen has the right to own a high-powered assault rifle for personal protection, so the government should do little to control their availability. However, this is a classic example of a solution that creates the very problem it aims to solve. By making automatic weapons widely and easily available to everyone, gun lovers also arm the people they fear, criminals.

To Digress/To Go Off on a Tangent

A digression is a brief discussion of something that only partially relates to the main idea. Digressions are more common in lectures than conversations; speakers try to limit or avoid them because digressions are off-topic. A student might be explaining that he must miss a test because of a family wedding, and briefly describe how beautiful his hometown is. The beauty of his town is a digression because it adds nothing to his explanation, but it is still partially related to it. A digression could involve some of the following vocabulary:

> That reminds me/makes me think about…
>
> Let's/I'll digress for just a minute and talk about/discuss…
>
> Let me/I'll briefly mention/discuss/describe/go off on a tangent about…
>
> Get sidetracked/get off track…
>
> Beside the point/not really relevant/irrelevant for now

To Conclude

A conclusion is the end of a conversation or lecture, although it is more than just an ending. It can also be a summary, final opinion, or judgment based on what has been discussed. Similar to a hook, a conclusion can include a description, quote, fact, anecdote, comparison, or contrast. However, unlike the hook, the conclusion includes some logical connection to the previous material because it is based on what was mentioned earlier. Also, a conclusion can state or suggest the topic of the next conversation or lecture. Although conclusions don't always follow key words, a speaker can use some of the following vocabulary in a conclusion:

> In conclusion/to conclude
>
> In a nutshell, overall
>
> Finally, final
>
> In summary/to sum up/summing up
>
> To wrap up/let's wrap up/wrapping up

Strategy 10: Be prepared to think about rhetorical structure in two related ways: "infinitive + object" and "by + gerund."

As you can see by the question forms below, rhetorical structure can be discussed in two ways. In the first, the question begins with "why" and quotes or paraphrases a statement, and then asks what the statement achieves. The correct answer is one of four infinitive phrases. Look at the definition below:

Professor: NASA's Space Shuttle is a manned, two-winged aircraft that is carried into outer space on rockets and then returns to Earth on its own power.

The first type of question asks, "Why does the professor mention a rocket traveling into outer space?" The answer could be "to define the Space Shuttle."

In the second way of discussing the above quote, the question begins with "how" and could instead ask, "How does the professor define the Space Shuttle?" The correct answer is one of four prepositional phrases: "by describing how it looks and how it travels to and from outer space."

Both questions discuss the same statement, but they discuss the statement in opposite ways.

Rhetorical Structure Question Forms

Rhetorical structure questions can be identified by their form. There are four different forms of this type in the Listening section. All are multiple-choice with four answer choices.

Form 1: The question paraphrases a statement or idea.

> Why does the professor say/mention/refer to X?
>
> To emphasize his point
>
> To refute the student's point
>
> Etc.

Form 2: The question mentions a rhetorical structure and asks for the method, or how it is achieved.

> How does the professor illustrate her point about X?
>
> By comparing X to Y
>
> By giving the dimensions of X
>
> Etc.

Form 3: An excerpt from a lecture is followed by a shorter quote from the excerpt and a question.

> Listen to a part of the conversation/lecture. Then answer the question.
>
> [A short excerpt of a lecture previously heard is played]
>
> Why does the professor say this?
>
> [A specific sentence from the excerpt is repeated]
>
> To point out a flaw
>
> To repeat a point
>
> Etc.

Form 4: The question quotes a longer excerpt and asks for the purpose of the entire quote.

> Why does the professor say this?
>
> [Short excerpt of a lecture previously heard is played]
>
> To define an important term
>
> To exemplify a key point
>
> Etc.

Distracters

Distracters are incorrect answers. Rhetorical structure questions have four answer choices; one is correct and the other three are distracters. The distracters for this type of rhetorical structure are the following:

Distracter 1: The answer choice uses an incorrect infinitive or incorrect gerund.
A speaker mentions the steps of a dying star, for example. You should understand that he mentions the process in order to explain how a star dies (correct infinitive = to explain the death of a star). Also, another way to describe it is that the speaker explains a star's death by giving the steps involved (correct gerund = by mentioning/by referring to the steps). This distracter might say that the speaker mentions the process "to define a star" or it might say that he explains the death of stars "by defining a star."

Distracter 2: The answer choice uses ideas from the passage in a way that is unrelated to the question.
In this type of distracter, the answer choice refers to ideas that are mentioned in the passage but are not connected to the question. The choice relates to other statements, not the one that the question asks about.

Distracter 3: The answer choice refers to ideas that are not mentioned in the passage.

Distracter 4: The answer choice refers to ideas that are untrue based on information in the passage.
This type of distracter uses words related to the question and uses the right infinitive or gerund, but the ideas are untrue. For example, a speaker explains Newton's law of inertia (a body will continue in the same state unless a force acts on it) by describing an apple resting on a table. If the question asks why the speaker mentions an apple on a table, the correct answer could be "to explain why a body cannot change its own state." However, this distracter will mention something like, "to explain why a body can change its own state" or "to explain why a body's state changes by itself," which are incorrect ideas (bodies must be acted on in order to change states).

Practice

Now, practice rhetorical structure—statement questions for a conversation and a lecture, and review questions for purpose, main idea, and rhetorical function. Using the strategies you've learned, take notes in the spaces provided and answer the multiple-choice questions.

Notes – Conversation

_____

_____

_____

_____

Play CD 2, Track 23 for the following conversation. Questions 36 to 37 for this conversation are Tracks 24 to 25 on the same CD. For each question, you must play the corresponding track.

33. How does the woman indicate her problems with her roommate?

○ By elaborating on one specific event

○ By describing a general pattern of events

○ By comparing her problems to those of other people

○ By describing how the problems have affected her studies

34. Why does the woman visit the Student Housing Office?

○ To request a refund on her housing deposit

○ To get advice on negotiating with her roommate

○ To ask to be reassigned to a different dorm room

○ To inquire about off-campus housing options

35. What problem does the woman have?

○ Her roommate makes noise at night.

○ Her roommate's belongings are taking up too much space in the room.

○ Her roommate wakes her up early in the morning.

○ Her roommate's friends visit too frequently.

36. Listen again to part of the conversation. Then answer the question.

Track 24
Why does the man say this?

     ◯ He wants the woman to clarify her earlier question.

     ◯ He wants to gain an overview of the situation.

     ◯ He wants the woman to repeat what her roommate said.

     ◯ He wants to introduce a possible solution to the woman's problem.

37. Listen again to part of the conversation. Then answer the question.

Track 25
Why does the woman say this?

     ◯ To show that she made an attempt to find a solution

     ◯ To explain that she already went to the housing office.

     ◯ To explain that she and her roommate were co-workers.

     ◯ To highlight the number of problems she has with her roommate

Notes – Lecture

_____

_____

_____

_____

Play CD 2, Track 26 for the following lecture. Question 42 for this lecture is Track 27 on the same CD. For this question, you must play the corresponding track.

38. The professor discusses the apprenticeship system in order to

     ◯ describe the philosophical foundations of technical training in the U.S.

     ◯ show how the factory schools used training techniques invented by the European guilds

     ◯ argue that 19th century technical training benefited from Colonial municipal administration

     ◯ illustrate how the need that had been served by the apprenticeship system came to be filled by other means

39. Why does the professor mention Lincoln's Land Grant Act of 1862?

     ○ To show how the debate over public training came to be resolved

     ○ To illustrate the role government played in funding the mechanics institutes

     ○ To provide an analogy to the earlier municipal administration of the apprenticeships

     ○ To argue that governmental support came only after private training had demonstrated its value

40. How does the professor explain the growth of technical training in the 19th century?

     ○ By demonstrating how new training methods resulted from new ideologies

     ○ By connecting it to changes in the nature of work brought about by the Industrial Revolution

     ○ By comparing its development to the simultaneous development of the American public school system

     ○ By contrasting the opposing views of people who supported and opposed the integration of technical training in the public schools

41. The professor mentions the companies Westinghouse, General Electric, and Goodyear to provide examples of companies that

     ○ spearheaded the factory schools movement

     ○ competed for workers with the technical societies

     ○ realized the importance of training workers in basic skills

     ○ opposed the integration of technical training in public schools

42. Listen again to part of the lecture. Then answer the question.

Track 27

Why does the professor say this?

     ○ He already discussed the reasons.

     ○ He plans to discuss the reasons later.

     ○ He doesn't plan to discuss the reasons.

     ○ He plans to discuss the reasons briefly.

43. What is the main point the professor makes about technical training in the United States?

○ The history of technical training begins with the rise of European industrialization

○ From the beginning, technical training in the United States combined job skills training with education in basic skills

○ Technical training in the United States has a history concurrent with, but separate from, the history of public school education

○ As the United States developed technologically, technical training helped it develop into the world's premier economic power

Content Relationship Questions: Drag-n-Drop and Tables

In content relationship questions, you must recognize key points from a passage, such as comparisons, contrasts, causes and effects, steps in a process, etc. The question requires you to relate information rather than simply remember what the speaker said. In other words, you must be able to perform an action, such as inference, deduction, prediction, or generalization, in order to choose the correct answer.

Use the following strategies to help prepare you for this task.

Strategy 1: Listen actively (see also General Listening Strategies).

Strategy 2: Take notes on major points, not minor details (see also General Listening Strategies and detail questions).

Strategy 3: Pay special attention to sequence.
Anytime you hear or recognize an order, you should anticipate a question about it. The order could involve historical events, a natural or scientific process, instructions, etc. Don't assume that the order in which they are mentioned is the correct order. For example, a speaker may mention something third, but it actually occurs second. Use logic and the main idea if you don't hear or forget the correct order.

Strategy 4: Pay special attention to categorization.
Categories are important because you may have to connect certain characteristics to a particular type or kind. This kind of description is always important, so record it in your notes.

Strategy 5: Pay special attention to multiple details.
You should recognize anytime that you have multiple actors, places, causes, effects, problems, solutions, and so on. Whenever you hear multiple details, you may have to combine them correctly in a table or sequence. In a lecture on global warming, the speaker could mention several problems and then a solution for each one. This should be a clue that you will have to match each problem with a possible solution later.

Strategy 6: Be prepared to infer.

As always, you may have to infer a connection that isn't stated explicitly. If a speaker describes Y as an effect of X, then logically Y must occur after X since effects follow causes. You must recognize that even if the speaker never states which happens first (see also Inference questions below).

Question Forms

Content relationship questions have two main forms: drag-n-drop and table completion (You may get an Inference question about content, which will be a standard multiple-choice question, but this is discussed below under inference questions). You can recognize drag-n-drop and table completion questions by the following question forms and related vocabulary.

Drag-n-Drop: Sequencing

As the name suggests, you must use your mouse to select an item and move, or drag, the item into its correct place.

> The student/professor mentions several events/steps/directions/instructions…? Put the following in the correct order.
>
> [The question is followed by empty spaces and below the spaces is a list.]
> [You must select each choice with your mouse and drag it into the proper order: the top space is first, and the bottom space is last.]

Table Completion

As the name suggests, you must use your mouse to click an X in the appropriate space in a table.

1. Which of the following is/indicates/demonstrates/refers to/shows…?

   [Question is followed by a table]
   [You must mark an X under Yes or No next to each choice in the table.]

2. The professor describes the features/traits/characteristics of…Which belongs to each type of…?

   [Question is followed by a table]
   [You must match each trait with the correct type by marking an X next to each trait and under the right type.]

Distracters

Obviously, these questions are not standard multiple-choice questions. Most don't have distracters because all choices must be put into order or chosen. Only yes/no table completion questions have distracters. When you must choose yes or no, certain questions are incorrect answers and must be marked under no. The distracters for yes/no Table questions are similar to those that you have seen already:

Distracter 1: The answer choice rearranges incorrectly several ideas repeated from the passage.

This distracter type is also found in many other question types, such detail questions and NOT/EXCEPT questions. As you've learned already, this distracter repeats ideas (nouns, actions, states, adjectives, etc.) from the passage and changes the relationship among the ideas (switch actor and object, etc.). This distracter is one reason why you should always relate ideas to each other as you listen and take notes. Pay close attention to actors, objects, starting points and destinations, sequence, and so on. For example, imagine that a student needs access to the rare books collection. The librarian tells the student to get a reference letter from his professor, pick up a copy of his course schedule from the registrar's office, fill out a library form, and submit everything to the librarian. This distracter could rearrange those details in a number of incorrect ways, such as pick up a course schedule from his professor, fill out a registrar's form, and bring his form and copy to the registrar.

Distracter 2: The answer choice refers to ideas that are not mentioned in the passage.

This kind of distracter may rearrange ideas in the same way as distracter 1, but this distracter also adds ideas that were not mentioned. Often, this unmentioned information could seem logical or possible based on the context, but the choice is still incorrect. For example, based on the librarian's instructions to a student mentioned in Distracter 1 (see above), this distracter could give the following incorrect choices: get a reference letter from his advisor (no advisor is mentioned above, but advisors are often needed) or obtain his parent's signature (also not mentioned but plausible).

Distracter 3: The answer choice adds or removes a negative, or uses antonyms for the ideas from the passage.

Like the distracters in many other question types, this distracter refers to the opposite of the information in the passage. For example, this distracter could say that the student in the example mentioned above doesn't need a reference letter or that the student has no reason to visit the registrar.

Practice

Now, practice content relationship questions for a conversation and two lectures. Using the strategies you've learned, take notes in the spaces provided and answer the multiple-choice questions for each passage.

Notes – Conversation

_____

_____

_____

_____

Play CD 2, Track 28 for the following conversation.

44. The administrative staff person gives the student directions to the campus post office. Place the following directions in the order in which the student is expected to follow them.

(On the actual TOEFL, you will use your mouse to drag each choice into the correct order. Here, you can only write them in the correct order by hand.)

○
○
○
○
○

· Passing the student life office	· Walking down stairs
· Going behind the cashier	· Crossing the graduate plaza
· Turning right	

Notes – Lecture 1

_____

_____

_____

_____

🎧 Play CD 2, Track 29 for the following lecture.

45. In the lecture, the professor mentions several facts about insulin and diabetes. Indicate whether the following facts are mentioned by clicking in the correct box for each phrase.

(On the actual TOEFL, you will mark an X with your mouse. Here you can only mark an X by hand.)

	Yes	No
Insulin allows glucose to move from the bloodstream and into muscle and fat cells.		
The Islets of Langerhans reduce insulin production when blood sugar is at a normal level.		
Type II diabetes is much more common than type I diabetes.		
Insulin resistance means that the body begins to destroy the Islets of Langerhans.		
Glucose levels rise in the bloodstream whenever food is consumed.		

Notes – Lecture 2

_____

_____

_____

_____

🎧 Play CD 2, Track 30 for the following lecture.

46. In the lecture, the professor describes features associated with end stages, thrust stages, and arena stages. Which feature belongs with which type of stage? Click in the correct box for each phrase.

(On the actual TOEFL, you will click beside each choice with your mouse. Here you can only mark an X by hand.)

	End Stage	Thrust Stage	Arena Stage
Performance space entirely surrounded by the audience			
May include a proscenium arch			
Surrounded by the audience on three sides			
Developed in order to hide scenery changes			
May take the form of an amphitheater			

Inference Questions

Inference is an important skill throughout the Listening section and throughout the TOEFL in general. In fact, you've already read about inference in main idea questions, attitude questions, rhetorical function questions, rhetorical structure questions, and content relationship questions. Therefore, you should be prepared to infer any of the above ideas (main idea, attitude, etc).

Inference is the understanding of unspoken ideas based on logic and facts. It is related to implication, which is the communication of unstated ideas. Speakers (and writers) imply, and listeners (and readers) infer. Look at the following conversation:

Professor:	If you won't be back to campus in time to take the bus with the rest of the class, do you have a vehicle of your own?
Student:	It's in the shop.
Professor:	I see, well…

The student here doesn't say directly that he can't take his own car on the trip, but he implies it by saying that it is being fixed. "It's in the shop" is an idiom that means "It's being fixed." Based on that fact, the professor (and you) can infer that the student can't drive.

(Also, notice the possibility for a rhetorical function question. The professor's question "do you have a vehicle…?" implies his reason for asking it: "do you have a vehicle of your own?" implies "are you able to drive yourself?" In other words, the professor asks the question not only to find out if the student has a vehicle, but also to find out if the student can drive himself. A question about the professor's purpose in asking that would be a rhetorical function question, but understanding the speaker's purpose also requires inference.)

Now, review the following strategies for making inferences.

Strategy 1: Listen actively.

Inference requires good attention to detail and good analytical skills. Therefore, as always, it's very important to think about what you hear during a conversation or lecture.

Strategy 2: Think critically about details.

Details are essential for inference because an inference must be based on them. Therefore, use the tenses, descriptions, numbers, comparisons, contrasts, times, places, and reasons and think about what else is true but not mentioned. Look at the following statement from an American History lecture:

> Once the colonies of New England and others were well established along the eastern seaboard, it didn't take long for explorers to begin looking for the Pacific.

You should infer that the explorers moved west because the Pacific is on the west coast of the United States.

Strategy 3: Use context to infer correctly.

Almost any part of speech can be used for an inference, but correct inference depends on context. For example, people going west are not always going toward the Pacific Ocean. Look at the reference to the Pacific in the following statement and compare it to the reference above in Strategy 2:

> After uniting their conquests north of modern day China, the Mongolian warriors soon left their homeland on the coast of the Pacific and steered their horses toward the riches of the Persian Empire.

In the above statement, the context is Mongolia. The Pacific is to the east of Mongolia (and China), so logically the Mongolians could only have ridden their horses west (if they had ridden east, they would have gone straight into the Pacific!). Also, the Persian Empire must have been to the west of the Mongolians based on the same logic.

In these examples, geographical detail helps you determine direction. However, detail can also give you clues about causes, intentions, times, origins, material, identity, and many other ideas.

Strategy 4: Recognize and use idioms.

An idiom is a set expression whose meaning has little or no relation to the literal meaning of the words in it. For example, "throw in the towel" means quit or give up, but the speaker does not have to be holding an actual towel for the meaning to be understood. You can often recognize idioms by this absence of a stated noun or action. The context will help you understand an idiom. For example, if someone will throw in the towel, the person will have a negative tone and use vocabulary related to quitting (can't go on, tired, take a break, try again later, etc). An inference question doesn't ask about the meaning of an idiom directly, but an idiom could be necessary to make an inference (see the example above about the vehicle).

Strategy 5: Think about details together, not individually.

Usually, an inference requires more than one piece of information, such as descriptions, questions, conditionals, suggestions, advice, reasons, and warnings, etc.

> **Student:**   Don't take Professor Martin's class. Bill almost failed his class last semester, and Bill's a genius!

In this statement, the information about Bill clarifies why Prof. Martin's class is a bad idea: it is difficult. However, this fact must be inferred since it isn't mentioned explicitly.

Strategy 6: Be prepared to perform some action to connect details correctly.

The correct answer to an inference question represents some understanding that you are expected to take from the passage. In order to recognize the correct answer, you will need to perform some action on the information you've heard. Look at the following chart:

Action	Meaning
To infer	To understand or recognize unstated meaning based on logic and stated detail.
To deduce	To draw a logical conclusion based on stated or implied information. (This is often used as a synonym for inference.)
To predict	To make a statement about future events or states based on current facts or implications.
To generalize	To make a statement about an entire category, species, group, class, or entity.

Question Form

Inference questions can be recognized by their form and the related vocabulary.

> What comparison/contrast does the student make?
> What comparison/contrast does the professor make?
> What does X demonstrate/show/indicate?
> What does the student/professor imply about X?
> What can be inferred about X?
> It can be inferred that...?

Distracters

Distracters are incorrect answers. Inference questions have four answer choices; one is correct and the other three are distracters. The distracters for inference questions are the following:

Distracter 1: The answer choice is illogical or impossible.
You must use your knowledge about the main idea, times, locations, sequence, results, to recognize that some idea is impossible. For example, a horse rider on the eastern seaboard of the U.S. cannot ride east because he would drown in the Atlantic.

Distracter 2: The answer choice is the opposite of the correct inference.
Like the same distracter type in other question types, this distracter often incorrectly adds or removes a negative (*no* or *not*), or it uses antonyms. For example, look at the following statement by a professor giving instructions to students on group projects:

> Now, the group work must be done in the library conference room, not in class. Although I won't be there to monitor your contribution or attendance, remember that you must sign in whenever you use the conference room.

If a question were to ask what the professor implies about attendance during group work, this distracter type would say the professor implies that she has no way of knowing their attendance. This is the opposite of the truth; the professor actually implies that she can check the sign-in sheet afterwards. The use of *although* emphasizes this by contrasting the professor's absence with the need to sign in.

Distracter 3: The answer choice is too extreme or takes an inference too far.
It's always important to pay attention to degree in vocabulary. For example, if a student is unhappy about her mark and wants to challenge the teacher's assessment, she may want a new grade but she probably doesn't want to get the teacher fired. Remember to use common sense when answering questions.

Distracter 4: The answer choice uses an incorrect meaning of a word or phrase.
If a student says that the enrollment has exploded for a particular course, meaning that many students have signed up for it, this distracter might say that the class has been destroyed by an explosion. Idioms are commonly used in this type.

Distracter 5: The answer choice rearranges detail from the passage.
This type of distracter is similar to a distracter in detail questions. It uses exact words from the conversation or lecture, but changes the actors, objects, actions, states, locations, and so on.

Practice

Now, practice inference questions for a conversation and a lecture, and review attitude questions for the same lecture. Using the strategies you've learned, take notes in the spaces provided and answer the multiple-choice questions for each passage.

Notes – Conversation

_____

_____

_____

_____

Play CD 2, Track 31 for the following conversation.

47. What can be inferred about the student?

    ◯ She no longer lives in New Orleans.

    ◯ She took another class in meteorology.

    ◯ She got a low grade on her previous paper.

    ◯ She lived through several hurricanes in the past.

Notes – Lecture

_____

_____

_____

_____

_____

🎧 Play CD 2, Track 32 for the following lecture. Questions 48, 50, 51, and 52 for this lecture are Tracks 33 to 36 on the same CD. For each question, you must play the corresponding track.

48. Listen to part of the conversation. Then answer the question

Track 33
What does the professor imply when he says this?

- ⃝ Marketers do not fully understand how consumers make their purchasing decisions.

- ⃝ Consumers base most of their purchasing decisions on their personal preferences.

- ⃝ Marketers study consumers' decision processes so they can influence buying behavior.

- ⃝ Personal factors and external factors have equal influence on consumers' decision making.

49. What does the professor imply about intelligent, or "smart," consumers?

- ⃝ They employ a variety of decision making styles.

- ⃝ They gather information before making a purchase.

- ⃝ They spend less money than other consumers.

- ⃝ They make fewer purchases than other consumers.

50. Listen again to part of the conversation. Then answer the question.

Track 34
What does the professor imply?

- ⃝ He is amused by the student's comment.

- ⃝ He is upset that the student interrupted him.

- ⃝ He is unhappy that the student made a joke.

- ⃝ He is concerned that the student misunderstood him.

51. Listen again to part of the conversation. Then answer the question.

Track 35
Why does the man say this?

- ⃝ He likes to watch the same television program.

- ⃝ He has a sister who watches the same TV program.

- ⃝ He thinks the woman has no reason to feel guilty.

- ⃝ He does not like the TV show the woman watches.

52. Listen again to part of the conversation. Then answer the question.

Track 36
What does the student imply?

- ○ She is excited to be learning about market segmentation.
- ○ She is confused about the professor's definition of the term.
- ○ She is surprised by how specifically markets are segmented.
- ○ She is sympathetic toward the people mentioned by the professor.

IMPROVE-YOUR-SCORE STRATEGIES

Use the following strategies to improve your score by studying on your own. Since the TOEFL is an English language test, these strategies all require practice in English, which is easiest in an English-speaking country.

Strategy 1: Expand your knowledge.
While the TOEFL is not a knowledge test, background knowledge does help. For example, it is easier to understand some issues of American history if one knows that the United States used to be a colony of England (Which other countries tried to establish colonies in the same part of the world?). Assess your own level of knowledge about TOEFL-related fields, and improve your weaknesses by focusing your practice in certain areas.

Strategy 2: Practice the strategies from this book.
Whenever you listen, listen actively and try to recognize the speaker's main idea, purpose, attitude, organization, and implications. Ask yourself questions: What are they talking about? Why is that important? How is this organized? What will be mentioned next? What can I infer based on that fact? The more you practice, the easier it becomes.

Strategy 3: Watch television.
Watch nonfiction programs on television, such as travel shows, the news, and history. Look for shows and documentaries on topics related to fields covered by the TOEFL.

Strategy 4: Join the library.
Rent documentaries and movies related to fields covered by the TOEFL. One obvious advantage of rentals is that you can pause and rewind difficult parts. Any big North American library has multivolume histories of the United States and biographies of notable people. Stores that rent movies also carry some non-fiction titles, but usually not as many as libraries do.

Strategy 5: Go to the movies.
Documentaries are also available in movie theaters. Look for new documentaries, and especially documentary festivals. In fact, reading the newspaper regularly will not only

improve your reading and vocabulary but also inform you about new television shows, movies and events.

Fictional movies are also useful listening activities, but focus on those that could give you useful background in history, art, and science. Look for movies based on novels by famous writers, such as Charles Dickens, Ernest Hemingway, or Jane Austen. And look for movies about famous artists such as Picasso or Mozart.

Strategy 6: Do a variety of activities.

All language skills are related. The vocabulary, structure and knowledge that you gain while you do one activity, such as reading, can help you when you listen, write, or speak. Therefore, don't think that you must only listen to become a better listener. Be an active learner and vary your activities.

Strategy 7: Communicate with others.

Create or join a study group or conversation club. Watch shows and movies together and then discuss them (in English!). Discussion is an excellent way to vary your practice. With a good group or tutor, you can practice listening and speaking together while improving your vocabulary and knowledge. It's hard to develop and improve without interaction and communication, so be social.

ANSWER KEY

On the actual TOEFL, the multiple-choice answers are *not* letters (A), (B), (C), (D), as you may have seen on other exams. Beside each choice is an empty circle; you select an answer by clicking on the circle with your mouse and darkening the circle. For that reason, the answer choices in this book have no letters in the practice.

For the purposes of identification here, however, the choices in the practice are identified in the answer key as letters. For each multiple-choice question in the practice:

   ◯ = (A)
   ◯ = (B)
   ◯ = (C)
   ◯ = (D)

Each question number in the answer key below is followed immediately by the correct answer, and then a brief explanation.

For questions that involve a table or list, either the correct order is given or the table is recreated with the correct additions.

1. (C)

(C) is correct because the student explains why he has missed some classes and asks for an extension on his paper. (A) is wrong; the student does not suggest that his illness was caused by stress. (B) is the opposite of the truth: in the conversation, the professor expresses willingness to grant the student an extension. (C) refers to a minor detail: "...a few more weeks before I'm back to normal."

2. (C)

(C) is correct based on the statement "Today I want to talk about the most appropriate methods to use for studying cougars." (A) is unmentioned; there is no discussion about how to divide the work that needs to be done. (B) is also unmentioned; no changes to an earlier plan are discussed. (D) refers to a minor detail: "...their solitary nature, their stealth..."

3. (B)

(B) is correct because the need to study cougars is framed within the context of the cougar management debate: "Our ignorance means we don't know how to manage them." (A) repeats only one detail ("Because of their solitary nature...") and cougars are not compared to other cats. (C) refers to minor detail: human interaction with cougars is not mentioned in the larger context of the future of wildlife conservation. (D) uses detail incorrectly: "...it sounds like several of you are leaning toward the approach that requires the least intervention."

4. (B)

(B) is correct; the first student describes the views of conservationists, developers, hunters, and animal rights activists, and the students briefly argue about the cougar-human conflict. (A) is wrong; there is no discussion of actual techniques for managing cougars. (C) rearranges detail: there is no mention of how the cougar debate influences the ability of researchers to study cougars. (D) is wrong; the cougar debate is not compared to other conservation issues.

5. (C)

(C) is correct: she inquires about outside organizations and is shown how to use the Internet to search for scholarships. (A) refers to unmentioned ideas: she is looking for help from outside organizations, not in-school scholarships. (C) is a distracter: "I need a little more help in covering tuition." (D) is too specific; she comes in looking for a list, and the Internet is only part of her search.

6. (C)

(C) is correct: Adams's Zone System influenced other photographers, and his pictorial calendars helped to popularize photography. (A) refers to minor detail: Adams's development of the Zone System is mentioned, but this is not the main focus of the lecture. (B) is incomplete; his modern popularity is mentioned, but Adams influenced his contemporaries as well. (D) refers to minor detail: The natural world is mentioned as an important influence on Adams's work, but this is not the main purpose of the lecture.

7. (A)

(A) is correct: "Oh, here. Fill in this form." (B) refers unmentioned ideas: he does not make a payment or receive a receipt during this conversation. (C) is wrong: he is asked to examine a brochure; he is not handed a contract. (D) is wrong; the university and insurance company Web sites are mentioned over the course of the conversation, but the actual address is not.

8. (A), (B)

(A) and (B) are correct: his question about dental coverage is what leads the staff person to mention the Vital Savings plan, and fitness is mentioned a couple of times by both speakers. (C) is wrong; the staff person is "not sure" if acupuncture is covered. (D) refers to unmentioned ideas: prescription drugs are not mentioned.

9. (B)

(B) correctly refers to the statement "...Moses' basic philosophy about cities, was that cities were made for driving, not living." (A) is the opposite of the truth: "Since Moses, many city planners have begun to favor pedestrian-friendly, intimate communities..." (C) refers to the wrong reason: the professor says that Moses believed people were meant to live in the suburbs, but not because cities were overpopulated. (D) is the opposite of the truth: according to the speaker, the routes were "...made for driving..."

10. (B), (D)

(B) and (D) are correct based on the following statements: "...blamed for...promoting suburban sprawl throughout the United States" and "...ruined hundreds of neighborhoods in the process by building highways right through them..." (A) might be known through outside knowledge, but is not mentioned by the professor. (C) rearranges detail: "He's been blamed for...killing public transportation systems..." Note that New York's public transportation system is not explicitly mentioned.

11. (D)

(D) is correct because the professor believes the wolf spider will be a better topic when the class studies entomology. (A) is plausible since he calls it interesting, but his main suggestion is that it is not appropriate for the current project, so (D) is better. (B) contradicts the passage: "...fill 15 minutes of presentation time." (C) also contradicts the passage: "...enough data to talk about..."

12. (B)

(B) is correct because the student expresses excitement and mentions what he feels to be an unusual fact about the wolf spider. (A) is wrong; the student makes no mention of other students. (C) distracts you with the professor's use of the word "but" in the previous comment ("...that's an interesting one but..."). (D) mentions the wrong cause for the emotion: his frustration is with his inability to more clearly focus his research, not with the information available about the wolf spider.

13. (D)

(D) is true because at no time does he indicate doubt about the theory. (A) is incorrect because he does not indicate any opinion about whether or not the theory is provable. (B) is wrong since he does not address any competing theories, not mention that they exist. (C) is wrong; he never suggests that anyone has trouble understanding the theory.

14. (A)

(A) is the best answer because he implies surprise that the "obvious" point was not scientifically presented until the early 1900s. (B) is wrong because his mention of the world map does not indicate that Wegener was a mapmaker. (C) is incorrect, for his comment does not address developments to Wegener's theory since it was proposed. (D) is also wrong; he does not imply an opinion about whether or not Wegener received enough credit.

15. (B)

(B) correctly reflects the use of the verb; "I think" can be used to soften a statement. He is indicating the woman's answer is incomplete, but in a polite manner. (A) is plausible in another context, but he is actually certain about his point, so (B) is better. (C) contradicts the passage; he does not indicate any trouble understanding the theory. (D) is plausible since he is amplifying her answer. But he does not actually express disagreement; he only wants to add an important detail.

16. (A)

Choice (A) correctly relates to the context. (B) is wrong because the professor's authority is not called into question and does not need to be established. (C) is the opposite of the truth: the professor expresses a willingness to be flexible regarding the assignment. (D) is unmentioned; the professor expresses no preference for one form of education over the other.

17. (A)

(A) fits the context and sequence of statements; the phrase expresses a request for the student to describe his thoughts. (B) is out of context; the student's opinion is not the subject of discussion. (C) also contradicts the context; it's not a test of his knowledge. (D) is too general; it is not an invitation to discuss homeschooling, only the student's assignment.

18. (A)

(A) is best because he wants to make sure he understood her correctly. (B) is out of context; the author's name is not under discussion. (C) is incorrect because he is not demonstrating his knowledge; he is asking for more information. (D) is incorrect, for since understands the relevance of the professor's point, which is why he wants to make sure he understands it clearly.

19. (C)

The best answer is (C); she mentions the name of a character from *Beowulf* to show she knows something about it. (A) is wrong because Grendel is one of the monsters from *Beowulf*; mentioning his name does not correct the professor's statement, it elaborates on it. (B) is incorrect; the professor already mentions that the events involving monsters and dragons could not have happened; the student has no dispute with this. (D) is wrong; her mention of Grendel does not invite an alternative interpretation of the events in *Beowulf*.

20. (B)

(B) accurately combines the key points; the professor primarily discusses the conflict between value systems, and how *Beowulf* represents a new worldview for the Anglo-Saxons. (A) is plausible, because some plot details are mentioned, but the bulk of the discussion is spent discussing *Beowulf's* themes. (C) is also plausible, because the historical accuracy of *Beowulf* is mentioned, but the bulk of the discussion is spent discussing *Beowulf's* themes. (D) uses detail with unmentioned ideas; the unusual language is mentioned, not its influence on later developments in the English language.

21. (C)

The best answer is (C) because *Beowulf* represents a break from the worldviews of the pre-migration Anglo-Saxons; the professor says, "It really is the foundation of English literature." (A) incorrectly relates a historical fact to the present; *Beowulf* was not appreciated until the 19th century; but clearly appreciated after that point. (B) is wrong; the moral choices in *Beowulf* are not presented as universal, but as specific to the 6th century Anglo-Saxons. (D)

is plausible, but the professor does emphasize the continuity with the Germanic traditions; she emphasizes how *Beowulf* is different from those traditions.

22. (A) and (D)

Choices (A) and (D) correctly relate to the following quote: "... two competing value systems—the heroic code of the Germanic people he's writing about, and the more docile Christian code..." (B) misuses details; the monsters and dragons in *Beowulf* are not presented as representative of a worldview. (C) also incorrectly presents details as a worldview; "In *Beowulf's* world, the people rally around strong kings who protect them from danger."

23. (D)

(D) matches the statement "The idea is to highlight the information on your resume that's most relevant for each employer you send it to." (A) is plausible because her suggestions relate to the student's intention to find a job, but this does not provide the unifying structure, but (D) is better. (B) refers to incorrect action; the student's resume is not compared with other resumes. (C) is unmentioned; general suggestions are offered, not a set of guidelines.

24. (B)

The best answer is (B) because the ice cream server job is not relevant to the Assistant Curator job, and so should be removed. (A) is wrong; she mentions the Assistant Curator job in order to strengthen her point, not to suggest the student apply for a job. (C) is incorrect, for she offers an example regarding a resume detail, not its overall structure. (D) is the opposite of the truth.

25. (C)

(C) is correct because it fits the context; the student's impending graduation suggests that he needs to improve his resume soon. (A) is wrong; his graduation is not presented as a job qualification. (B) is unmentioned; there is no mention of how many jobs are on his resume. (D) is plausible, but he's not really explaining why he's never visited Career Services, he's asking for help. (C) is still the best.

26. (A)

(A) is correct; the primary focus of the conversation is material contained within the student's resume. (B) contradicts the passage; "...so, if I'm applying for jobs related to art..." (C) is out of context; the student has not yet applied for any jobs because he is still preparing his resume. (D) refers to a minor detail; the detail is subordinate to the main idea: how to improve the student's resume.

27. (D)

(D) correctly fits the main topic; he is visiting Career Services because he needs help preparing his resume. (A) is unmentioned; visiting Career Services is not mentioned as a graduation requirement. (B) is also unmentioned; there is no mention that the student was invited to speak with the advisor. (C) is minor; career options are discussed, but the main reason the student visits the advisor is to improve his resume.

28. (B)

(B) is correct; David Copperfield, Pocahontas, and Ben Franklin's autobiography are all provided as examples of classic texts that do not suffer from being simplified. These texts provide the rhetorical framework for the professor's talk. (A) is unmentioned; she does not provide any specific methodology for using simplified texts. (C) is minor; Krashen provides scientific validity to her point, but this is not the central device she uses to structure her argument. (D) refers to incorrect action; the lecturer does not compare the benefits of reading texts in their simplified or original forms.

29. (C)

Choice (C) reflects the thesis: "In my opinion, the educational advantages of using simplified graded readers far outweigh the philosophical concerns about art." This opinion forms the core of her argument. (A) rearranges ideas: "Extensive reading is the opposite of intensive reading." (B) is unmentioned; she makes no statement about people who express the opinion that simplifying texts devalues them—she just disagrees with the opinion. (D) rearranges ideas; the culture and history are interesting parts of the books, but learning about them is not the reason for using simplified texts.

30. (B)

(the roommate—"I have to get up early." (D) omits the time; "...brings friends over late at night, almost every night."

31. (D)

(D) is correct; she immediately cuts off this point and returns to her central argument. (A) is wrong; the quote does not introduce her central thesis; it is a tangent. (B) is wrong; the quote does not support the benefits of reading classic texts. (C) is the wrong action; the statement does not present a comparison between two points of view.

32. (B)

(B) is correct; asking the question leads her into the point she wants to make. (A) doesn't fit the context; she does not suggest that the terms are poorly defined. (C) also contradicts the context; she does not invite answers from the class. (D) confuses rhetorical and interrogative questions; she immediately answers the question she asks, so she clearly does not want students to answer the question for themselves.

33. (B)

(B) reflects the correct organization; she describes a pattern of events: her roommate keeps the lights on, brings friends over late at night, and plays music. (A) is too specific; she does not go deeply into any single event; she describes a general pattern of events. (C) refers to incorrect action; she does not compare her roommate problems to those of other people. (D) is minor; she implies that being unable to sleep influences her 8 am classes, but this is not how she structures her argument.

34. (C)

Choice (C) correctly relates to the student's statement "I'd like some information about changing dorm rooms." (A) is minor; the idea of getting her housing money back comes up in the flow of the conversation; it is not the reason she visits the office. Choice (B) is also minor; "Ok…but have you tried negotiating with her?" (D) is minor; living off-campus is mentioned, but this is not why the woman has gone to the office.

35. (A)

(A) accurately reflects the key points; her roommate keeps lights on and brings friends over nearly every night, and it is interrupting the woman's sleep. (B) is unmentioned; the roommate's possessions are not mentioned. (C) rearranges detail; the woman gets up, not the roommate—"I have to get up early." (D) omits the time; "…brings friends over late at night, almost every night."

36. (B)

(B) is correct because he wants her to explain her situation calmly, so he can understand what her problem is. (A) is wrong because the woman has not asked a question. (C) is also wrong; the woman has not claimed that her roommate said anything. (D) refers to later ideas; at this point, he has not yet begun to suggest solutions.

37. (A)

(A) correctly reflects the woman's previous experience; "work with" means "cooperate with" or "help." (B) uses another meaning for "been there." (C) uses another meaning for "work with." (D) doesn't refer to problems; it refers to an attempt at solving them.

38. (D)

(D) is the best answer because the apprenticeship system, mechanics and factory schools served the same need: to train people to perform the jobs of their day. (A) is wrong since no philosophy is mentioned in connection with the apprenticeship system. (B) is unmentioned; the actual techniques that were used to train people are not mentioned. (C) is also unmentioned; the professor does not state that colonial municipal administration led to any benefits of any kind.

39. (A)

(A) is correct because the speaker concludes that the act "did much to legitimize the value of technical training in the public eye." (B) is factually wrong; the mechanics institutes were privately, not government, funded. (C) mentions something the speaker never does; no analogy is made to the municipal administration of the apprenticeships. (D) is the wrong order since government involvement came first.

40. (B)

(B) is the correct choice since the speaker says, "New methods were needed for training workers for industrial labor." (A) is wrong as neither training methods nor ideologies are mentioned. (C) is incorrect; because the development of the public school system is not covered in detail, no comparison is really possible. (D) is a detail, but does not provide the lecture's unifying structure.

41. (A)

(A) is right; according to the lecture, "By the turn of the 20th century, several of the nation's major corporations...had their own factory schools." (B) is unmentioned; there is no mention of competition between the technical societies and these companies. (C) is also unmentioned; there is no mention that the factory skills trained workers in basic skills. (D) is also wrong because the speaker never mentions where these companies stood on the question of public technical training.

42. (B)

(B) is the best answer because "shortly" implies that he will discuss it in the near future, or soon. (A) is wrong since no reasons have been mentioned. (C) is untrue; He discusses the reasons in detail later in the lecture. (D) incorrectly uses "shortly" to mean "briefly," which isn't true.

43. (C)

(C) is the best choice. The municipal administration of the apprenticeship system evolved into the public school system, while technical training was privately administered until the late 19th century. (A) is out of context; Technical training is placed in the context of American, not European, industrialization. (B) is the opposite of the truth; the professor makes no mention of basic skills education in the context of the apprenticeship system. (D) is unmentioned; there is no mention of how technical training helped the U.S. compete with other economies.

44. Correct order

> Crossing the graduate plaza
> Going behind the cashier
> Walking down stairs
> Passing the student life office
> Turning right

45.

	Yes	No
Insulin allows glucose to move from the bloodstream and into muscle and fat cells.	X	
The Islets of Langerhans reduce insulin production when blood sugar is at a normal level.	X	
Type II diabetes is much more common than type I diabetes.		X
Insulin resistance means that the body begins to destroy the Islets of Langerhans.		X
Glucose levels rise in the bloodstream whenever food is consumed.	X	

Yes. "...acts like a key that opens cells—muscle and fat cells, mainly—and allows glucose to enter them."

Yes. "Once your blood sugar returns to normal, the Islets of Langerhans reduce the output of insulin."

No. This is a fact that might be known through outside knowledge, but it is not mentioned in the lecture.

No. Insulin resistance is described as feature of Type II diabetes. Type I diabetes is when the body destroys the Islets of Langerhans, so they can no longer produce insulin.

Yes. "...when you eat a meal, all this glucose suddenly enters your bloodstream."

46.

	End Stage	Thrust Stage	Arena Stage
Performance space entirely surrounded by the audience			X
May include a proscenium arch	X		
Surrounded by the audience on three sides		X	
Developed in order to hide scenery changes	X		
May take the form of an amphitheater			X

Answer: (End Stage = Column A; Thrust Stage = Column B; Arena Stage = Column C)

Column C: "The Arena stage is a performing space that's totally surrounded by the audience."

Column A: "Since the Renaissance, Western theater has been dominated by a type of end stage known as the proscenium theater."

Column B: "…the thrust stage, also known as a three-quarter round. The thrust stage is a platform surrounded by the audience on all but one side."

Column A: "…was developed in response to performers' desire to hide scenery changes as well as…their offstage exits and entrances from the audience.."

Column C: "…the amphitheater…which is an open-air variation on the theater-in-the-round, meaning the audience watches from all sides."

47. (A)

(A) correctly uses the key words of location; She says she's "from" New Orleans, and the professor refers to New Orleans as "there." (B) distracts you with a statement: "…this is a science class…" (C) is based on unmentioned ideas; the only paper the speakers discuss is the paper the student is working on. (D) also relies on unmentioned detail; the only hurricane mentioned is Katrina.

48. (C)

(C) is the best answer because the focus of the comment is on how understanding consumer decision processes helps marketers influence them. (A) is illogical based on the detail; her description of the decision-making process is quite detailed, indicating that there is some agreement in marketing about how consumers make their decisions. (B) is plausible, since the factors each consumer considers important are somewhat personal, but this is not implied by the current statement, so (C) is still a better answer. (D) refers to an unrelated action; the statement makes no comparison between personal and external factors.

49. (B)

(B) correctly infers a sequence of actions based on the following idea: "smart consumers gather all the information they can about the cost and benefits of each possible solution." (A) is the opposite of the truth; based on the detail, one can assume the five-step process described by the professor is shared by intelligent consumers (unintelligent consumers may not gather product information). (C) uses unmentioned ideas; the professor never mentions levels of spending. (D) also uses unmentioned ideas; the professor does not discuss the number or frequency of purchases.

50. (A)

(A) correctly fits the context; the continued laughter of the class indicates that the professor is responding humorously to the student's joke. (B) contradicts the professor's actions; he responds to the joke with another joke, indicating that he is not upset. (C) also doesn't fit the context; there is no clue that he is unhappy about being interrupted. (D) might be true in another context; he corrects the student, but only to set up his joke, not because he is concerned.

51. (D)

(D) is correct based on the speaker's intonation; his tone implies that he finds it hard to believe that the woman watches that program. (A) is the opposite of the correct inference; he teases the woman because he apparently does not like the program. (B) misuses an idiom; "Oh brother" is an idiomatic expression often used to indicate humorous disbelief or disagreement. (C) is the opposite of the truth; his words and tone imply that he is teasing the woman because she likes the program.

52. (C)

(C) is correct because she mentions the level of specificity as the factor she considers surprising. (A) is plausible, since she is presumably somewhat excited by the subject matter, but her comment primarily indicates surprise. (B) is out of context; she is surprised, not confused. (D) refers to unmentioned ideas.

LISTENING TRANSCRIPTS

Conversation: Office Hours

CD1, Track 1

<u>Thesis Extension</u>

Narrator:	Listen to a conversation between a professor and a student.
Student (male):	Professor Grey, may I talk to you for a minute?
Professor (female):	Sure. You haven't been in class for a while.
Student:	Yeah, I've been sick.
Professor:	That's no good. Is everything ok?
Student:	Yeah, now I'm ok.
Professor:	Well that's good…so have you been able to start your term paper?
Student:	Actually, that's what I wanted to talk to you about. So, well it started off as a sore throat, but I ended up having tonsillitis, and then I had to have my tonsils taken out.
Professor:	I'm sorry to hear that. Are you feeling better?
Student:	I feel a lot better, but I'm still a little run down. The doctor said it will take a few more weeks before I'm back to normal. He said I should take it easy, but I have so much work to do.
Professor:	Well, the doctor is right. You should try not to get too stressed out over school. Your health should come first.
Student:	So…I was wondering if I could get an extension on the paper.
Professor:	Okay…well, have you started it yet?
Student:	Um…I have a thesis and I'm about halfway finished with the research…so I'm wondering…I think I need two more weeks if that's okay with you.
Professor:	I don't see why not. I'd like to take a look at your thesis and outline, though, and I will also need a doctor's note from you…
Student:	Okay, no problem. I'll bring everything tomorrow. Thanks a lot.
Professor:	You're welcome. I'm glad you're feeling better.

Lecture

CD1, Track 2

Cougars

Narrator (female): Listen to a professor and a group of students discussing a proposed study of cougars.

Professor (male): Well, here we are again. Today I want to talk about the most appropriate method to use for our cougar study. Before we look at the various options, let's just make sure everyone is familiar with the background to this discussion. Why do we need a cougar study?…Yes?

Student 1 (male): Umm…well, cougars are a pretty hot topic for lots of people, and managing cougars is really controversial. On the one hand are the conservationists who want to protect the whole ecosystem, including this large predator. On the other hand is the development lobby that sees the wilderness as a potentially lucrative building site. Then there are the hunters, who hate cougars because they think, umm, they think the cougars kill too many of "their" deer. Oh, and at the opposite end of that spectrum are the animal rights people who don't believe in hunting or killing anything. The entire debate is overshadowed by concern about cougar-human conflict. There have been several instances over the past few years of cougars attacking, and umm, sometimes even killing and eating humans.

Student 2 (male): Yeah, but the cougars were there first! Can you blame them if people start invading their space?

Student 3 (female): That's ridiculous! People should be able to live in their homes without fear of getting eaten up by a dangerous predator! We need to kill them, especially if they hang around areas where humans live or play.

Professor: All right, everyone, those are all interesting points, but I don't want to go into that debate just now. All we're going to do today is try to decide what methods would be best for studying cougars. Joe mentioned that cougar management is a delicate issue. One of the reasons it is so controversial is that we know very little about these cats. Because of their solitary nature, their stealth, and perfect adaptation to their habitat, they have always been elusive research subjects. As biologists, we rarely get a chance to observe them in the wild to find out about their ecology. We don't know how far they range, what or how much they eat, how they interact with each other, how young adults establish new home-ranges, how they relate to other predators like bears and wolves….We also don't know much about how they respond to anthropogenic influences. I mean, umm…impact from the presence of humans. Our ignorance means we don't know how to manage them. We don't know how much land…in what areas we need to set aside for healthy cougar populations to survive; we don't know how deer hunting affects their prey base; we don't know how a cougar hunt might affect their population. Lots of questions. What do you guys think of the available research options?

Student 3: I like the idea of training sniffer-dogs to find cougar feces and then doing DNA analysis on the feces. We'd be able to identify individuals, so we'd find out how they roam by mapping the locations where we find the feces. Umm, we could also identify familial relationships between individuals, which would tell us about intra-species interaction. Oh, yeah…of course we'd also be able to tell what the cats have been eating by analyzing whatever hairs and bones we find in the feces.

Student 1: What about GPS collars? You know, the kind of radio-collars that use satellites, like the thing you can get in cars nowadays to tell you where you are. I propose we capture and place GPS collars on as many cats as we can. That technology is so advanced now we'd get extremely detailed location data.

Professor: You're right about GPS technology, but do you know how much those collars cost? I don't think our project budget will allow us to buy enough to get the data we need. Also, you need to remember that capture rates tend to be biased toward males and for our purposes it is crucial that we get a representative sample of the population.

Student 2: I support Pat's suggestion of using dogs to locate feces and then doing food studies and DNA analysis. That way, we wouldn't need to lay hands on the cats at all. There's always a risk when you capture an animal in the wild and we don't want to damage these creatures if we can help it.

Professor: Well…it sounds like several of you are leaning toward the approach that requires the least intervention. That pleases me, because your suggestions are in line with what many biologists are currently advocating in the literature.

Conversation: Service Encounter

CD1, Track 3

<u>Scholarships</u>

Narrator: Listen to a conversation between a student and a Financial Aid officer.

Student (female): Hi…. is there someone here who can help students find scholarships?

Financial Aid (male): Um…is there something in particular you're looking for?

Student: Well…I got a loan from the university but…uh…I need a little more help covering tuition, and I heard from some students that they got money from outside organizations. Do you have a…uh…list of these types of scholarships?

Officer: Ah, well….unfortunately, since there're so many different kinds of scholarships, there's no comprehensive list.

Student: Oh…..

Officer: But…what we can do…is search the Internet for scholarships by subject.

Student: Oh, ok…sounds good.

Officer: What's your major?

Student: Mechanical engineering, but…are you going to check out different companies' Web sites for scholarships? 'Cause I already tried that…

Officer: Actually…I'm just going to enter "Mechanical Engineering Scholarships" right into the Internet browser and see what comes up. (short pause-waiting for page to load) Here's one—from the State Institute of Technology.

Student: Wow, that was quick! But… what happens if I don't qualify—or if the deadline's past already?

Officer: Well, there're actually some other things to explore when searching for scholarships—like ethnicity. What's your cultural background?

Student: Uh, well…I was born here, but…my family comes from Australia. I don't know if that matters…

Officer: Well, some scholarships are set up to…uh…help students from specific backgrounds. There may even be a scholarship out there specifically for a woman, your age, of Australian descent, studying mechanical engineering!

Student: So…you mean that my gender, even my age could help me get a scholarship? How about extracurricular activities?

Officer: Sure! Anything could be a factor…I suggest you make a list of all the things that make you, you—and use it to search the Internet.

Lecture

CD1, Track 4

<u>Ansel Adams</u>

Narrator: Listen to a professor give a lecture about a famous artist.

Professor: Ansel Adams has contributed significantly to the field of photography. To many, his photographs are representative of the definitive pictorial statement on the American western landscape. Adams was born and grew up in San Francisco. He was influenced by the beauty of nature that surrounded his home. The marvelous views of the ocean and rocky beaches helped him to appreciate the natural wonders around him.

His love of photography really began in 1916, when on a family trip to Yosemite, his parents gave him his first camera, a Kodak Box Brownie. This enabled him to record the magnificent views he saw. He was so profoundly moved by the experience of capturing the beauty of his surroundings that he viewed it as the beginning of his lifelong inspiration. Throughout his career he continued to record the natural world around him.

The sheer beauty of his images is enhanced by their technical perfection, which are a result of Adams's determination to maintain total control of the photographic process. In his images he aimed to capture the full range of tones—from the deepest black to the purest white—to achieve perfect photographic clarity and to convey the beauty and power of the scenes. Later, Adams codified his approach to exposure, processing, and printing so others would be able to gain control over the photographic process and the characteristics of black and white film. This system, known as the Zone System, is a set of techniques that aims to produces the best possible negative to print the image as the photographer previsualized it. Once mastered, the Zone System allows photographers to consistently control the range of tones in the negative and, consequently, in the print.

Adams also believed that his work should be seen and enjoyed by everyone—not only those with enough money to purchase original prints so he chose three images and arranged for them to be printed as affordable posters. This was so successful that the mass production of pictorial calendars was the next step. As you know, these calendars are still popular today. Adams's images are immediately recognizable and still available in many forms, including books, posters, calendars, and prints. His goal of making his work accessible to all is still being realized today.

Conversation: Service Encounter

CD1, Track 5

Student Health Insurance

Narrator:	Listen to a conversation between a student and a staff member.
Student (female):	Hi...um, do you have a moment? I have a question about health insurance.
Staff (female):	Okay, shoot.
Student:	So the fee I pay as part of my tuition covers me if I need to use the university health clinic right?
Staff:	So you're a full-time student, then?
Student:	Right. Full-time.
Staff:	Yes, that's right. That fee covers you for health clinic visits, as part of the basic plan.
Student:	Okay, and let me see that I understand...I'm married, so I just want to check...so that fee covers me and my wife for any nonemergency care, right?
Staff:	No, not exactly.
Student:	No? I'm pretty sure I read it on the university Web site that my spouse is covered...I mean, she's been complaining of headaches and I need her to go to the clinic but so I just needed to make sure she's covered...we're both covered...
Staff:	Okay. I'm not saying your wife can't be covered, I'm just saying you'll have to sign up for an additional plan...
Student:	(getting anxious) But I remember on the original form I had to fill out. I had to check a box to say whether I was married or not and...
Staff:	(trying to diffuse the situation) Yes, but...here, let me show you the brochure...Ok, so see where it says, "Coverage can also be obtained for a spouse or dependent children?"
Student:	Yeah but I thought that was for part-time students 'cause the health insurance fee isn't part of their tuition.
Staff:	Mm... I can see how you could interpret it that way but I'm afraid it's a separate plan that you would need to apply for separately and pay a separate premium for.
Student:	Okay, so what's the premium?

Staff: Well, let's see. I believe it's around $150 or thereabouts... Yes, here. It's $145.

Student: Okay...Okay...well that's not horrendous. And will that cover her for dental?

Staff: Oh, no. I'm afraid not. The basic doesn't cover you for dental either.

Student: I'm not covered for dental?

Staff: Nope. You can only get dental if you sign up for the Vital Savings plan. It's a separate plan you sign up for in addition to your basic plan.

Student: In addition, huh?....Hmm...So is it expensive? I mean, I'm a student and she's, um, not really working.

Staff: Hmm, let me see. I have information on that somewhere... (pause). Students get a pretty big discount. Um...yeah, here: it's $25 annually.

Student: So $25 a year? That's definitely doable. Would that be for both or just per person?

Staff: It's per person but um... if it's a student plus one...um, for your wife...um, then the fee is $44. And the Vital Savings plan includes vision, as well as fitness and alternative health care. It's actually a pretty good deal.

Student: What kinds of alternative health care does it cover?

Staff: Let me see, Vital Savings covers...um, vision, fitness, as I said...I think it covers massage...definitely chiropractor, and it might even include acupuncture, but I'm not sure.

Student: Great, my wife will be thrilled. Okay...um, so I guess I have to fill out a form?

Staff: Well you have to enroll directly via the insurance company Web site...here take this brochure with you. The Web site address is on the bottom.

Student: So, if I enroll directly through the insurance company's Web site, will I still get my student discount?

Staff: Yes, there's a section where you input your student info.

Student: Ok, great, thanks. So I sign up for the Vital Savings plan through the insurance site, and it'll get me and my wife all this additional coverage: dental, fitness, all that.

Staff: Right.

Student: But I still have to sign my wife up as an addition to my regular basic plan, right? How do I do that?

Staff: Oh, here. Fill in this form. If you put it in today, I can start the process and she'll be covered within a couple of weeks.

Student:	Still a couple of weeks?
Staff:	Yeah, I'm afraid so.
Student:	Ok, well thanks. I actually have to get to class but you've been a great help. I really appreciate it.
Staff:	My pleasure.

Lecture

CD1, Track 6

<u>Robert Moses</u>

Narrator:	Listen to a professor and some students in a sociology class discussing the urban planner, Robert Moses.
Professor (female):	Robert Moses may well be the most polarizing person in the history of urban planning. He's been blamed for destroying urban communities, creating ghettoes, killing public transportation systems, and promoting suburban sprawl throughout the United States. His supporters, on the other hand, point out that he built some of New York's most important landmarks and made it easy for people to commute between the city and the suburbs. So who was Robert Moses, and what's his legacy?
	Moses has been called the "master builder" of New York City. Interestingly enough, he was never elected to public office—though there were times in his career when he was politically more powerful than the city's mayors. He came into power in the early 1920s, with the election of Governor Al Smith. Smith put Moses in charge of parks and other public works projects, and Moses turned around and built several prominent parks: attractive, on time, and under budget—Jones Beach is considered his finest example. Shortly after, Moses applied his development talents to highway construction, and he built virtually all of Long Island's major parkways. Because of this, Robert Moses is often considered the father of the New York State Parkway System.
	Moses became extremely powerful in the 1930s when he took over as chairman of the Triborough Bridge Authority. This bridge brought in millions of dollars a year in tolls, and Moses' control over this bridge gave him the ability to fund nearly any public works project he wanted. One of the first places where he clashed with the public was in the 1930s, when there was a major controversy over the building of the Battery Park Tunnel...that, uh, currently links Brooklyn and Manhattan. Business owners, politicians, unions—everybody favored the tunnel, but Moses wanted to build a bridge—and he was probably right that it would've carried more traffic and been cheaper to build, but it also would've destroyed a lot of Wall Street's

historical architecture. And since Moses controlled the Triborough Bridge money, he was able to hold up the project for years. As you've probably figured out, Moses eventually lost the battle over the Battery Park Tunnel, but not before the federal government was forced to step in and settle the dispute.

If this were Moses' only controversial act, it would be a footnote. But what stamped his legacy was his commitment over the years to building a city that, to put it bluntly, favored cars over people. Let me give you an example of what I'm talking about: from the 1930s to the 1960s, Moses was responsible for building several of New York's major bridges and highways—his projects included the Brooklyn-Queens and Staten Island expressways, the Belt Parkway, the Verrazano Narrows Bridge. But he also ruined hundreds of neighborhoods in the process by building highways right through them, and he displaced thousands of people. So this brings up the question: why did he do it? What was his reasoning? Well, Moses' basic philosophy about cities, was that cities were made for driving, not living. To him, cities were supposed to be entertainment centers ringed by highways, and people were meant to live in the suburbs. And unfortunately, his influence grew outside of New York as well. Several smaller cities hired him to design their freeway systems—many of which were never built, incidentally, due to public opposition. Eventually, Moses' desire to build an expressway right through the heart of New York's Greenwich Village led to one of his great failures—there was so much public opposition that the city rejected it in 1964. And the failure of the Greenwich Village project signaled the start of his political decline.

Since Moses, many city planners have begun to favor pedestrian-friendly, intimate communities, and see value in preserving historic architecture. Maybe the kindest thing we can do is to try to see him in a historical perspective. He helped put thousands of people to work during the Great Depression and built some of New York's most famous landmarks. And his projects never had technical failures, which speaks to his skills as a planner. One has to wonder what he might have done if he'd had compassion to go with his talent.

Conversation: Office Hours

CD1, Track 7

Topic

Narrator:	Listen to a conversation between a student and a professor.
Student (female):	Hey Professor Yang!
Professor (female):	Oh, hi, Tahir…How's that habitat project coming along?
Student:	Uh…actually that's what I'm here to talk to you about…
Professor:	I see, well…you won't be the first one to have questions. Have a seat.
Student:	Thanks….so….I'm uh, trying to make my topic a little more focused…
Professor:	Ok…
Student:	Um….for the final presentation, you said we have to make a class presentation about a unique plant or animal habitat, right?
Professor:	That's right…
Student:	So…I think I'd like to work with the idea of animals who survive in harsh conditions, like the penguins chapter we read for yesterday's class.
Professor:	Sounds like it would make for a fascinating presentation. What types of animals are you thinking of?
Student:	Um…I'm thinking of using the wolf spider that lives in the deserts of Arizona.
Professor:	Um…well, that's an interesting one but…
Student:	(excited interrupting) Did you know that the male wolf spider performs a drumming serenade to get the female's attention?
Professor:	Uh…no, I…hadn't heard that. But…you know, we're going to be doing a unit specifically on entomology in a few weeks. Maybe you can use the wolf spider for a project or paper when you study bugs and spiders in entomology?
Student:	Uh…okay, well…another idea I had is the Viper Fish, which lives 500 to 2,500 meters below the ocean surface, in total darkness.
Professor:	Mmm…yes…that is an amazing fish, and it certainly fits your harsh environment criteria…
Student:	Yeah, and it has huge teeth…and these weird internal organs that produce light! So interesting…but, um…would you recommend using mainly journal articles for that one? Or the Internet?

KAPLAN)

Professor: Yes, well…that's one problem with researching some of these more obscure species…Since only a little has been learned about them due to their hard-to-reach location, it might be difficult for you find uh…enough information on the Viper fish to fill 15 minutes of presentation time.

Student: Right…well…what do you suggest?

Professor: Maybe…you could choose a plant and an animal that share a harsh environment and then focus half the presentation on each one. That might…uh, give you more to work with…

Student: Hmm…uh, so…I guess I should pick a habitat first and then find a plant and an animal who both live there?

Professor: That's the idea…that way, even if there's not a lot of information available about them, you should have enough data to talk for about seven minutes on each one.

Student: Yeah…I see your point. Great…well, back to the drawing board!

CD1, Track 8

Professor: Um…well, that's an interesting one but…

Student: (excited interrupting) Did you know that the male wolf spider performs a drumming serenade to get the female's attention?

Narrator: Question 12

Narrator: What can be inferred about the student when he says this:

Student: Did you know that the male wolf spider performs a drumming serenade to get the female's attention?

Lecture

CD1, Track 9

<u>Plate Tectonics</u>

Narrator:	Listen to a professor and his students discussing plate tectonics.
Professor (male):	So…today, we'll begin to discuss the chapter you've read on plate tectonics. When we look at a map of the world…there are the seven continents, right? Can anyone tell us what they are? Yes…?
Student 1 (female):	The seven continents are, um…Europe, Asia, North America, South America, Africa, Australia, and, um…oh, Antarctica.
Professor:	Good. And…some of the continents are separated by oceans. America and Europe are separated by the Atlantic Ocean; Africa and Australia are separated by the Indian Ocean, and so forth. But, if you look closely at the coastlines of some continents, it almost looks like they used to fit together, like a puzzle. This is very noticeable along the Eastern coastline of South America, when you compare it with the western coastline of Africa, for example.

While it may seem an obvious point when we look at the world map, this idea wasn't scientifically presented until the early 1900s, when a meteorologist named Alfred Lothar Wegener proposed the theory of continental drift, stating that Earth's continents were once all one big land mass. The theory suggested for the first time, that the continents came to their present positions only after moving for millions of years. This one original land mass is known as Pangaea. Can anyone tell me more about Pangaea?

Student 2 (male):	Pangaea means "all Lands" in Greek, which is why the word was chosen to describe the landmass. And…didn't another scientist divided Pangaea into regions?
Professor:	Yes, that's true, it was divided into an area called "Laurasia" in the north, and …"Gondwanaland" in the South. Very good. Now, on to the theory of plate tectonics, which was presented scientifically only about 30 years ago. In geology, a plate is a large, slab of solid rock. The word tectonics also comes from Greek, from a word meaning "to build." Together, plate tectonics explains that the earth's surface is built of rock plates. This means…that the outer layer of the earth, beneath all of the continents and all of the oceans, is fragmented into large and small plates that are moving in various directions, relative to one another's positions. Can anyone tell us about the place where two plates meet?

Student 1 (female):	Yes, Um…between two plates, there's a boundary, called a fault line. There're different kinds of boundaries.
Professor:	Right. Do you know what they're called?
Student 1:	Um…Let's see…there's uh, divergent boundaries, where the plates are pulling away from each other, kicking up new crust…and then there's convergent boundaries, where one plate slides underneath another plate.
Student 1:	I think there's another one…the uh…transform boundaries?
Professor:	Good. Yes, you've got them all now. Divergent boundaries are where the plates are pulling away from each other…and convergent boundaries occur where one plate slides underneath another plate. This one's sometimes called a "very slow collision," because some of earth's crust is destroyed in the process.
	The uh…transform boundaries occur where two plates slide past one another, without creating or destroying any crust. Then there's places where the boundaries between plates aren't clearly defined, and we don't know how those plates interact…we refer to those as plate boundary zones.
Professor:	Very good, both of you…yes, there are four types of boundaries between plates. Divergent, convergent, transform, and plate boundary zones, and all of this occurs far beneath the surface of the earth. Now let's talk about some features of plate tectonics that we can see in the surface landscape. What are some natural features of the earth that are formed by plates moving at their boundaries? Yes…?
Student 2:	So…it's mountains, volcanoes, and…I think earthquakes too, right?
Professor:	Yes, that's right. So let's discuss all three of these…mountains first.
	Plate tectonics formed some of the most impressive mountain ranges on earth. The Himalayas were formed at a convergent boundary, when the Eurasian Plate crumpled up and slid over part of the Indian Plate, bringing India and Asia together around 50 million years ago. After the collision, the slow continuous convergence of the two plates over millions of years pushed up the Himalayas up even higher, to their present heights.
	Now, for volcanoes…We have to go back to plates for a minute. Plates are…actually part of earth's mantle, and below the mantle is another layer called the *asthenosphere*, made up of hot, partially molten—that is liquified—rock. You can think of a volcano as a vent, which when opened, allows the hot molten rock from the asthenosphere layer to leak out.

Speaker's Attitude Questions

Narrator: Question 14

CD1, Track 10

Narrator: What can be inferred about the professor when he says this:

Professor: While it may seem an obvious point when we look at the world map, this idea wasn't scientifically presented until the early 1900s, when a meteorologist named Alfred Lothar Wegener proposed the theory of continental drift, stating that earth's continents were once all one big land mass.

CD1, Track 11

Narrator: Listen again to part of the conversation. Then answer the question.

Student 1: Um…Let's see…there's uh, divergent boundaries, where the plates are pulling away from each other, kicking up new crust…and then there's convergent boundaries, where one plate slides underneath another plate.

Student 1: I think there's another one…the uh…transform boundaries?

Narrator: Question 15

Narrator: What does the man mean?

Conversation: Office Hours

CD1, Track 12

<u>Writing Assignment</u>

Narrator: Listen to a conversation between a student and a professor.

Student (female): Uh…professor…Could we have a word?

Professor: Yes, Brian?

Student: I just wanted to say that I really appreciated your lecture on homeschooling… My…uh…parents actually chose to teach me and my brother at home until 11th grade.

Professor: Really?(excited) I've been teaching this course for three years and never actually had a homeschooled student in the class before. I'd like to hear about your experiences.

Student: Oh (surprised)…uh…sure, anytime. Right now though, I'm a little worried about the writing assignment on the effect of homeschooling families on the public school system…

Professor: Hmm, yes…I can see why that might be a tricky subject for you. What do you suggest?

KAPLAN

Student:	Well…I don't know much about the public school system, since I only went for one year…So…I was wondering…uh…Would be all right if I did a slightly different assignment?
Professor:	What do you have in mind?
Student:	…Like maybe one about what it's like to be homeschooled kid coming into a normal school environment?
Professor:	I think that would be very interesting paper. Go ahead and use that topic.. uh…
Student:	(interrupting) Really? That's a big relief—thank you…
Professor:	…But, as part of your paper, I would still like you to address the topic of effects on the public school system because, while I understand that you may not be as familiar with institutional schooling as many of our students, I'm sure we could all learn from your opinion.

Rhetorical Function Questions

CD1, Track 13

Narrator:	Question 17
Narrator:	Why does the professor ask this:
Professor:	What do you have in mind?

Lecture with Comments

CD1, Track 14

Beowulf

Narrator:	Listen to part of a lecture from an English literature class.
Professor (female):	The first story we're going to read this term is the epic poem, *Beowulf.* It's often referred to as the first important work of literature written in English… though I caution you, if we were to read it in its original form, there's not one among us who would understand what in the world it was saying.
Student 1 (male):	When was it written?
Professor:	It was written by an unknown Anglo-Saxon poet around 700 A.D., though elements of the story had probably been in the oral tradition for hundreds of years. The story refers to events that had happened hundreds of years earlier—it's essentially a record of historic deeds. What do you guys know about the Anglo-Saxons? Anybody?

Student 2 (female):	They migrated from northern Europe to the British Isles, like two thousand years ago, am I right?
Professor:	Close, but you're off by a few hundred years. They first appear in Britain around the fourth century A.D.
Student 1:	They were from northern Europe, is that what you said?
Student 2:	Germany...Scandinavia...
Professor:	Exactly. And in its original form, *Beowulf* represents a period of old English in which the language is still heavily Germanic, showing none of the Latin influence that came into the language after the Norman Conquest.
	One of the interesting things about *Beowulf*...it tells the story of actual figures from history. The characters in the story are recognizable from the historical record as Swedish and Danish kings from around 500 A.D. Although once we get into it, you'll recognize that most of the events—battling monsters and dragons and whatnot—couldn't have actually happened.
Student 2:	Grendel...
Professor:	Oh, so you've read it, then?
Student 2:	No, but I read John Gardner's novel, Grendel, which tells the *Beowulf* story from the perspective of the monster. I'm looking forward to reading the original...
Professor:	Terrific! Just so the rest of you know what we're talking about, the central plot element in *Beowulf* is the battle between *Beowulf* and the monster Grendel—and Grendel's mother, and later a dragon. These battles take place over a period of 50 years...
Student 1:	So *Beowulf* is somebody's name...
Professor:	Yes, *Beowulf* is a warrior of the Danish kingdom of Geatland. And the story gives us tremendous insight into the, I would say, the heroic code that permeated not only Saxon civilization, but most of Europe throughout the Middle Ages. In *Beowulf's* world, the people rally around strong kings who protect them from danger. There is constant fear of invasion from neighboring kingdoms, and as we see in the story, from monsters as well.
Student 1:	So the person who wrote *Beowulf* was Scandinavian?
Professor:	No, he's clearly not, and I'll explain why. The story is written around 700 A.D., as I said, and refers to events that had occurred about 200 years earlier. In the interim, most of the Scandinavian people had converted to Christianity, and in this respect, the author of *Beowulf* is quite different from the people he's writing about. One of the tensions in the story, one of the reasons it's so

important for our understanding of English poetry and literature, is that, in *Beowulf*, the author seems at pains to reconcile two competing value systems—the heroic code of the Germanic people he's writing about, and the more docile Christian code to which he and most of his people had converted by the time the story was written.

Student 2: So the obvious question is, how were these value systems different?

Professor: Lots of ways. So while the warrior culture states that it is better to mourn than to grieve, the Christian value system emphasizes forgiveness toward one's enemies. While the warrior code states that a person acquires prestige through his actions in life, Christian doctrine emphasizes that real glory is to be found in the afterlife. Characters in *Beowulf* frequently face difficult moral choices, and are given very little direction on how to proceed. And in this regard, *Beowulf* represents the beginning of a very different worldview, one that is uniquely English, in that it represents a significant break from the attitudes of the Germanic peoples before their migration to Britain. In many ways it really is the foundation of English literature.

Now before we get too far into it, I want to clarify one possible misconception. I've called *Beowulf* the foundation of English literature because it represents a different sensibility from that of the people who had migrated to Britain. This is not to say, however, that this story in any way influenced the evolution of English literature. In fact, there is nothing to indicate that Chaucer, Shakespeare, Shelley, Keats, or any other English authors were familiar with it. *Beowulf* really wasn't widely read until the 19th century. You can find references to it in the works of 20th century English writers, but historically, though it is an early example of English poetry, its influence on English poetry has been limited.

CD1, Track 15

Rhetorical Function Questions

Narrator: Listen again to part of the conversation. Then answer the question.

Professor: Terrific! Just so the rest of you know what we're talking about, the central plot element in *Beowulf* is the battle between *Beowulf* and the monster Grendel—and Grendel's mother, and later a dragon. These battles take place over a period of 50 years...

Student 1: So *Beowulf* is somebody's name...

Narrator: Question 18

Narrator: Why does the student say this

Student 1: So *Beowulf* is somebody's name...

CD1, Track 16

Narrator: Listen again to part of the conversation. Then answer the question.

Professor: Although once we get into it, you'll recognize that most of the events—battling monsters and dragons and whatnot—couldn't have actually happened.

Student 2: Grendel...

Narrator: Question 19

Narrator: What does the student mean to say?

Conversation: Service Encounter

CD1, Track 17

Career Services Office

Narrator: Listen to a conversation between a student and a professor.

Advisor (woman): Hi Stefan. So...is this is your first time at Career Services?

Student (male): Yes.... I'll be graduating next month, and I could really use some help with my resume....

Advisor: Sure...that's a good place to start. Let's take a look. (pause) Well, this is good...but you may want to leave out some of your summer jobs like the ice cream server job, the pool lifeguard job...

Student: (interrupting) Really? ...I thought they made my resume look longer.

Advisor: Well, they do...but longer isn't always better. The idea is to highlight the information on your resume that's most relevant for each employer you send it to.

Student: I see.....huh....so, if I'm applying for jobs related to art, maybe I should take out the job as the receptionist for the Theater Department office?

Advisor: Um, actually, I would keep that one, since it shows you've learned some office skills—which can be applied to many jobs. What you want to cut is stuff that doesn't really apply. Like, for example, would knowing you're a good ice cream server help the director of the City Museum decide if you'd make a good assistant curator?

Student: Hmm...I guess not. So...maybe I would keep the ice cream shop experience on my resume if I was trying to get a job in a restaurant?

Advisor:	Right, right….exactly. So for our assistant curator example, you'd want to highlight the gallery assistant job you had last summer at the Bernard Sculpture Museum, and the internship you're doing now at the modern art gallery.

CD1, Track 18

Rhetorical Function Questions

Narrator:	Question 24
Narrator:	Why does the advisor say this:
Advisor:	Like—for example, would knowing you're a good ice cream server help the director of the City Museum decide if you'd make a good assistant curator?

CD1, Track 19

Narrator:	Listen again to part of the conversation. Then answer the question.
Advisor (woman):	Hi Stefan. So…is this is your first time at Career Services?
Student (male):	Yes…. I'll be graduating next month, and I could really use some help with my resume….
Narrator:	Question 25
Narrator:	Why does the student mention his graduation date?

Lecture

CD1, Track 20

<u>Teacher Training</u>

Narrator
(male): Listen to part of a lecture in a teacher training class.

Professor
(female): Although the use of simplified texts in teaching, especially in EFL, uh, that is, English as a Foreign Language teaching, has increased dramatically over the past decade…and the publishing industry is bursting with graded readers of every kind…heated debate continues about the artistic ethics of umm diluting classic works of literature. There are those who argue that great literature is great because of the words, phrases, and structures chosen by the author. Therefore, they say, it is a crime to change those words, phrases, and structure in any way. By doing so, the very essence is removed and thus the work of art becomes entirely devalued.

Well, this debate brings up good questions regarding the definition of great literature. What makes a novel great? What makes it a classic? Is it the plot, the characters, the descriptions, the dialogues, the prose itself? Or does greatness lie in some kind of originality; does a work need to break barriers to be considered "art"? Or is it just a question of popularity? I don't believe there is a simple answer; in fact, I think…well anyway, I mustn't get sidetracked.

The question for us is, do we use simplified classics in teaching? In my opinion, the educational advantages of using simplified graded readers far outweigh the philosophical concerns about art. I believe great literature is great because of a combination of things. Characters and plot can be crucial; for me, it is these elements that take my imagination beyond the mundane, outside of ordinary life. Have you ever been asked: "If you were going to be stranded on a desert island and could only take three books, which three would you take?" I'll tell you one title that would be on my list: David Copperfield. There's a book that makes me laugh out loud, moves me to tears, and never fails to absorb me completely. The young hero experiences death and hardships, meets amazing characters, and survives everything thrown his way. It is the events, the characters, the story (emphatically) that make this book so rich. Those can be very effectively conveyed in a simplified text.

Another example is Pocahontas. Unlike David Copperfield who is 100-percent fiction, we know that Pocahontas was a real person. However, as is the case with many legendary people, no one can be certain that all the stories written about her are completely true. She did indeed marry John Rolfe in

1614 and traveled with him to London in 1616. Regardless of any potential embellishments, this is a memorable story of Native American culture and the impact white settlers had when landing on the new continent. Because of the characters and the way their interactions are portrayed, it is not only interesting historical insight; it's a page-turner.

A nonfiction classic is Benjamin Franklin's autobiography. So much can be learned about the way the world worked in the late 1800s from this absorbing and intimately written book, even though it centers on just one life. Now, Benjamin Franklin used grammar, vocabulary and especially punctuation that would baffle even the most accomplished modern day reader! Any version of his autobiography that is published today has been simplified to a certain extent. Why not simplify it far enough to enable EFL learners to access it?

I could go on. But aside from the fact that reading offers escape, entertainment, and windows on unfamiliar or bygone worlds, there are many reasons from a pedagogical,—that is, instructive—point of view to encourage reading anything at all!

Stephen Krashen demonstrated way back that extensive reading improves all aspects of language learning. This includes vocabulary, speaking skills, fluency, and writing skills. In other words, learners who read in English learn more English, more quickly than people who don't read. And, the more you read, the better you read; the better you read, the more you enjoy it; the more you enjoy it, the more you read…get my drift?

Extensive reading is the opposite of intensive reading. Okay…but what does that mean? It means not reading for detail and not needing to understand or even read (emphatically) every single word. Competent readers are fast readers. Stopping to use a dictionary all the time slows you down. Okay, that's all very well, but if you don't understand any of what you are reading, it won't be a very positive experience. So, how to encourage extensive reading among English-language learners? Provide stimulating materials at a level that is accessible. Many of the classics are great stories; great stories are stimulating. So, if simplified classics work for your learners, then for goodness sakes, use them! Don't let the purists put you off.

CD1, Track 21

Rhetorical Function Question

Narrator:	Listen again to part of the lecture. Then answer the question.
Narrator:	Question 31
Narrator:	Why does the professor say this?
Professor:	What makes a novel great? What makes it a classic? Is it the plot, the characters, the descriptions, the dialogues, the prose itself? Or does greatness lie in some kind of originality; does a work need to break barriers to be considered "art"? Or is it just a question of popularity?

CD1, Track 22

Narrator:	Listen again to a part of the lecture. Then answer the question.
Professor:	Extensive reading is the opposite of intensive reading. Okay…but what does that mean?
Narrator:	Question 32
Narrator:	What does the professor suggest?

Conversation: Service Encounter

CD2, Track 23

Roommate Trouble

Narrator (male):	Listen to a conversation between a student and a Housing Office employee.
Student (female):	Um…Hi there…is this the Student Housing Office?
Admin Staff (male):	Yes, can I help you?
Student:	I really hope so. I'd like some information about changing dorm rooms.
Admin staff:	Oh…I'm sorry, but…it's almost midsemester you know. You had a whole month at the beginning of the semester to request a room change, but now it's really too late.
Student:	No way…there has to be something we can do here…I can't stay in that room.
Admin staff:	All right, ok, calm down. Why don't you tell me, what's the problem?

KAPLAN

Student:	Oh…ok, um… Well…everything was ok at first, which is why I didn't…well anyway my roommate keeps the lights on and brings friends over late at night, almost every night. I'm not getting any sleep!
Admin staff:	Have you tried talking with your roommate, letting her know this is a problem for you?
Student:	Yeah, I told her… I have to get up early. I have 8 A.M. classes, and that I'm studying all day in the library, so when I finally come home around 10 or 11, I need to crash. So yes, yes…I told her.
Admin staff:	Ok…but have you tried negotiating with her? Like, working out a schedule of when each of you gets some private time in the room? I know this works for some students who…
Student:	(interrupting) Oh, been there. I tried to work with her. Like, I said the room's all hers if she wants to bring friends over during the day or evening, after class, as long as I can have it quiet at night.
Admin staff:	And did she think it was a good idea?
Student:	Yea, but then she didn't keep her end of the bargain. She goes out and then comes in late and turns on music at night…ugh. I'm really tired of this—I'm ready to move out! I've had it. I don't care if I have to move off campus for a little while. So…I guess my question for you is can I get at least part of my housing money back?
Admin staff:	Look, I understand how you feel, but like I said, it's too late to get your housing money returned to you.
Student:	It's just too much money for me to just throw it away, but I'm really at the end of my rope…
Admin staff:	The only other thing you could try is to find someone willing to swap with you for the remainder of the semester.
Student:	Swap?
Admin staff:	Yes—if you can find another student who is also dissatisfied with their room assignment, sometimes students arrange a swap, and move into each others' rooms.
Student:	Hmmm…so how do I find them?
Admin staff:	I think there's a notice board in the dining hall—you could try looking there.
Student:	All right then, I guess that's worth a try. It's a gamble, but it's better than staying where I am.

CD2, Track 24

Rhetorical Function Questions

Narrator: Listen again to part of the conversation. Then answer the question.

Narrator: Question 36

Narrator: Why does the man say this:

Admin staff: All right, ok, calm down. Why don't you tell me, what's the problem?

CD2, Track 25

Narrator: Listen again to part of the conversation:

Oh, been there. I tried to work with her. Like, I said the room's all hers if she wants to bring friends over during the day or evening, after class, as long as I can have it quiet at night.

Narrator: Question 37

Narrator: Why does the woman say this:

Oh, been there. I tried to work with her

Lecture

CD2, Track 26

Technical Training in the United States

Narrator: Listen to a workforce education professor giving a lecture on technical training in the United States.

Professor: Technical training has a rich and interesting history in the United States. For centuries in Europe, crafts guilds had allowed master craftsmen to pass their trade to apprentices, and the Europeans who settled in North America brought the apprenticeship system with them. But as the colonies lacked the strong guild organizations of Europe, apprenticeships came to be administered by municipal authorities. Later, the apprentice system came to be replaced by the school system, but in the eyes of the colonists, the apprenticeship system was seen as an important method for training the people who would build a new nation.

America's shift from an agrarian to an industrial society in the 18th century led to a permanent decline in the apprenticeship system—the system was just unable to keep pace with the growing demand for factory workers. New methods were needed for training workers for industrial labor.

Put simply, the changes in the workplace required a change in the preparation of workers.

Now, free public schools for elementary education had been established early on in the colonial era. Yet the means for training people to work in the nation's expanding factories, as we said, didn't exist. To us now, combining technical with academic courses in the schools seems like an obvious option, but for reasons we'll get into shortly, this was not seriously explored until the late 1800s. So, separately from the system of public education, technical societies of mechanics and tradesmen—precursors of today's unions—formed what were called "mechanics institutes." These mechanics schools provided technical training in a particular trade, as well as basic education in reading, mathematics, and other subjects.

At the same time as these technical societies were opening their mechanics schools, businesses began realizing the importance of training workers not just for general trades, but for factory-specific jobs. And so you get, in the late 19th century, the opening of factory schools—that is, training classes that were held on company grounds. By the turn of the 20th century, several of the nation's major corporations—companies like Westinghouse, General Electric, Goodyear—had their own factory schools. And these schools became the precursors of today's corporate training departments.

Now, even though privately sponsored technical training was popular, there was strong resistance to the integration of technical training within the public schools. Opposition came from educators who were concerned that technical training would lower academic standards, and who felt that moral training and instruction in the basic subjects would provide the best preparation of students for the outside world. Simultaneously, however, the public schools were coming under increasing criticism from people who felt that public school education lacked relevance. It wasn't until the late 19th century that several state legislatures—seeing the success that the mechanics schools and factory schools had in developing skilled workers—began funding technical training curricula within public schools. At the national level also, Lincoln's Land Grant Act of 1862 established college-level training in agricultural and industrial education, and did much to legitimize the value of technical training in the public eye.

CD2, Track 27

Rhetorical Function Question

Narrator: Listen again to part of the lecture.

Yet the means for training people to work in the nation's expanding factories, as we said, didn't exist. To us now, combining technical with academic courses in the schools seems like an obvious option, but for reasons we'll get into shortly, this was not seriously explored until the late 1800s.

Narrator: Question 42

Narrator: Why does the professor say this:

...but for reasons we'll get into shortly,

Conversation: Service Encounter

CD2, Track 28

<u>Asking Directions</u>

Narrator: Listen to a conversation between a student and an administrator.

Admin staff: Can I help you with something?

Student: Uh, no...actually, I came in here because I guess I'm a little lost. Isn't the campus post office supposed to be near here?

Admin staff: Um...no...it's across the graduate plaza...it's actually kind of far from here.

Student: Uh-oh...its getting late and I have to send something by express mail. I'm afraid they're probably going to close soon, don't you think?

Admin staff: Yeah, it's close to five. Maybe, if you run, you might still make it...It's actually in the basement of the cafeteria.

Student: Oh no! I just came from the cafeteria...I didn't even know it was in the same building!

Admin staff: Yeah, it's, uh...sort of hidden. You need to take the stairs behind the cashiers...

Student: Wait a minute...what stairs?

Admin. Staff: Okay, look, go back across the graduate plaza and enter the cafeteria through the main doors.

Student: Okay.

Admin. staff: When you first enter the cafeteria, if you go all the way through to the back, where the cashier is, you'll see some stairs.

Student:	Okay, I go in, I see the cashier, and there are some stairs. Oh, you mean the stairs to the bathroom.
Admin staff:	Exactly. Go down those stairs as if you were heading to the bathroom…
Student:	Okay, I'm with you.
Admin staff:	Walk past the bathrooms, and you'll see the office of student life and take a…um…a right.
Student:	After the student life office? Isn't that part of the health center?
Admin staff:	Um, it used to be, but they moved some offices around last summer.
Student:	Oh….so that's why I was so confused when I went down there.
Admin staff:	Yeah, they should have kept the signs up longer…they were only there for a few days at the start of the semester. Everyone's confused.
Student:	Yeah, I obviously am. So at any rate, I turn right after I pass the student life office...
Admin. Staff:	Correct, you turn right. And there you are. It's right at the end of the hall.
Student:	I wish I'd known that when I was over there. Okay, well, now I know. Thank you so much for your help with the directions. I really appreciate it.
Admin. Staff:	All right, you'd better hurry! It's almost five!
Student:	[laughing] Okay, I'm going…I'm going!

Lecture

CD2, Track 29

<u>Insulin and Diabetes</u>

Narrator:	Listen to a professor giving a lecture on a medical condition.
Professor:	(female) Let's briefly discuss what diabetes is and how it's caused.
	The standard definition of diabetes is excessive glucose in the bloodstream. What is glucose? It's simply a sugar, one of your body's main sources of energy. Your body produces a hormone called insulin, which controls the amount of glucose in your blood. The way insulin works—it travels through your bloodstream, and acts like a key that opens cells—muscle and fat cells, mainly—and allows glucose to enter them. If glucose doesn't enter the cell, your cell basically starves because it has no fuel to burn. So insulin is pretty important.
	How is insulin produced? Within your pancreas you have these cells called the Islets of Langerhans. Their job is to sense the amount of glucose in the

blood and produce enough insulin to move it out. For example, when you eat a meal, all this glucose suddenly enters your bloodstream. The Islets of Langerhans detect this, produce insulin to move that glucose out of your bloodstream and into your cells, and reduce your blood sugar to its normal levels. Once your blood sugar returns to normal, the Islets of Langerhans reduce the output of insulin. In this way, your pancreas adjusts the amount of insulin it produces on a minute-by-minute basis, always in proportion to the amount of glucose in the bloodstream.

Blood sugar can begin to rise, either because your body fails to produce enough insulin, or the insulin stops working. And if this remains undetected, it can lead to one of two types of diabetes. Type I diabetes occurs when your body begins to destroy the Islets of Langerhans so they can no longer produce insulin. Without insulin, your blood sugar rises, and your muscles starve, because, as we said, there's no insulin to open them so glucose can enter. People with type II diabetes, on the other hand, have plenty of insulin in their bodies, but their bodies become insulin-resistant, meaning the insulin doesn't do what it's designed to do. The pancreas becomes unable to keep up with the body's insulin demands, and this can lead to a number of other problems. If the warning signs are diagnosed before the onset of diabetes, many potential diabetics respond to lifestyle changes, such as improved diet and exercise. Once people have diabetes, their health risks become much greater.

Lecture

CD2, Track 30

<u>Types of Theaters</u>

Narrator: Listen to part of a lecture in a drama class.

Professor (male): Theater is all about seeing and being seen—so the word theater, from the Greek word *theatron*, is quite fitting—it means, "the seeing place." There're four essentials of a theater: the actor, the audience, the performance, and the space. We call this empty space the *stage*. Today we'll examine three types of theaters that have been used all throughout western theatrical history: the end stage, the thrust stage, and the arena stage—and then we'll cover a few variations of those basic set ups.

So we'll start with the end stage. An end stage is basically a raised platform, and it's usually placed at one end of a rectangular space, facing the audience. Since the Renaissance, Western theater has been dominated by a type of end stage known as the *proscenium theater*.

In that type of theater, there's the proscenium arch, which is the opening in a wall at the front of the stage through which the audience can see the performance. This, uh, design was developed in response to performers' desire to hide scenery changes as well as their offstage exits and entrances from the audience. The end result is a proscenium, with its separating curtain, which actually enhances the illusion of the theater by eliminating all sights that aren't part of the actual scene.

Next we have the thrust stage, also known as a *three-quarter round*. The thrust stage is a platform surrounded by the audience on all but one side. The ancient Greeks, the Elizabethans, and also classical Spanish performers often used this type of stage. One notable feature of the thrust stage is a wall, which backs the stage on the side that's not occupied by the audience. The stage is usually bare except for some scenic elements and props.

The arena stage, or theater-in-the-round, takes exposure to the audience one step further than the thrust stage. The arena stage is a performing space that's totally surrounded by the audience. So, you can imagine that providing a good view for all spectators is a real challenge that puts limitations on the type of scenery that can be used in this type of theater. And on the same note, the arena stage severely limits the movements of the actors because, at any given time, the performer will show his back to one section of the audience.

Finally, we come to one of the oldest types of theater, the amphitheater which is an open-air variation on the theater-in-the-round, meaning the audience watches from all sides. Used to show gladiatorial events and even mock sea battles, the amphitheater is akin to the modern-day sports stadium. One example that you all might recognize is the Colosseum in Rome, completed back in 80 A.D.

Conversation: Office Hours

CD2, Track 31

<u>Hurricane Homework</u>

Narrator:	Listen to a conversation between a student and a professor.
Student (female):	Hey Professor Gould?
Professor (male):	Oh hi, Gina how are you?
Student:	Not bad, do you have a minute to talk?
Professor:	Uh…well I'm teaching in 10 minutes, but go ahead, shoot.

Student:	Oh, great. Ok, so I'm working on the meteorology paper due next week…
Professor:	Uh-huh…
Student:	So…I think, since I'm from New Orleans, I'd like to write about hurricane Katrina.
Professor:	Oh! I didn't know you were from there…it should be a pretty interesting paper then…
Student:	Yea…so…I'm wondering…well, this is a science class, so everything should be really factual, right?
Professor:	Well, yes…I'd like your paper to focus mainly on meteorological data…
Student:	Right…so…I guess pictures of my family's house and my town after the storm would be a little…too, uh…personal for this assignment?
Professor:	Oh…I see. No, uh…actually I think that would just make your paper that much more effective. Instead of just grabbing photos from the Internet or newspapers like most other students probably will….
Student:	Really? Ok…in that case, is it ok to also include some…uh…quotes from my parents or neighbors about their experiences during Katrina?
Professor:	Um…perhaps, as long as that part isn't too long, and it supports the scientific content of your paper.
Student:	I see. (pause) I…uh…think I can manage that. Thanks professor!

Lecture

CD2, Track 32

Consumer Decision-Making

Narrator:	Listen to a professor and his students discussing consumer decisions.
Professor (male):	So I want you to take a few minutes and put yourselves into the mind of an average shopper about to make a moderately important purchase. If you're a smart shopper, you'll go through a five-step consumer decision-making process before reaching into your wallet.
	The first step results from having a want, need, or goal that you'd like to satisfy. For example, you and a roommate have moved into an apartment off campus that doesn't have any laundry facilities. You want or need to find a way to wash and clean your clothes. This is the first step in consumer decision-making: recognizing and defining your want, need, or goal. Since you're intelligent college students on a limited income you want to solve your problem in both an effective and economical way, right? This leads us

to the next step in the process. Anyone want to take a stab at what that could be?

Student 1 (male): How about…um take 'em home to my parents on the weekends?

(group laughter)

Professor: (jokingly) (A) I said it was a five-step process, and (B) I said you were intelligent students…

(group laughter)

Student 2 (female): Can I answer? (waiting for confirmation) Well, this is off the top of my head but um well, I guess you'd start thinking about your options, right? I mean, what are my choices? I can yeah, go home to my parents, or um, say check out local laundromats...um...yeah, I dunno…

Professor: Exactly. So this stage is when you start gathering information. During this stage, you gather as much information as you feel you need, and you probably think about alternative solutions. So Lisa already gave us some solutions to our laundry problem. Anyone got any other ideas?

Student 1: Ok, seriously. I'd probably look at buying a washer and dryer.

Student 2: You or your parents?

Professor: (Ignoring woman) Ok, James. (encouraging) That's right. You might find it worth your while to buy a washer and dryer rather than dealing with the inconvenience of a Laundromat. You might also find it's actually cost-effective. So then you have the choice of buying either a new washer or dryer, or a used washer and dryer, right? Now what do you do? Considering this is a fairly important purchase in terms of cost and value to you…

Student 1: Um…Well, I'm a bit of a research nut so I'd research anything and everything about the product to find the best product at the best price.

Professor: Precisely. Smart consumers gather all the information they can about the cost and benefits of each possible solution. You probably talk to friends, consult the consumers' report, and look in the classified and advertising sections of the newspaper to find information on the best brands of washers and dryers, sales, and so on. Next, you want to compare these three alternatives in terms of money, time, and convenience. This is the third step. Finally you move into step four where you actually select one of these alternatives. Last, but not least, you enter step five and reach into your wallet. This five-step process varies in length depending on the consumer and the importance of the purchase.

Now, the better a company understands its customers and potential customers, the better it can serve them. Marketers wanting to influence consumers' buying behavior must understand what goes on in the minds of consumers during the decision process, and the factors, both personal and outside, that affect consumers' preferences, choices, and spending.

Let's look at some of these personal factors. A consumer's demographic characteristics—which include age, gender, income, education, race, marital status, housing, and number of children—affect a person's buying behavior and purchasing power. Take a minute and think about how your buying behavior differs from that of your parents. Think about how your mother's buying behavior differs from your father's....Certainly, college students like yourselves spend differently than the CEO of a top American company and certainly television programs, products, or ads that appeal to you would never appeal to your parents. Can you give me any examples? (pause) Naomi?

Student 2: Yeah, like…(a little embarrassed) well um…like, my parents watch the news every night before dinner and following the news they watch more kinds of current affairs type shows. Now I guess I should be keeping up with what's going on in the world but I hate watching the news. It's just too depressing.

Professor: Ok, perfect. So what do you watch?

Student 2: No! I can't say!

Student 1: Go on!

Student 2: Um. Ok. The Gibson Girls (confessionally) It's my guilty pleasure.

Student 1: Oh brother!

Professor: Ok. No, it's good. So companies use these demographic characteristics to divide the market into segments in order to position a product. A demographic segmentation might include single 18-25 year old women, working mothers with professional degrees and two children or more, white male senior citizens, or professional single Latino-Americans grossing $40,000 or more annually…

Student 2: Wow, that's pretty specific, huh?

Professor: Yes, it is…

CD2, Track 33

Inference Questions

Narrator: Question 48

Narrator: What does the professor imply when he says this:

Professor: Marketers wanting to influence consumers' buying behavior must understand what goes on in the minds of consumers during the decision process, and the factors, both personal and outside, that affect consumers' preferences, choices, and spending.

Speaker's Attitude Questions

Narrator: Listen again to part of the conversation. Then answer the question.

CD2, Track 34

Student 1: How about…um take 'em home to my parents on the weekends?

(group laughter)

Professor: (A) I said it was a five-step process, and (B) I said you were intelligent students...

Narrator: Question 50

Narrator: What does the professor imply?

CD2, Track 35

Narrator: Listen again to part of the conversation. Then answer the question.

Professor: Ok, perfect. So what do you watch?

Student 1: No! I can't say!

Student 2: Go on!

Student 1: Um. Ok. (confessionally) The Gibson Girls. It's my guilty pleasure.

Student 2: Oh brother!

Narrator: Question 51

Narrator: Why does the man say this:

Student 2: Oh brother!

CD2, Track 36

Narrator: Listen again to part of the conversation. Then answer the question.

Professor: A demographic segmentation might include single 18–25 year-old women, working mothers with professional degrees and two children or more, white male senior citizens, or professional single Latino-Americans grossing $40,000 or more annually…

Student 2: Wow, that's pretty specific, huh?

Professor: Yes, it is…

Narrator: Question 52

Narrator: What does the student imply?

Chapter 3: **Speaking**

The Speaking section of the TOEFL is the third section of the exam. It follows a mandatory 10-minute break after the Listening section, so you should be relaxed when you get to this portion. Instructions appear on the screen at the beginning of this section, telling you to put on your headphones. The microphone that you speak into is attached to the headphones, and you also need the headphones to hear the narrator's instructions and listen to the various conversations and lectures. So you should keep your headphones on throughout the section.

There are six tasks in the Speaking section, and the entire section takes 20 minutes to complete.

First two tasks: Independent (which means that you speak about familiar topics without reading or listening to any passages beforehand)

Last four tasks: Integrated (which means that you must first read and/or listen to a passage, and then speak about what you've read and heard)

To perform well on the Speaking section of the TOEFL, you must be able to generate ideas quickly, listen and read actively, record notes efficiently, speak clearly from fragmented notes, and accurately summarize passages in your own words. The following strategies review the necessary skills for each Speaking task on the exam. After reviewing each type, you can practice a question similar to the one you will get on the actual TOEFL and then read a sample response.

GENERAL SPEAKING STRATEGIES: INDEPENDENT AND INTEGRATED

The following strategies are useful for all six tasks in the Speaking section. You should keep them in mind whenever you are speaking.

Strategy 1: Be prepared to speak from notes, not a transcript.
You won't have enough time on the actual exam to write out a transcript of your answer. Even for the independent tasks, you will only be able to record your ideas in note form:

single words, fragments, and abbreviations. Also, the integrated tasks include some complex ideas that you must record or remember while you read and listen (see also General Speaking Strategies: Integrated below). Therefore, you must be able to speak clearly and correctly with only notes as a guide.

Don't practice by writing out whole sentences and then speaking them out loud. That is reading aloud, which is useful for practicing pronunciation, stress, and rhythm, but it doesn't completely prepare you to perform well on the TOEFL. Your notes should only be clues about what you want to say, and you must be able to speak with only a few words as clues.

In order to be properly prepared for the exam, you should practice by writing single words or phrases and then using those ideas in complete sentences. Avoid speaking from a transcript even when you are practicing in class, in a study group, or with a tutor. For example, the first task could have a question like "Describe the qualities of a good parent." You should practice speaking about this with notes such as the following:

Good—patient, responsible, helpful

Patient—problems, not angry easily, calm

Resp—duties, always there, don't forget/run away

Help—advice, explanation

Strategy 2: Organize your thoughts and speech as a paragraph, not a standard essay.
Your answers in the Speaking section are not spoken essays, but they must still be organized. Your answer must be relatively brief (45 seconds or one minute maximum), so it should be organized like a typical body paragraph. You need to make your point quickly, so don't bother with an extended hook like those used in lectures and essays. There is no need, or time, for an introductory paragraph like the one in an essay or lecture.

Instead, your first sentence should be your thesis statement or topic sentence, and it should give your main idea and possibly a clue to your supporting points. The next sentences should explain your main idea with details. This is the longest part of your answer, but don't think of it as separate paragraphs. Likewise, your conclusion can't be a separate paragraph. Because of the limited amount of time, your conclusion should be one or two sentences at most.

Strategy 3: Use familiar vocabulary.
Be yourself and be conversational. Use the vocabulary that you would normally use in a serious but friendly conversation about an academic topic. Although you may have to learn and use new vocabulary from a notice, conversation, or lecture, don't try to sound like a professor by thinking of long, unfamiliar words and sentences unnecessarily. You will waste time and make more errors if you try that.

Strategy 4: Use short, simple sentences.

Comfortable, natural speech in English doesn't sound like written English. Spoken English must still be grammatically correct, organized, and detailed. Yet native speakers use a more casual, relaxed style when speaking, even when they discuss academic topics. If you listen to them, most English speech is made up of sentences with one or two clauses. For example:

> My grandmother has been a very important person in my life. She's always given me excellent advice and encouraged me. For example, when I was in my senior year of high school, I was unsure about university. I couldn't decide between chemistry and math. My math marks were higher, but I loved using the test tubes and mixing chemicals in the lab. My grandmother told me…

Notice how the sentences above are not very complex. The longest sentences have only two clauses, yet the answer includes detailed description (senior year, chemistry, math, test tubes, etc.). Of course, academic lectures are more formal and structured with complex sentences, but you are not expected to recreate or give a lecture on the TOEFL. Therefore, don't try to speak in a formal, written style with long, complex sentences of three or more clauses.

Strategy 5: Vary your vocabulary and sentence structure.

Speaking naturally does not mean speaking repetitively. Use different word forms when you discuss the same idea or related ideas. It is not only more interesting but also demonstrates your vocabulary. Look at part of a summary about a student complaining about her study partner:

> The student is frustrated with her partner. She is frustrated because she feels she does more work than…Because she's frustrated about her partner, she wants...

Though this is not terrible English, it is repetitive and suggests that the student knows only one way to express the idea (subject + be + adjective). Instead, the speaker should vary, or change, the words and structure:

> The student expresses a lot of frustration about her partner. She feels frustrated because she believes she does more work than…She is so upset about this that she wants…

There is always more than one way to say something. In fact, often there are dozens of ways, but variety requires a broad vocabulary and good grammar skills. One way to develop these is to use the same set of notes and try expressing the same ideas in a variety of ways (see also Improve-Your-Score Strategies).

Strategy 6: Use cohesive devices effectively.

Cohesion means unity, and it refers to the number of connections among the ideas in a sentence and paragraph. There are many ways to add unity to your answer. Some of the most common cohesive devices are personal pronouns (*he, she, they,* etc.), demonstrative pronouns (*that, this,* etc.), adjectives (*next, another,* etc.), articles (*a/an, the*), synonyms, transitions (*first, however, in fact,* etc.), and word form (ex: *inform, information, informative,* etc.).

Your listener (and reader) requires clues to follow your answer easily. Moreover, using cohesive devices is an excellent way to add variety to your speech (and writing).

> I disagree for several reasons that the city should build a highway through my neighborhood. First, the equipment will create noise and dirt. For example, the dump trucks and jackhammers are loud, and they will disturb residents. Second, the highway will be dangerous. The highway will be very close to some homes, so children might be killed if they try to cross it. Also, people's pets might be killed as well.

The speaker here connects ideas with articles (*a highway/the highway*) and pronouns (*highway/it, children/they*), and the speaker guides the listener with transition devices: *first, second, for example, so, also.* See also Reading Strategies, Listening Strategies, and Writing Strategies.

Strategy 7: Use transitions appropriately.

Throughout your speech, use transitions to indicate sequence (*first, next,* etc.), connection/similarity (*moreover, furthermore,* etc.), contrast (*however, but,* etc.), examples (*for example, for instance, like,* etc.), or explanation (*therefore, because, due to,* etc.). However, don't overuse them. Transitions can guide a listener when they are necessary, but they can waste time and confuse the listener if they are used excessively. See also General Reading Strategies, Reading Strategies: Referent Questions, and Coherence Questions.

Strategy 8: Be concise.

A concise answer gives a lot of information in a short amount of time. In other words, it is brief but informative. Many students make the mistake of speaking faster in order to say more. Although your answer should not be slow, it also should not be too quick. A natural, comfortable speed is best. Speaking too quickly creates confusion and unnecessary errors. The real way to say more is to be concise.

Concision is often the difference between a good answer and an excellent one. This is true not only in speech but also in writing. High level answers include a lot of detail: examples, descriptions, reasons, steps, etc. This is a difficult ability to master, however, because it requires a broad vocabulary and excellent grammar skills. Therefore, you should expand your vocabulary and improve your grammar in order to shorten your statements without losing meaning.

Using cohesive devices is a good way to be concise. Pronouns, for instance, avoid unnecessary repetition and also save time because they are often shorter. Also, avoiding unnecessary dependent clauses and using parallel structure are two common ways to shorten your statements.

Original	Parallel	Parallel with Fewer Clauses
A good leader is a person who is confident. Also, a good leader is someone who is educated. Workers are inspired by this kind of leader. For example… (The first two sentences repeat the same basic structure: Subject + verb + noun + adjective clause.)	A good leader is a person who is confident and educated. Workers are… (The two adjectives "confident" and "educated" are parallel.)	A good leader is confident and educated. Workers are…

There is nothing grammatically wrong with the original sentence. However, notice how the parallel structure removes one unnecessary sentence (Also, a good leader is someone…). Always try to use parallel structure instead of repeating the same structure. Moreover, notice how the adjective clause (who is confident and educated) is unnecessary. Describing a leader as "a person who" is unnecessary because that fact can be assumed; leaders are always people. Mentioning the qualities right away is more concise.

There are too many other ways to review here. Consult a good grammar exercise book and review compound nouns, appositives, reduced adjective clauses, and reduced adverb clauses.

Strategy 9: Use idioms in moderation, but not slang.
Don't be afraid to use idioms in any of your answers even if the speakers don't use them. As you learned earlier, an idiom is a fixed, or set, expression whose meaning is different from the literal meaning of the words in the expression. For example, if someone passes a test by a very wide margin or with very high marks, the person passes with flying colors. The verb *pass* and prepositional phrase *with flying colors* form a useful idiom that you could use in your answer.

Using an idiom appropriately shows good language skill, which can raise your mark. However, be careful: using the wrong idiom or using an idiom out of context can lower your mark. Also, don't overuse idioms. In academic discussions and lectures, native speakers use idioms sparingly because idioms don't illustrate or define concepts effectively. Even in friendly conversations, a situation normally requires only one or two idioms, not several.

Finally, an idiom is not slang, which is nonacademic language from popular subcultures and youth movements in music, sports, and urban culture. Slang is inappropriate for the TOEFL.

Strategy 10: Pay attention to your pronunciation.

Pronunciation is a vital part of any spoken language, and it is often the most difficult skill to improve. Unfortunately, some TOEFL students focus more on *what* to say rather than on *how* to say it. However, your answers in the Speaking section are not only marked on the content but also on your speaking ability. You should prepare for the exam by focusing on the pronunciation patterns of English words. Consult a good exercise book and practice pronunciation with a partner, tutor, or class.

The proper pronunciation of vowels and consonants in English depends on manipulating the mouth, lips, tongue, and breath in specific but subtle ways. Just as a dancer or athlete practices the same essential movements over and over, you should practice the actions necessary for various sounds. Identify your weaknesses and improve them before your exam.

Vowel and consonant sounds are put together in words as syllables, and each word is stressed on a particular syllable. English words have particular stress patterns, depending on the part of speech (verb, phrasal verb, noun, compound noun, adjective, etc.), the number of syllables, and the presence of prefixes or suffixes. It is common for students to pronounce new or academic words incorrectly by emphasizing the wrong syllable, so learn the stress patterns of English words. If you cannot stress a word properly, then you also cannot reduce vowels.

Reduced vowels are vowels that are not fully pronounced and sound different than they do when fully pronounced. Not all vowels can be reduced, but reduced vowels are very common in spoken English. They rhyme with the vowels in *up* or *it*. Generally, a stressed vowel is fully pronounced and longer, but unstressed vowels are often reduced according to set patterns. For example, a vowel is reduced when it comes before or after a stress. The verb *be* is only one syllable, so the only vowel is stressed and fully pronounced. The vowel *e* rhymes with the vowels in *see* or *knee*. However, the verb *become* has two syllables with a stress on the second syllable (*be'come*). Because the first syllable (*be-*) is unstressed, it is reduced and rhymes with the vowel in *but* or *bun*.

Depending on their place in a sentence, prepositions, pronouns, and articles are also reduced. Because reduction plays such a big role in English, over-pronunciation is a common problem for students. Reduced vowels are easier to pronounce, so they are also faster to say. Therefore, pronouncing words without reduction slows your speech and makes it difficult to speak with a natural pace and rhythm (see also Improve-Your-Score Strategies).

Strategy 11: Speak at an even pace and with a regular rhythm.

Good English is spoken at the same overall pace, or speed, throughout. Although a speaker may speed up or slow down slightly, the fluctuation is minor. It is common for students to speak too quickly at first due to nervousness, which is why practice is so important. The more familiar you are with the exam, the more comfortable you are. At the same time, don't speak too slowly. Sometimes, students speak too slowly because they want to be clear and understood. This is risky because of the time limits. An answer that is clear but incomplete or too short will still receive a low mark. Therefore, practice speaking at a constant speed that is comfortable for you and easy for a listener to understand.

Rhythm involves the proper stress of words and proper pauses between phrases and clauses. Although each word is stressed on a particular syllable, certain words in a sentence are also stressed more than others. This stress, or emphasis, of syllables and words creates part of the rhythm, or beat, of English speech. While syllables are stressed according to certain patters (part of speech, prefixes, etc.), the words in a sentence are also stressed according to type. Content words provide most of the meaning of a sentence, so they are usually stressed more heavily on the appropriate syllable. Function words fulfill a particular grammatical role, and they are usually unstressed. Look at the following chart:

Function Words (Unstressed)	Content Words (Stressed)
Articles: a, an, the, some	Nouns: person, place, or thing
Prepositions: in, at, on, to, for, etc.	Verbs and participles (not auxiliary verbs)
Pronouns: I, you, he, she, it, etc.	Negatives: no, not, don't, aren't, can't, etc.
Conjunctions: and, but, yet, for, that, since, etc.	Adjectives: big, expensive, thorough, etc.
Auxiliary verbs: be (passive voice), has/have (perfect tense), will, modals, etc.	Adverbs: quickly, finally, too, very, etc.
	Numbers
	Interrogative pronouns: who, what, where, etc.
	Demonstratives: this, that, those, these.

Based on the general stress patterns above, English sentences have a natural rhythm depending on the mixture of content and function words. Look at the following examples (stress is marked '):

> I en'joy 'volleyball because I 'like 'outdoor 'sports and 'group ac'tivities.
>
> The 'woman disa'grees with the 'new 'fees. She 'thinks that they are 'too 'high and un'necessary.

Strategy 12: Group words and pause appropriately.
Most students already know that they must pause for breath at a comma or a period, but this advice is difficult to follow without a transcript. You should understand that English words are spoken in groups, not individually. The groups generally follow the phrase and clause structure of a sentence, and speakers rarely pause in the middle of a phrase or clause. Natural pauses occur between phrases and clauses, and between sentences. This is why speakers pause at a period; it is between one sentence and the next. Common phrases include the following:

Prepositional phrases (prep + object):   in a class, for an hour, by a student, on a field, etc.

Infinitive phrases (inf + object):   to buy a book, to pass the test, to be late, etc.

Gerund phrases (gerund + object):   studying math, skipping class, getting a loan, etc.

Common independent clauses include:

Subject + verb + adverb: The man's partner works too slowly.

Subject + verb + complement: The woman is a dancer…;The class is full…

Subject + verb + object: The man failed the test…;The teacher canceled the class…

Subject + verb + object + complement: Her marks made her happy…He called his partner lazy…

Common dependent clauses include:

Noun clauses: The teacher asked *what he said*; The woman didn't know *why he left*, etc.

Adjective clauses: She wanted a class *that interested her*, etc.

Adverb clauses: *Because he has no money*, the man can't go on the trip, etc.

You should group your words into phrases and clauses, and pause between them:

> The woman wants to be a dancer [pause] and she has to work after school [pause] to pay for her tutorials. [pause] Although she's on a scholarship, [pause] it doesn't pay for everything. [pause] Dancing is very demanding [pause], which is why she needs a tutor. [pause] Unfortunately,…

There are two main exceptions to the clause rule:

First exception: When a noun clause is the object of a verb of argument/belief: subject + main verb + noun clause. In this case, there is rarely a pause after the main verb:

> The man wants to drop the class [pause] because he thinks that it's too hard. (no pause after *thinks*)

Second exception: When an adjective clause identifies a specific noun or idea. In this case there is no pause between the noun and the identifying adjective clause:

> I think the student should take a class that interests her. (no pause after *class*)

Strategy 13: Link and glide smoothly between pauses.

Linking and gliding are two main ways of grouping English words smoothly. Between pauses, speakers link consonants and vowels and glide between vowels. Linking is the combination of the final consonant of one word with the first vowel of the next word. When one word ends with a consonant and the next word begins with a vowel, it is easy to combine the two words together and say them without a pause. Try the following examples (stress marked '):

> He 'needs a 'book on the 'Civil 'War. = He 'needs-a 'book-on-the 'Civil 'War. (no pauses between *needs* and *a* and between *book* and *the*)
>
> The 'woman is a 'dancer. = The 'woman-is-a 'dancer. (no pauses between *woman* and *dancer*)

On the other hand, when a word ends with a vowel and the next word begins with a vowel, the technique is called *gliding*. A glide is an extra consonant added between each word. The glide could be an extra *w* sound as in *wet*, or it could be a *y* sound as in *yes* or *you*. The choice depends on the final vowel of the first word and the shape of your mouth.

If you end a word with a round, circular shape to your mouth, then the *w* glide is easier, but if you end a word with your mouth stretched to the sides, like a smile, then the *y* is easier. Look at the following:

> 'Information is e'ssential to in'vestors. = 'Information-is-e'ssential to-*w*-in'vestors.
>
> 'Time should be im'portant as 'well. = 'Time should be-*y*-im'portant-as 'well.

In the first sentence, a speaker can link the first three words, and then glide between the last two with a *w* sound. In the second sentence, a speaker can glide between *be* and important with a *y* sound, and then link important and as.

Both of these skills are common because they make English easier to say. With practice, they should feel natural and shouldn't be forced or exaggerated (see also Improve-Your-Score Strategies).

GENERAL SPEAKING STRATEGIES: INTEGRATED

There are four integrated tasks in the Speaking section. Because these tasks ask you to read and/or listen before you then speak about the reading and listening, there are some unique strategies to learn for this type of task. The following general strategies cover the basic skills, and you can review specific strategies for each task later in this section.

Strategy 1: Take notes in fragments, symbols, abbreviations, and acronyms.
You must be able to take good notes while you read and/or listen for the last four Speaking questions. Since it is impossible to write down everything and you need to record several key points and details, you must record efficient, clear, and detailed notes. The best way to do this is by using fragments (pieces of incomplete sentences), symbols (drawings that represent ideas), abbreviations (shortened forms of words), initials (the first letter of each part of a name), and acronyms (an artificial word made up of the first letter of each word in a phrase or compound).

Fragments, or incomplete sentences, usually omit articles, unnecessary prepositions, and even subjects, verbs, or objects if they're assumed or implied. In conversations, there are usually only two speakers. If they speak about themselves most of the time, you don't always have to record them as subjects or objects in your notes. You can and should rely on your memory to reduce your note-taking.

Abbreviations are really only necessary for words with more than three letters. An abbreviation can be the first few letters of a word (government = gov), the first and last letters of a word (want = wt) or the word without vowels (believe = blv). In your notes, try not to write abbreviations longer than four letters.

Initials and acronyms are similar since both use the first letters of several words: Isaac Newton's initials are IN and the acronym for the Federal Bureau of Investigation is FBI. Finally, symbols represent words in shapes and pictures: math (+,=,<,>), numbers (to/too = 2; before = b4), and many others are useful (see also General Listening Strategies).

Look at the following examples of notes on part of a conversation:

	Original Passage	Notes
Man:	Professor Miller's biology class moves too quickly. I'm always lost, and I'm not the only one.	Pr M bio = 2 fast, alw lost, +
Woman:	Why don't you ask him to explain more?	Ask expl?
Man:	Well, it is an advanced class, and nobody made me take it. I think I was too ambitious.	Adv cl, not req, 2 amb
Woman:	What about Professor Taylor's class? It's more of introductory type thing, and I think there's room.	Pr T ? = intro + space
Man:	That sounds like my cup of tea. I'll…	Pr T yes

As you learned in the Listening section earlier, you should ignore correct spelling, grammar and punctuation when taking notes. Remember that you are never marked on your notes; they are only clues to remind you and guide you while you speak. In order to be efficient and avoid confusing yourself, develop a personal system and be consistent. Many words are similar and could be confusing in abbreviations, such as example, explain, expand, exam, exact, extract, and extra.

Strategy 2: Paraphrase while taking notes.
Some of the reading and listening passages in the Speaking section are academic, so they have a formal style, which includes more words, a higher level of vocabulary, and more dependent clauses than spoken English. However, you don't have to repeat or mimic that

style. As you've learned above, spoken English can be more casual and relaxed than lectures and formal, written English. In fact, your mark depends on your ability to paraphrase what you read and hear.

Since you must express the same ideas in different words and structures, it is an advantage if you can understand the meaning but record it in your own words in your notes right away. Therefore, don't always copy the speaker's exact words. Look at the following chart:

Original Passage: Intro and Key Points	Notes
Though trains saw widespread use over the course of the 19th century, the 20th century saw nothing but a decline in the use of railroads in the United States and a rapid growth in American car culture. This progressive loss of interest in trains and the growing love affair with the automobile can be explained by convenience, versatility and variety…	Trains use ↑ 19th 20th trains ↓ + cars ↑ conv, vers, var
If one's objective is to conserve energy, neither mode of transportation offers any real advantage…	If save power, both X
On the other hand, the safety of train travel is a very real advantage…There are relatively few deaths on America's tracks while US roads claim an average of over one hundred lives every day…	But, trains = safer Dead / train < cars
It is never necessary for an owner to wait for her car…The reach of American roads is wider and farther than that of US rail road tracks.	0 wait car Rds go ↑ places
Cars can be customized, and with enough resources they can be altered again and again to suit any need or taste…However, trains don't allow riders to make any such statement of personal style.	cars style with $ trains not

Strategy 3: Recognize the basic organization of the passage.
Conversations are organized according to turns. A turn is made up of the statement or question of one speaker and the other speaker's reaction or answer. Be aware of delayed or extended turns, when one turn is incomplete until one or more other turns have been completed.

Lectures are organized according to paragraphs. The first paragraph is the introduction, which may or may not include a hook and/or thesis. It is followed by two or more body paragraphs, each of which begins with a topic sentence and is followed by detail. The final paragraph is the conclusion, which may or may not include summary, comment and/or prediction (see also General Reading Strategies, General Listening Strategies, and Writing Strategies).

Strategy 4: Read and listen for the main idea and key points.

The main idea of a passage is not only the topic but also the message, opinion, or perspective of the speaker. In the Speaking section of the TOEFL, one conversation is about an announcement and you must paraphrase the announcement and the student's opinion of it. The other conversation is about one student's problem and possible solutions. You must paraphrase the opinion and also explain your reasons for preferring one solution.

The lecture is about an academic topic from a field common to North American universities and colleges. You must paraphrase the lecturer's main idea and the supporting points. A common mistake for students is to repeat some details (*names, numbers, dates, examples,* etc.) without explaining how the details connect together. The main idea connects the details.

In the notes above from the conversation, the details (*moves quickly, always lost, advanced, too ambitious*) relate to the student's decision to take an easier biology class. Also, in the notes from the lecture on trains and cars, the details (*energy use, safety of trains, waiting time, roads*) are all connected to the speaker's main idea, which is that cars have replaced trains because of their benefits.

Strategy 5: Connect details to the main idea.

Details support a main idea, and you must recognize, record, and explain that connection. If you hear a student referring to a teacher, why is the teacher mentioned? What has the teacher done or not done? What has happened? Always ask yourself, "Why is s/he saying this?" or "Why is this mentioned?"

It isn't enough to say that the professor mentions that cars are dangerous. You must identify this as a problem, disadvantage, or drawback, and explain that cars have become more popular than trains despite the danger; that the greater danger of cars hasn't hurt their popularity; that people enjoy/use/prefer cars regardless of the danger.

SPEAKING STRATEGIES: TASK TYPES

In the next section, you can review strategies for each Speaking activity, or task, on the actual exam. After you review the strategies for a task, you can practice a sample activity similar to the one you will encounter on the TOEFL.

Speaking Task 1: Independent Description

The first task in the Speaking section of the TOEFL is independent, which means you must speak about a familiar topic without reading or listening to anything beforehand. In this task, there are two types of question. The first type involves a noun of personal importance, for example "Describe a place that is important to you." The second type involves the ideal qualities of a noun, for example "Describe the qualities of a good parent." Although the second type doesn't mention you or your personal experience, both questions can be supported by details from your own knowledge and experience.

After you read the task, you have 15 seconds to generate and organize your ideas, and 45 seconds to speak. Your answer should be expository, which means that you must describe and explain your topic.

Use the following strategies to help prepare you for this task.

Strategy 1: Be prepared to describe a person, place, thing, or an event/activity.
Before your exam, you should practice describing as many types of nouns as possible with the strategies discussed below. Look at the following chart with some, but not all, of the possibilities:

A Person	A Place	A Thing	An Event/Activity
A family member	A room	A class/course	A hobby
A teacher	A house/building	An instrument/tool	A sport
A coach	A park	A skill/ability	A festival/holiday
A friend	A lake/river		A custom
A boss	A mountain		
A neighbor	A city		
	A country (this is broad and tough to describe in 45 sec.)		

Strategy 2: Focus your answer and identify a specific noun.

The prompt or task for the first Speaking task is intentionally general or vague. For example, the prompt uses nouns such as event, person, place, or holiday. The prompt does not specify any qualities or type because you are expected to focus the topic yourself. A common mistake for a lot of students is to omit any identification, such as a name, adjective or adjective clause that identifies how the noun relates to you. For example, look at the following answer to the prompt "Describe a holiday that is important to you:"

> During this holiday, nobody works or goes to school. I try to relax and also visit my family and friends. I always sleep in on a day off. Normally I wake up early for school, so on this holiday I sleep longer. It is common in my country to see your family…

The student incorrectly continues to use the same general noun *holiday* or the synonym *day off*. Instead, the student should name the holiday, specify its meaning, and describe food, clothing, music, activities, and so on.

Strategy 3: Organize your ideas before you speak.

As you learned above in General Speaking Strategies, your answer should be organized as a paragraph. Your first sentence should be your thesis or main idea sentence, which introduces the specific noun, and the two or three traits you will discuss. The rest of your answer should give details about each trait. Therefore, prepare this kind of organization in your notes. In the 15 seconds you have to prepare, quickly choose a noun, traits, and details. Arrange them so that their organization is clear for when you speak.

Strategy 4: Use concrete imagery.

Your goal in the first task is to help the listener imagine what, where, or who you are talking about. The best way to do this is to use concrete imagery. Concrete means tangible, or perceivable. It is the opposite of abstract, which means intangible or not perceivable. A cat or a car is concrete because you can recognize one in real life with one of your five senses (you can touch, hear, and smell a cat, but you wouldn't want to taste one!). Therefore, a cat or car is easy to imagine, but you would need more description than a name.

However, love is abstract because you cannot perceive it on its own. Therefore, it is difficult to imagine alone. Abstract nouns are usually understood as ideas with the mind or intellect. However, because they are intangible, abstractions must be connected with something concrete. For example, if you imagine love, you probably also picture two people who are in love. Also, love is easiest to recognize when there is some concrete evidence, such as hand-holding or a wedding ring. Thinking of an intangible idea requires concrete imagery.

Imagery refers to descriptive detail that helps a listener (or reader) imagine the topic. When you describe a noun, use concrete detail that gives the listener something to imagine. Avoid using too many abstract ideas without concrete support. For example, look at the following description of a friend, Bill:

> My friend Bill is a patient and kind man. This is why I like him so much. His patience makes him a good friend because patient people are comfortable. Also, he is kind. Kind means nice, and I like nice people. Kindness is an important quality.

In the above example, the speaker uses very abstract description. Although there is some detail (*because* is explanation, and *means* is definition), the detail is all very abstract. What does *nice* look like, sound like, smell like? It is difficult to imagine the quality without thinking of something concrete: a person, a place, a thing. Look at the following answer to the same question:

> My friend Bill is a patient and kind man. This is why I like him so much. Whenever we go somewhere together, like the movies, he never worries about being late and never asks me to hurry up or to ignore anything. One time Bill stood calmly by the door while I looked frantically for my car keys. I threw books and cushions in the air, and Bill never complained. His patience makes him a good friend, and also his kindness…

Here, the detail is more concrete because it includes actions and objects that can be imagined, or pictured, easily. Concrete detail is an important part of not only speech but writing as well (see also Writing Strategies).

Strategy 5: Describe more than appearance.
Many students focus on looks, or appearance, when they describe something. However, humans have five senses, so looks aren't everything. Smell, touch, taste, and sound are also useful qualities, but some might be irrelevant depending on the topic. Try to be creative, but don't include unnecessary senses. Make sure any detail adds to your explanation. Look at the following partial list:

Smell:	pungent, dank (wet and stale)
Taste:	sweet, salty, bitter, sour (4 main tastes), fresh, stale/old/rotten, juicy, creamy, hot/spicy
Touch:	dry, wet, smooth, greasy/oily, sticky, sandy, rough
Sound:	soft/faint/distant, lilting, pleasant/pleasing, harmonious/inharmonious, jarring, startling, loud/booming/thunderous
Size:	tiny, miniscule, small, short, average, moderate, tall, big, huge, gargantuan, humungous
Shape:	round/oval/circular, square, rectangular, triangular
Color:	green, red

Too often, students don't know many adjectives for the senses, so they say things like "it tastes good" or "it sounds bad." Try to be more creative and use similes (see Strategy 7 below).

Strategy 6: Use actions and events to make ideas more concrete.

Students sometimes have a hard time describing personal qualities, such as patience, enthusiasm, generosity, or kindness. People and their characteristics are always best explained by their actions. A common mistake is that students just define the quality or use synonyms. Look at the following description of a favorite coach:

> My old basketball coach from high school was very supportive, and that's why I liked her so much. I mean she supported us a lot. We always got a lot of help from her, and this made her a good teacher…

The speaker never gives an example of the support or help and only repeatedly explains the idea of support. The easiest way to give an example would be to describe an actual event or supportive action by the teacher:

> My old basketball coach from high school was very supportive, and that's why I liked her so much. She supported us a lot. For example, she would always comment on our throwing or ball handling. If we missed the net a lot, she demonstrated techniques and worked with each player. When I first started, I was terrible at defense, but she taught me how to block shots and steal the ball. Also, she gave us emotional support as well. My friend's parents died during our last year, and the coach talked and listened to my friend…

Strategy 7: Use metaphors and similes.

Even native speakers run out of specific vocabulary for certain ideas. Especially when you are speaking on an exam, it can be difficult to come up with the exact word quickly, so you must get creative by using a metaphor or simile to expand your description.

A simile is when you compare or contrast two things using *like* or *as*. "My friend is as fast as a cheetah" is a simile that refers to a friend's incredible speed. Also, "my sister is like an angel" is a simile that describes a sister's beauty and innocence. Similes are an excellent way to capture a lot of detail quickly.

A metaphor is a simile without *like* or *as*. "My sister is an angel" is a metaphor that implies the same comparison between the sister and an angel. Metaphors do not have to use the same sentence structure. "I flew down the stairs" is metaphorical because the speaker did not literally fly down the stairs. The speaker uses *flew* metaphorically to mean "moved quickly" or "ran quickly."

Metaphors and similes are excellent ways to be concise and creative. A sentence such as "I was drowning in homework" not only reflects the overwhelming amount of work but also the feeling of desperation. Also, similes are extremely useful for all five senses. If something smells like old garbage, you know it smells very bad. If something smells like flowers blooming in spring, you know it smells very good. (See also Listening Strategies: Rhetorical Structure.)

Strategy 8: Don't be too emotional.
Emotions are useful because how we feel is often a big reason for our likes and dislikes. However, emotions are very abstract, so don't rely on emotions to explain everything. Look at the following description of a person's favorite place, the beach:

> I love the beach because it is so relaxing. When I feel stressed during the day, I always go to the beach after work. It feels calm and comfortable. I go to the beach because I can relax…

The sentences are grammatically correct, and the description includes much emotional detail, but it is too emotional, and therefore too abstract. Compare that answer with the following, which uses more concrete detail and a simile:

> I love the beach because it is so relaxing. When I feel stressed during the day, I always go down to the water, and stretch my blanket on the warm, yellow sand. The sound of the waves hitting the shore relaxes me, but sometimes the smell of the barbeques makes me hungry. Also, I always bring sunscreen so that I don't fry like bacon! When I get too hot, I take a quick dip in the water. It feels refreshing…

Strategy 9: Remember to explain the noun's importance or your choice of qualities.
If the task asks you to describe something that is important to you, you should choose details that explain this importance. A noun can be important for a variety of reasons. Usually, importance involves usefulness/practicality, knowledge/learning, help/support, pleasure/enjoyment, or culture/history. You should mention details that help you explain one or more of these ideas.

A description that doesn't relate to importance is incomplete or irrelevant. Many students focus on physical description and personality when describing a person, but this is not always relevant to the question. Look at the following example of an answer to the prompt "Describe an important person in your life":

> My friend Bill is a very tall man and also very funny. He has to buy his clothes at a special store because most stores don't have his size. Sometimes he even hits his head on the ceiling. Also, he tells a lot jokes. He makes people laugh and always has funny stories. Bill has a great sense of humor, and...

This answer is grammatically correct and has good detail (special store, hitting head, jokes), but it doesn't really explain why Bill is an important friend. His size and humor don't explain his importance to the speaker. Compare the above quote with the following:

> My friend Bill is important to me because of his patience and generosity. First, Bill is always patient. He almost never gets angry and he has saved the day more than once. One time, while we were camping, we had difficulty putting up our tent. Bill never got too frustrated, and he managed to finish just before the sun set. That was a close call! Also, Bill is very generous. For example, he bought the tent we used to go camping, and he never asked me to pay him back. Bill's qualities make him a great friend, and I'll always want him in my life.

In the second answer, the detail (patience, generosity, camping, tent, sunset, etc.) explains the speaker's positive attitude and respect for Bill. The importance is part of the main idea of the answer, so the detail should support it.

Likewise, if the question asks you to describe the ideal qualities of something, you must explain your reasons for choosing those qualities. If the task says, "Describe the qualities of a good parent," you must not only describe a good parent, you must also explain the importance of those qualities.

Strategy 10: Be logical but subjective.
Your description and explanation must make sense, but it can be very subjective. In this task, you must explain a preference, not defend an argument. Therefore, your answer can be subjective, or based on your personal likes, dislikes, and preferences.

Subjective means personal or related to a certain personality, and your answer can be based on what you like, enjoy and prefer. You might describe a favorite hobby by saying that you enjoy volleyball because it is a group activity and a non-contact sport. This is subjective because not everyone likes group activities and non-contact sports. The preference only has to be true for you, the speaker.

Although a preference can be explained objectively, it doesn't have to be. Objective means impersonal or not specific to the speaker or writer. An objective idea is not based on personal likes and tastes; it uses logic that anyone can agree with. However, preferences don't have to be explained this way. If you said that you enjoyed volleyball because the cardiovascular exercise reduces your chances of a heart attack, the answer makes sense but seems overly logical and impersonal. The reason isn't wrong, but seems awkward and unnecessary for a preference.

Question Forms

Independent description is always the first task in the Speaking section. Once the Speaking section begins during the actual test, directions will appear on the screen, telling you to put on your headphones, which have the microphone that you speak into. You hear the instruction in your headphones and also read it on the screen. An electronic beep tells you when to start speaking. You can recognize this type by one of the following forms:

> Describe an important X.
> Describe a/an X that is important to you.
> Describe a/an X and explain why X was important to you.
> Describe the qualities/traits/characteristics of a good X.

Practice

Now, let's practice what you have just learned. So that you can review your answer and monitor your progress, record your answer on a cassette player or computer. As mentioned, during the actual test, you can listen to the narrator's question and also read it on the screen.

 Play CD 2, Track 37 and practice by answering the following:

Choose a person from your childhood. State why that person was significant to you. Give details and examples to support your choice.

Preparation: 15 seconds
Answer: 45 seconds

You can read a sample response in the speaking transcript at the end of this section.

Speaking Task 2: Independent Opinion

The second task in the Speaking section of the TOEFL is also independent, so you don't read or listen to anything before you speak. Yet unlike the first, it requires an opinionated or argumentative response. In your answer, you must support or oppose an argument or plan with logically defended reasons. After you read the task, you have 15 seconds to generate and organize your ideas, and 45 seconds to speak.

Strategy 1: Be familiar with the possible topics.
The second Speaking task may include topics from a range of possible areas. Although there is no set list of topics, the topics must be familiar to most college-age students. You only need to use your own experience to defend your position. Therefore, the possible topics must be limited to a range of common experience.

In order to prepare for this task, practice by defending positions for or against a variety of familiar topics. Your goal is not to predict the topic on the exam, but instead to practice a range of vocabulary. This way, you have a better chance of speaking successfully regardless of the topic on the actual test. Look at the following list of some, but not all, of the possibilities:

Parents/Children	Schools/ Education	Civic Projects	Government
Should parents be strict or not? Should parents raise children in the city or the country? Is television bad for children or not? Should parents force lazy children to exercise?	Should female and male students live together or separately? Is it better for a student to live on campus or off? Should all education be free? Should students study many subjects or just their major?	Can the city build a new community center near your home? Can the city build a highway near your home? The city will reduce bus service but make it free. Do you agree or disagree? Are zoos good or bad?	Is government censorship good or bad? Do government workers need privacy? Should government officials be paid or not?

Strategy 2: Organize your ideas before you speak.
Your answer should be organized as a paragraph. Sentence One should be your thesis, which clearly states your opinion and the two or three reasons that you will discuss. The rest of your answer should explain your reasons in detail. Therefore, prepare this kind of organization in your notes. In the 15 seconds you have to prepare, quickly choose a position, reasons, and details. Arrange them so that their organization is clear for when you speak:

> Thesis: opinion + two reasons
> First reason + detail
> Second reason + detail

Strategy 3: Give a clear, definite opinion.
While you may not always feel strongly about every issue, it is easier to defend a position if you choose only one side. Students often have mixed or uncertain opinions about a topic, and they try to be completely honest and explain exactly what they think. Unfortunately, this approach makes the task more difficult, might take too long, and could confuse the listener.

The purpose of the task is to test your ability to defend an opinion in speech. The purpose is not to learn what your opinion is. As long as you defend your opinion clearly and logically, your opinion itself doesn't matter. In order to make it easier for you to answer and to avoid confusion, decide to either support or oppose and then do so strongly.

Avoid neutral or mixed opinions. Don't say things like "I'm not sure one way or the other" or "Both sides seem good to me." In other words, don't mention or suggest that the opposing view is also correct or good in certain circumstances or in certain ways. This weakens your argument and could confuse the listener. Be clear about which side you choose, and only defend your position. A strong opinion is not only easier to defend but also easier to understand.

Strategy 4: Use recognizable language of opinion.

Opinions don't always have to be introduced as opinions: "The planned highway is a waste of money" is clearly opinionated and argumentative. However, you should demonstrate your vocabulary and grammar skills by using phrases and clauses to identify your opinions. Look at the following list of some possibilities:

> I believe/think/feel/argue that…
>
> It's my strong/firm opinion that…
>
> It's impossible to deny that…
>
> It's clear that..
>
> From my point of view,…
>
> There's no doubt that…
>
> I have no doubt that…
>
> What I think/believe is that…
>
> I'd have to say that…
>
> I agree/disagree that…
>
> I oppose/support…

Don't overuse the expressions above. For example, don't say "It's impossible to deny that I firmly believe…" or "There's no doubt I strongly think that…" These statements are redundant, or unnecessarily repetitive, and sound awkward.

Strategy 5: Organize your reasons with cohesive devices.

> First…/One reason is…/My first reason is…
>
> The best thing about X is…/The worst part of X is…/The most important issue is…
>
> Second…/Another reason is…/My second reason is…/Another advantage/benefit is…
>
> This is because…
>
> That is why I…
>
> Etc.

Strategy 6: Be specific.

Not every word in your answer must be specific, and not every word requires an example. However, your answer should not be completely general or vague, so you should give at least a few examples. A common problem for students is that they can't recognize when their ideas are general and examples are needed. This is primarily a vocabulary problem.

Understanding the specificity of a word is a key part of vocabulary. You must be able to categorize or rate words on a scale of general to specific. One important quality of specificity is the number of possible types of a word. The word *equipment,* for example, can have many types: construction equipment, exercise equipment, medical equipment, camping equipment. So *equipment* is a very general noun that requires an example.

Strategy 7: Defend your opinion logically and objectively.

In this task, you must defend an opinion that is for or against a plan or argument. Unlike a preference, an opinion must be defended objectively. This means that your reasons should not be based on your own tastes or preferences. In order to convince other people that you are right, you must mention more than just what you like or don't like. An argument that is too subjective, or based on personal tastes, is weak. Look at the following thesis:

> The city should build a new movie theater in my area because I enjoy movies and I don't like to travel far.

Although the above thesis is reasonable and makes sense, it is a weak argument because it is based only on what the speaker likes. The fact that the speaker likes movies but not travel is not a bad idea, but it isn't enough to defend the position. There is no mention of the economy, traffic, environment, children, or other citizens. Look at the following improvement:

> The city should build a new movie theater in my area because everyone needs more activities and children will be safer if they don't have to travel far.

The second thesis is more objective because it focuses on the benefits to other people, not just the speaker, and this strengthens the argument.

Question Forms

The second speaking task asks for your personal opinion about a topic of general interest and experience. At this point during the actual test, you should still be wearing your headphones. As you did for Task 1, you will hear the narrator's question as well as read it on the screen. The question can have one of the following forms:

> Do you agree or disagree that [opinion]?
>
> Do you agree or disagree with the following: [opinion]
>
> Some people think [opinion X]. Others believe [opinion Y]. Which opinion do you agree with/Which side do you support/What do you think?
>
> [Opinion] Do you agree or disagree?
>
> X plans/wants/intends/hopes to [proposed action]…Do you agree or disagree with [proposed action]?

Practice

Now, practice the strategies you've just learned. So that you can review your answer and monitor your progress, record your answer on a cassette player or computer.

 Play CD 2, Track 38 and practice by answering the following sample:

> In many countries, the government practices censorship of television programs. Do you support this practice or not? Give specific reasons and examples to support your answer.

Preparation: 15 seconds
Answer: 45 seconds

You can read a sample response in the speaking transcript at the end of this section.

Speaking Task 3: Reading and Conversation Integrated

In the third speaking task, you must first read a school notice, listen to a conversation between two students about the notice, and then summarize the notice and one student's reaction to it. You have 45 seconds to read the notice and take notes. After you listen to the conversation, you have 30 seconds to organize your answer and then 60 seconds to speak.

Use the following strategies to help prepare you for this task.

Strategy 1: Be familiar with possible topics and related vocabulary.
Although it is impossible to predict exactly what the topic of the notice will be, it is possible to anticipate a range of possible topics. The notice will not discuss or explain topics or ideas from a lecture or textbook. The notice will be campus-related, and it will mention news that a college or university would likely announce to students. The news may affect a particular group of students, several groups, or the entire school.

To prepare for the test, you can think about what a college or university might start, cancel, or change during a semester. Some, but not all, of the possibilities are listed below:

> Curriculum: add/drop a course, new textbooks, etc.
> Mid-term/final exams
> Dormitories: gender mixing/segregation, visitors, curfew, etc.
> Activities: field trips, galleries, movies, concerts, student associations/committees, etc.
> Visiting professors/guest speakers
> School services: cafeteria, councilors/advisors, gym, lounge, media, computer lab, etc.
> Majors/minors: deadline for announcing, course requirements, etc.
> Graduation requirements: minimum credits, types of credits, maximum time limit, etc.
> Security: guards, lighting, phone installation, volunteers to accompany people at night, required identification (pictures, biometrics), visitor restrictions, etc.

All of the above may also be changed according to schedule, location, cost, eligibility, or capacity.

Strategy 2: Take notes on just the key points of the notice.
Read the entire notice carefully, but you don't have to mention every detail from the notice. The key points are the change, reasons, and the people most affected by the change. This last detail is important for the conversation, which involves at least one person who is affected by the announcement. Record this information in note form on the paper provided for you.

The notice includes some extra or minor detail that makes the notice seem more authentic. This can include contact information, such as names, phone numbers, room numbers, as well as office hours, times, and so on. You don't automatically need to repeat this specific information in your spoken summary.

You should determine which extra detail is relevant based on the conversation. As a rule, a detail is important if the speakers discuss it in the conversation. If a speaker never mentions a number, name or schedule as a benefit or drawback, then you can omit that detail in your summary. Look at the following example:

> Starting next Monday, only drivers with passengers can park for free in the school's parking lot. Drivers without any passengers will now be charged $3.00 per day upon arrival. Students will be employed to collect fees and enforce the new rule. The Student Council wants to encourage car pooling and the use of public transit as a way to control over-crowding and reduce car emissions. As always, parking is available off-campus on various side streets, but be aware that parking meters are in effect between 6 A.M. – 8 P.M.

A student's notes on the notice above might look like this:

> 0 pass = $3 /day
> 1+ pass = free
> Students collect $
> For car pool, bus, train, less pollution

References to the Student Council, off-campus parking, and parking meters may become relevant if a speaker mentions them in the conversation.

Strategy 3: Put notes on each speaker in separate columns.
After you read the notice for 45 seconds, an instruction will appear on the screen, telling you to listen to a conversation with your headphones. The two speakers in the conversation discuss the announcement, and you should make separate columns for each speaker in order to distinguish each person's statements. Usually one speaker is male and the other female, but it is possible that both speakers are either male or female.

Strategy 4: Focus on the speaker with the pronounced or clear opinion.
One speaker will have a stronger opinion than the other. You should pay special attention to

that speaker's attitude, opinion, and his or her reasons. Usually one speaker expresses at least two main reasons for supporting or opposing the change.

You are not expected to summarize the entire conversation or everything that both speakers say. The question after the conversation will ask about one speaker, either the man or woman. Look at the following conversation about the announcement concerning parking fees:

Man:	Hi Julie. Can you believe the new parking rules?
Woman:	Hey, Mark. Yeah, I just read the notice in the school paper. It doesn't affect me too much since I live on campus, but it sounds like a good idea.
Man:	Well, I'm not so lucky, and I think the Student Council is being too heavy-handed. It's essentially forcing us to be more responsible.
Woman:	But, if it cleans the air and reduces traffic, it's worth it. I mean, the cars are packed like sardines in there.
Man:	Sure, I'll admit that the crowding has gotten pretty bad lately, but not much will change. Only students on limited budgets will be affected. We rely on free parking at school to avoid the prices downtown. I can barely make ends meet, but a lot of students have plenty of extra cash. I bet you'll still see a full parking lot tomorrow.
Woman:	You're not being forced to pay the fee. You can pick up some passengers along the way. Remember, it's free if you have passengers.
Man:	But that should be my choice. I shouldn't be forced to drive other people to school. Besides, nobody I know lives near me.
Woman:	Well, I guess it's time to make some new friends!

In this conversation, the man's opinion is the stronger one. You should focus on him while listening, but always take notes on both speakers. If you have enough time, you can mention something about the woman. The notes for the conversation above might look like this:

<u>Man</u>	<u>Woman</u>
not happy, extr, force resp	not driver, yes
crowd, > change, poor/ not rich, still full	↓ pollu + traff = good, cars = tight
wants choice	no force, pick up = 0$
none close	find pass

Strategy 5: Be prepared to infer meaning.
To *infer* means to understand unstated ideas that are implied by a speaker. When a speaker implies something, he is communicating information without mentioning it directly. Pay attention to any implications or unstated ideas. As always, use the context and surrounding statements as clues because any inference must be based on something that is said.

In the above conversation, there are several possible inferences. When the woman says that she isn't affected by the new rules because she lives on campus, you can infer that she probably doesn't own a car and doesn't drive. Immediately after that, the man says that he isn't so lucky, which implies that he probably drives. In order to properly understand his reference to luck, you must relate the man's statement to the woman's previous comment about being unaffected. Of course, the man's complete explanation of his opinion also explains why he feels unlucky.

Later in the conversation, the man complains about the fact that only poor students will be affected, and then uses the pronoun *we*. Because the choice of pronoun includes the man himself, he implies that he also has limited financial resources.

Strategy 6: Pay attention to rhetorical function.
Inference is also an important part of recognizing the rhetorical function of a statement. From the Listening strategies, you've already learned that rhetorical function refers to a speaker's reaction (comment, acceptance, rejection, etc.), the speaker's attempt to influence the listener (instruction, correction, encouragement, etc.), or the speaker's intentions or plans. A rhetorical question is a common method for performing any of those actions. For example, the man's first statement ("Can you believe the new parking rules?") is a rhetorical question that implies the man's negative attitude toward the news and the man's surprise or shock about the news.

Strategy 7: Be aware of sarcasm.
Although there is no sarcasm in the example above, it is possible in any conversation on the test, so you should be prepared for it. As you've already learned in Listening, sarcasm is when a speaker means the opposite of what he or she literally says. Sarcasm is a rhetorical device that a speaker may use to express an attitude, to show acceptance or refusal as well as imply intention. Speakers on the TOEFL are only momentarily sarcastic, so the sarcasm is limited to one or two statements only.

You have to infer a person's sarcastic meaning since speakers never explicitly say that they are being sarcastic. Also, you must use context in order to recognize sarcasm because the vocabulary contrasts with the nature of the speaker's situation and other statements. For example, imagine that a man already has a lot of homework, and then he learns that one big assignment is actually due sooner than he had thought. When he hears this bad news, the man says, "Well that's fantastic!" Although this adjective sounds positive, it is sarcastic because it contrasts with the man's situation.

Sarcasm can also be used to comment on positive as well as negative contexts. For example, imagine that a student has multiple scholarship offers, but she can't decide which one to choose. This is a positive situation for the woman because she is has plenty of financial support for her studies. However, a friend of hers might comment by saying, "Well, what a terrible dilemma!" Although *terrible* normally means bad or unfortunate, the friend's comment is sarcastic since it contrasts with the woman's positive situation.

Strategy 8: Identify the key speaker in the prompt.

After the conversation, the narrator will ask a question about one of the speakers. For example, based on the following prompt, you should focus on the man:

> The man expresses his opinion about the new rules. Explain his opinion and the reasons he gives for holding it.

Strategy 9: Organize your ideas into a paragraph.

After you hear the prompt, you will have 15 seconds to organize your thoughts. Based on the prompt, quickly identify the relevant opinions in your notes. Plan to speak about the conversation in one paragraph.

Begin by briefly summarizing the notice and the announced changes in one or two sentences. Then give the speaker's attitude/opinion and reasons with some detail. Finally, mention the second speaker only if you have time at the end.

Strategy 10: Focus on the opinion, not the notice.

The focus of your answer is one student's opinion, not the notice. The notice is context; it is the reason or cause of the conversation. The notice is certainly important, but it isn't the focus. It is only part of the topic.

Therefore, don't spend too much time on the notice. Many students make the mistake of trying to summarize the notice completely first, and then discussing the student's opinion. This is wrong because there isn't enough time. Remember that you only need to mention detail that the speakers discuss; you don't need to mention everything from the notice.

Paraphrase the key points from the notice in your first one or two sentences only. The rest of your answer should discuss the conversation.

Strategy 11: Summarize the student's opinion in your own words.

A summary is a restatement of the key points in different words. Therefore, you shouldn't try to paraphrase the conversation as it was spoken: "First, the man said…Then, the woman said…The man answered that…" This isn't a summary because a summary focuses on the key points only.

Also, you are asked about only one speaker, so that speaker's reaction to the notice should be your main topic, not everything that was said. Organize your answer in a paragraph as explained in Strategy 9.

Strategy 12: Use reported speech, not quotes.

You are expected to paraphrase the ideas in the conversation, not repeat them. Therefore, you must use reported speech in your answer. Reported speech is speech that has been paraphrased by someone who didn't say it. Because you must refer to someone else's statements, you must use reported speech effectively to perform well on this task.

When you report the speech of another person, you must change the pronouns, verb tense, modals, and question order. Reported questions follow the same rules as those for statements, but they also require certain unique changes. Review the following charts:

Reported Statements

Rules	Quote	Reported Speech
Change the present tense to the past tense. Change the past tense to the past perfect. Change the present perfect to the past perfect. Change pronouns accordingly. Change "can" to "could," and "will" to "would." Change "this/that/these" to "the/those."	1. Juan mentioned, "My class is full." 2. Paul said, "I studied at Harvard." 3. Susan argued, "I can do it myself." 4. Some students stated, "We won't go on this trip." 5. Someone pointed out, "Nobody has done this before."	1. Juan mentioned that his class was full. 2. Paul said that he had studied at Harvard. 3. Susan argued that she could do it herself. 4. Some students stated that they would not go on the trip." 5. Someone pointed out that nobody had done that before.

Reported Questions

Rules (in addition to those for statements)	Quote	Reported Speech
Remove "do/does/did " Change question order to regular subject-verb order. Change yes/no questions to "if/whether" noun clause. Change "w-questions" to regular subject-verb order, and keep the pronoun. If the pronoun is the subject of a question, do not change the word order.	1. Jen asked, "Did I pass?" 2. A teacher asked me, "Do you study here?" 3. Billy Bob wondered, "Where did my wife go?" 4. The teacher asked, "Who wrote this essay?" 5. The director asked, "What is happening here?"	1. Jen asked if she had passed. 2. A teacher asked me whether I studied there. 3. Billy Bob wondered where his wife had gone. 4. The teacher asked who had written the essay. 5. The director asked what was happening there.

Strategy 13: Use a variety of language in your summary.
Although reported speech is important, it isn't the only structure necessary for your answer. Therefore, don't just mention a sequence of reported statements and questions: "First, the

man said X…Then the man said Y…" As you've already learned, you should vary your vocabulary and structure. Besides reported speech, your answer requires language of attitude, opinion/argument, and explanation:

> The notice/announcement states that…
> According to the notice/announcement…
>
> The man sounded + adjective + about the news/change…
> The woman had a negative/positive attitude about the notice…
>
> First, the man argued/felt/believed that…
> The man's first reason…
> This is because…
> Etc.

A summary of the notice and conversation above could sound like the following:

> According to the notice, the school will begin new parking rules to limit pollution and the number of cars. Unaccompanied drivers will have to pay $3 to park at the school, but drivers with passengers won't have to pay anything, starting the following Monday. In the conversation, the man opposed the new fee and disagreed with the woman, who was supportive but unaffected by the new rule. First, the man said that they were being forced by the school to act responsibly. Second, the man agreed that the parking lot was too crowded, but he insisted that the fee wouldn't change that. According to the man, only poor students like him would be affected because wealthy students could just pay the fee. Finally, the man complained that he didn't have anyone to drive to school, so he couldn't avoid the charge, but the woman suggested that he try to meet new people.

Strategy 14: Don't give your personal opinion.
For Task 3, you are only required to summarize a student's response to the notice. You should not comment on the student's opinion, state any preferences, or make any suggestions.

Question Form

Since this is an integrated task, there are three stages and three sets of instructions. At this point during the actual test, you should still be wearing your headphones. First, you must read the announcement, and a narrator will give you instructions that you can hear but not read:

> Now, you will read an announcement/notice from a school…You have 45 seconds to read the notice… Please begin now.

After you read the notice, you must listen to a conversation. Again, the narrator will instruct you:

> Now, listen to a conversation between two students while they discuss…

Finally, after you listen to the conversation, you can both hear and read the final prompt, indicating the topic of your answer and the time. As always, a beep tells you when to start speaking:

> The man states/gives/presents his opinion about the announcement/ notice/news/ change…. State his opinion and explain his reasons…

Practice

Now, practice the strategies you've just learned. So that you can review your answer and monitor your progress, record your answer on a cassette player or computer.

 Play CD 2, Track 39

Practice by answering the following (remember, you can only hear the narrator's first three statements; you can hear and read the last):

Narrator: For this task you will read a short passage about a campus situation and then listen to a talk on the same topic. You will then answer a question using information from both the reading passage and the talk. After you hear the question, you will have 30 seconds to prepare your response and 60 seconds to speak.

Narrator: The University newspaper is looking for students to join its staff. Read the advertisement from the faculty advisor. You will have 45 seconds to read the advertisement. Begin reading now.

Reading: 45 seconds

The Campus Voice

The university newspaper, The Campus Voice, invites applications from students interested in joining its staff. The newspaper is a weekly publication written by students for students.

Although previous experience in journalism is helpful, it is not required. Openings are available for students interested in writing, photography, illustration, and advertising. Students will earn one academic credit for each semester they work on the paper. Please direct any questions to the faculty advisor, John Moore.

In order to work on the paper, students must maintain a minimum grade-point average of 3.0.

An application is available online at www.campusvoice.com.

Narrator: Listen to two students as they discuss the announcement.

Narrator: The woman expresses her opinion of the announcement. State her opinion and explain the reasons she gives for holding that opinion.

Prepare: 30 seconds
Answer: 60 seconds

You can read a sample response in the speaking transcript at the end of this section.

Speaking Task 4: Academic Reading and Lecture Integrated

In the fourth task, you must first read a relatively short academic passage on a common topic in North American colleges and universities and then listen to a relatively brief academic lecture related to the reading. Both passages have the same basic main topic but different main ideas. In other words, they both discuss the same thing but they do it in a different way. Also. both are relatively short.

In your answer, you must summarize both passages and synthesize, or combine, information from both. The reading explains an abstract concept, and the listening illustrates and expands on that concept with concrete details. In your summary of both passages, you must mention key points from both passages and discuss how the examples in the listening relate to the concept(s) in the reading.

You are given 45 seconds to read the first passage and take notes. After you listen to the lecture, you have 30 seconds to organize your notes from both passages and 60 seconds to speak.

Use the following strategies to help prepare you for this task.

Strategy 1: Focus on the main concept(s) and any definition in the reading passage.
The purpose of the reading is to give you a general understanding of a particular concept or theory. You must identify it, understand it, and record it in your notes. One way to do that is to identify how the idea is defined. Language of definition includes the following:

> X is…
> X means…
> X refers to…
> X involves/includes/relates to
>
> X is defined as…
> X is called…
> X is understood as…
> X is thought of as…
> X is referred to as…
>
> [Person/people] define/think of/understand/refer to X as…

The reading passage may involve multiple ideas or facts, and it may involve multiple definitions. Look at the following example about business and culture:

> With the onset of globalization, cultural sensitivity has become an increasingly important part of doing business overseas. In international business, cultural sensitivity is more than just cultural awareness or knowledge; it also refers to an ability to behave accordingly. Being culturally sensitive is like having good manners with people from another culture. It is crucial for business people to familiarize themselves with certain aspects of the local culture, such as greetings, social customs, and body language. In business, one must not only avoid offending clients and customers but also attract new ones. Therefore, a lack of cultural sensitivity is a major barrier to success in the new global economy.

In the above example, the verb *refers* to is a clue to the definition of cultural sensitivity.

Strategy 2: Pay attention to comparison and contrast.
Comparison and contrast are common ways of explaining a complex or confusing idea. Sometimes, a lecturer compares an idea to a similar one in order to use the similarity as a way to help listeners understand it, or a lecturer may contrast the idea with something different and explain the idea by showing what it is not like. Similes and metaphors are commonly used:

X is like/unlike Y…
X, like/unlike Y, is…
Like/Unlike Y, X is…

X and Y are alike in that…
X and Y are not alike in that…

X is similar to Y…
Similar to Y, X is…
X + comparative + than Y…

In the passage from Strategy 1, the concept of cultural sensitivity is compared to good manners, and the clue word is "like."

Finally, be aware that contrast is occasionally also used to reject or correct a commonly held misconception. Well-known ideas or phenomena are often misunderstood or only partially understood, so a writer might anticipate this misunderstanding and correct it.

Strategy 3: Note any details unique to the concept.
The idea or concept explained in the reading passage might involve a particular type of person, group, institution, or it could take place at a particular place or time, and so on. This could be important information for the short lecture that follows the reading because you will have to recognize that detail in the lecture.

For example, the passage above connects cultural sensitivity to successful business practices. Also, the passage mentions certain general ideas such as greetings and body language. All of these are clues to what you will hear in the lecture, which will expand on the general discussion in more concrete terms.

Strategy 4: Take notes on the reading.
You won't have time to reread the passage after you listen to the lecture. After the lecture, you need to organize your notes from both passages and prepare to speak. Therefore, you should write notes on the reading, and speak from those notes.

Don't plan on using the full text when you give your final summary. Some students make the mistake of reading from the passage while they summarize, but this is a mistake. You might quote too much from the reading, and you get no marks for quoting the writer's words. Instead, you must paraphrase the reading passage. Taking notes is the best way to avoid quoting by mistake. Notes on the above reading passage might look like the following:

> Intern bus need cult sens
>
> Cult sens = good manners
>
> Greetings, social, body lang
>
> no offense + new clients = ↑$
>
> no sens – X

Strategy 5: Listen for related vocabulary in the lecture.
The lecturer comments and expands on the ideas that you read about, so you can use the key words from the reading as clues to anticipate what you will hear in the lecture. However, don't rely on repeated vocabulary. Some words might be repeated, but many will be examples, illustrations, explanations, or descriptions of those in the reading. So you must recognize words that are more specific, that illustrate, and that give reasons.

Look at the following transcript of a lecture about the reading passage from Strategy 1 (You will only hear the lecture, and you will not see a transcript during the exam):

> When it comes to doing business in India, it is important to bear in mind that the business environment tends to be much more time flexible than it is here in the US. Punctuality, especially in government offices, is simply not a very high priority. You may find yourself waiting for hours to meet someone even though you had fixed an appointment days in advance. To avoid frustration, it's a good idea to call ahead and make sure the person you need to see is actually there.
>
> Greetings can be source of embarrassment, so be aware that Indian women prefer not to shake hands with men. By the same token, western women should not initiate a handshake with a man. Between men, on the other hand, a handshake is fine. The traditional greeting, which consists of hands pressed together, like this, is always acceptable.
>
> Talking of hands, avoid passing or receiving anything with your left hand. It is not polite. Also do not use a finger to point. Use your chin or your whole hand, held open.

If your associate is older than you are, make sure you find out the proper form of address. Indian elders are never addressed by their first name. If you happen to be older, do not be surprised if your associates are more comfortable addressing you with a title; don't insist on being called your first name, they may not be comfortable with such casual interaction.

Some vocabulary here is repeated, such as *greetings*, while other words are synonyms, such as *associates*, which is a synonym for client. However, most vocabulary is related to that in the reading by the main topic, culture and business, and the main idea: doing business in a foreign country requires cultural sensitivity. Moreover, note how the lecture is more specific and concrete than the reading. The reading defines and explains cultural sensitivity while the lecture discusses doing business in India, which is a concrete example of a situation that would require it. Also, the lecture includes specific ideas, such as punctuality, shaking hands, pointing, titles, and first names.

Strategy 6: Use your knowledge of lecture organization to take good notes.
As you do for longer Listening lectures, listen for the main idea by focusing on the opening. Try to recognize a thesis, but remember that lectures don't always have one, especially short lectures, such as the one in Task 4.

Strategy 7: Listen actively for the main idea, purpose, and key points.
While you listen, think about what the speaker says about the topic. In the example above, the speaker says that Indian customs affect how people do business. Related to that, think about what the speaker tries to accomplish in the lecture. In the lecture above, the speaker instructs the listener about the cultural practices in India and warns the listener about possible errors. The purpose is didactic and cautionary, not argumentative or persuasive. Finally, the key points are the specific customs related to punctuality, shaking hands, etc.

Strategy 8: Listen actively for a connection to the reading.
Don't just record names, dates, places, and times. Think about how the ideas in the lecture relate to those in the reading. Your answer must show some understanding. It cannot simply be a list of points.

Strategy 9: Write notes in fragments.
As always, you cannot write a transcript of the lecture, so do not try to write complete sentences as you listen. Focus on key words (see also General Listening Strategies). Notes on the lecture above might look like the following:

India

Time ↓ imp than US

may wait, call

women X shake / men do

greet ( )

no left pass

no finger →, open hand

no first name < titles

Strategy 10: Organize your response as a paragraph.
After you listen to the lecture, a narrator will ask you to synthesize, or combine, information from both passages. Perhaps you would be asked the following question about the reading and listening passages above:

> The professor describes some social differences between India and America. Explain how these differences relate to doing business.

You have 30 seconds to organize your notes. Begin by briefly paraphrasing the definition or explanation from the reading in order to show your understanding of the basic concept. Then, relate the key points of the lecture to the main idea of the reading.

Strategy 11: Paraphrase both passages; don't quote or copy.
Although not all language can easily be paraphrased, you must paraphrase most of the ideas from both passages. Some words, such as business, are hard to paraphrase. However, you should use synonyms and change the parts of speech and structure in your summary. For example, use reported speech, not quotes, when you discuss the lecture (see also Integrated Task 3: Strategy 9).

Strategy 12: Vary your expressions.
Don't repeat the same sentence structure too often. Avoid saying, "The writer says that…, and the speaker says that…Moreover, the speaker says that…" Grammar is not the only element of good language. Variety is also a part of good speech (and writing), so alternate your sentence structure. Some possibilities are:

> The writer mentions/states/defines/explains…
> The speaker mentions/states/explains/illustrates/demonstrates…
>
> The speaker/writer refers to + noun…
> The speaker/writer refers to the fact that + subject + verb…
>
> According to the writer,…/According to the lecturer,…
> According to the reading,…/According to the lecture,…
> Based on the reading,… /Based on the lecture,…

Strategy 13: Use the present tense for facts.
Reported speech is useful, but you shouldn't always change the present tense to the past tense. When an idea is a fact, it can remain in the present tense. For example, both the reading and the lecture discuss ideas related to business and culture. These facts remain true in the present, so you don't have to change the present tense to the past tense. Look at the following key point from the lecture:

Punctuality, especially in government offices, is simply not a very high priority.

As long as this fact remains true, you can paraphrase it in the present tense:

The speaker mentions that punctuality is not very important in India, even for government employees.

The exception to this is history. Historical facts occurred in the past, so their tenses must remain in the past.

Look at the following sample answer for the reading and listening above:

According to the reading, you are culturally sensitive when you behave well in another culture, and cultural sensitivity is very important when you do business internationally. The writer states that business people must meet new people and be polite with everyone if they want to be successful. The speaker discusses India, and explains specific customs that could be a problem for an American business person. In India, for example, it is rude to point with your finger, and instead you should point with your hand. In America, people use first names a lot, but in India you should use titles for older people. Also, Americans shouldn't get upset about some differences, such as the importance of schedules. In America, people are always in a hurry and mustn't be late, but Indians are more relaxed about being on time. Therefore, based on the information in both passages, a business person should be sensitive to these customs in order to be successful in India.

Question Form

Since Task 4 is integrated, there are two sets of instructions. At this point of the actual exam, you should still have your headphones on. First, the narrator instructs you to read the academic passage. On the screen, you see only the reading passage and hear the prompt:

Now, read the passage about/on…Start reading now/immediately.

After you read the passage, you hear the narrator instruct you to listen to a lecture, and the instruction also appears on the screen:

Now, listen to a professor give a lecture about/on…
Now, listen to part of a lecture about/on…

After you listen to the lecture, you can hear and read the narrator's final prompt for your spoken answer. As always, a beep indicates when you should begin speaking:

The professor describes/discusses/explains [lecture topic]… Show how [lecture topic] relates to/expands on [reading topic].

Practice

Now, practice the strategies you've just learned by summarizing a reading passage and lecture. So that you can review your answer, record your answer on a cassette player or computer.

Play CD 2, Track 40 and practice the following:

Narrator:	For this task you will read a short passage on an academic subject and then listen to a talk on the same topic. You will then answer a question using information from both the reading passage and the talk. After you hear the question, you will have 30 seconds to prepare your response and 60 seconds to speak.
Narrator:	Now read the passage about Impressionism. You have 45 seconds to read the passage. Begin reading now.

Reading: 45 seconds

Impressionism

Impressionism was a major movement in painting that developed in France during the latter half of the 19th century. Impressionism represented a radical break from the realism-inspired painting that had dominated Europe since the Renaissance. Impressionist painting was characterized by the use of short brushstroke techniques and bold colors to capture the effects of light on various subjects through different times of the day. Impressionists often placed strokes of bright, unmixed color next to each other on the canvas, letting the viewer's eye blend the colors, in order to capture the shimmering effects of light. This blending of color and form by the viewer's eyes was the effect, or impression, that the painters wanted to create.

Narrator:	Now you will listen to part of a lecture on the topic you just read about. Take notes while you listen.
Narrator:	Summarize the points made in the lecture you just heard, explaining how the lecture supports ideas mentioned in the reading.

Prepare: 30 seconds

Answer: 60 seconds

You can read a sample response in the speaking transcript at the end of this section.

Speaking Task 5: Conversation and Opinion Integrated

The fifth task in the Speaking section is also an integrated task. You must first listen to a conversation between two students about one student's campus-related problem or difficulty. In your spoken answer, you must then summarize the student's problem, options and final decision as well as give your own opinion of the student's choices.

After you listen to the conversation, you have 20 seconds to organize your notes and 60 seconds to speak.

Strategy 1: Be familiar with common problems for college and university students.
Before the exam, review the vocabulary related to common frustrations for college and university students. If you can't discuss these problems on your own, then you won't be able to discuss them on the actual exam. The following areas are some, but not all, of the possibilities (see also Listening Strategies: Main Idea Questions):

Workload: working/studying, too much homework and/or extracurricular activities, etc.

Schedule: two classes at the same time, class starts too early/finishes too late, etc.

Costs: increase in fees, rent, tuition, and/or a drop in salary, grants, scholarship, etc.

Relationships: pleasing parents, a conflict with a roommate, study partner, or friend, etc.

Travel: commuting, vacation, studying abroad, packing, method (car, train, plane), etc.

Look at the transcript of a conversation (remember that you will only hear the conversation on the actual exam, not read it):

Student 1 (female):	Hi Tony, what's up?
Student 2 (male):	Not much. What's going on with you?
Student 1:	Not much. How's your media studies project going?
Student 2:	Great, we are having so much fun in our group. Our presentation is going to be hilarious.
Student 1:	You're lucky. My group is having a terrible time.
Student 2:	Why is that? I thought all the assignments were interesting. What's the problem?
Student 1:	Well, the assignment is fine. It's just that we're not working well as a group at all. In fact, Frank never even shows up when we schedule a planning session. Out of the four meetings that we've had, I think I've seen him once. And even then he was 10 minutes late.
Student 2:	Well, that's no good. Doesn't he realize that this presentation is going to count toward 60% of his grade? You should remind him of that.
Student 1:	He knows, but he doesn't care. He's here on a football scholarship so he doesn't need good grades. He just needs to scrape by. That's why he picked a "soft subject" like media studies. But the problem is that if he doesn't pull his weight, our presentation will be lousy and the rest of us will also get lower grades.
Student 2:	You need to let him know he can't do that. Just tell him: "Frank, you are being a selfish little brat if you don't participate."
Student 1:	That's easy enough for you to say. This guy plays line backer! He's bigger than my refrigerator.

Student 2:	Hmm. Well, if you're nervous about confronting him then you need to let Professor King know that there's a problem. I know he told us that part of the challenge was to work well as a group, but this situation is unusual. Most students care about their grades, I don't know how you can persuade Frank to participate if he doesn't care about his grade.
Student 1:	Yeah. That's what I am beginning to think too. I hate to be a snitch, but I don't know what else to do.

One student has a problem with a study partner, who isn't participating and contributing to a group project.

Strategy 2: Be familiar with the organization of a conversation.

As you should already have learned in Listening, a conversation is organized according to turns. Each turn begins with one speaker's statement or question and ends with the other speaker's response or answer. Therefore, a statement or question is an automatic clue to what follows, since they must be related. In the above conversation, the man asks several questions (How's your media studies group going? Why is that?), which are clues to the next statements (description, explanation).

You must be able to follow delayed turns as well, when a response or answer comes after one or more other turns are completed first (see also General Listening Strategies: Conversations).

Strategy 3: Record notes for each speaker in a separate column.

Strategy 4: Write notes in fragments.

Notes for the conversation above could look like the following:

Man	Woman
happy, funny proj	bad time
all good?	proj good, Frank bad
know = 60%?	¼ meet, late
say you're lazy!	not care, football
see prof, not usual, most care	too big, scary
	no choice

Strategy 5: Focus on the speaker with the problem.

Although the conversation involves two speakers, you must focus on only one of them in your answer: one student's problem, options and decision. The other student is important only for suggestions and comments. Therefore, identify which student has a difficulty right away, and record the problem.

Although it is possible, it is unlikely that the speaker will explicitly say, "My problem is X" since this would be too obvious. Instead, listen closely to the opening statements because the problem, which is essentially the main topic, will be mentioned early.

Also, listen for negative vocabulary and clues about attitude to recognize the problem. In the conversation above, the woman uses negative vocabulary when discussing Frank: a terrible time, never even shows up, doesn't care, etc. Clue words may also be the following:

> My problem/difficulty/frustration is…
>
> I'm having difficulty/a problem with…
>
> I'm annoyed by/frustrated with/angry about/upset about…
>
> I can't/won't/am unable to…
>
> I don't know…
>
> I was hoping to/planning on…but…
>
> X is stopping me from/holding me back from/keeping me from…
>
> If it weren't for X, I could/would…

Strategy 6: Record the options and any reactions.

Usually, the student will have two choices. One or both might come from the other student. Neither student will be explicit or obvious about the suggestions. You almost certainly won't here statements like "You're first option is…and your second choice is…"

Instead, listen for language of opinion (argue, suggest, think, feel, believe, etc.), preference (want, prefer, choose, like, etc.) and suggestion. For instance, the man uses similar language in the conversation: You should remind him…You need to let him know…Just tell him… you need to let Professor King know…

Other vocabulary includes the following:

> Why don't you…?
>
> How about…?
>
> What do you think about…?
>
> Have you thought about…?
>
> Are you aware that…?
>
> Do you realize/remember that…?

> Think about/Consider X.
>
> Remember/Don't forget…
>
> It would be…if you…
>
> Maybe you could…
>
> You should/ought to/must…
>
> X is/would be…

> I recommend/suggest/advise that…
>
> My recommendation/suggestion/advice is…

Strategy 7: Be prepared to infer.
A speaker's attitude, opinion or intentions may not always be stated explicitly, so you must be able to infer meaning (see also Reading Strategies: Inference Questions, Listening Strategies: Inference Questions, and Speaking Strategies: Integrated Task 3).

Strategy 8: Pay attention to rhetorical function.
A speaker's statement may have a rhetorical purpose such as comment, acceptance/refusal, or suggestion, among many others. Remember to use inference and context, and always ask yourself why a speaker says something (see also Reading Strategies: Rhetorical Function Questions, Listening Strategies: Rhetorical Function Questions).

Strategy 9: Be aware of sarcasm.
Speakers may be sarcastic occasionally, so always relate any vocabulary to the context of the conversation (see also Speaking Strategies: Integrated Task 3).

Strategy 10: Record any final decision(s).
The student with the problem chooses one option and gives one or more reasons for the choice. Listen for why one choice is bad and the other choice is better. Often the choice involves limiting risk or cost and/or exploiting opportunity, but there are too many possible reasons to list here. As in Strategy 6, language of opinion or preference could be a clue, such as the woman's final decision in the conversation: "Yeah. That's what I am beginning to think too. I hate to be a snitch, but…"

Strategy 11: Review your notes and form your own opinion quickly.
After you listen to the conversation, you have 20 seconds to organize your ideas before you speak. In that time, you must identify the student's problem, choices and decision in your notes. Decide on your own opinion before you speak because you will have so little time to speak but a lot of information to summarize.

Don't wait until you speak to decide what you think. It takes at least a few seconds to form an opinion and reason, but that's all the time you will have after you summarize the conversation.

Strategy 12: Don't agree or disagree without an explanation.
Expressing an opinion about something you've just heard is an important part of the task. Therefore, it is important that you explain your opinion. Don't simply say something like "I agree with the man's opinion for the same reasons" or "I like what the woman says and I support her reasons." This isn't enough.

Even if you agree with one speaker's opinion and reasons, you should paraphrase those reasons and give your own reason. You must do more than just accept other people's ideas. You have to express those ideas as well as your own.

Strategy 13: Add your own ideas to those mentioned.
The conversation includes two options and the students discuss several reasons for choosing one or the other. Although you must defend one of the choices, you can also mention your own ideas. You don't have to choose an option for the same reasons as the speakers.

In fact, you can improve your mark by adding ideas to your explanation. This strengthens your argument and shows your creativity and vocabulary.

Strategy 14: Organize your answer as a paragraph.
Be prepared to summarize the conversation and give your opinion as one paragraph. Don't try to recreate the original order of the conversation ("First, he said…Then the woman said…And the man answered…"). This is unnecessary and impossible to do in the allotted time. Use the following general outline:

> Topic sentence: a brief but detailed description of the student's problem.
>
> Middle sentences: the student's two options, attitude, and final choice.
>
> Final sentences: your opinion and at least one reason.

Look at the following answer:

> The man and the woman are discussing preparations for, well…they are discussing progress on preparations for presentations that they will be making in a media studies class. The woman has a problem because one of the members of her group (I think his name is Frank) is not participating in the preparation. So, the woman is worried that the presentation won't be good and that her group will get a low grade. The man suggests that the woman talk about it, and tell him—Frank, that is—that he is being selfish. But the woman doesn't want to do that, I guess because that guy, Frank, the one who isn't participating, is a football player, so he is probably really big and strong. I think the man's suggestion is a good one. It doesn't matter how big and strong the guy is; he needs to know that he should contribute to the group. If Frank's behavior doesn't change, then I think the woman and other members of her group should also talk to the professor about the problem.

Practice

Now let's summarize a conversation similar to the one on the actual TOEFL. So that you can review your answer and monitor your progress, record your answer on a cassette player or computer.

 Play CD 2, Track 41 and practice the following:

Narrator: For this task you will listen to a conversation. You will then be asked to talk about the information in the conversation and to give your opinion about the ideas presented. After you hear the question, you will have 20 seconds to prepare your response and 60 seconds to speak.

Narrator: Now listen to a conversation between two students.

Narrator: The students discuss a problem faced by the man. Summarize the problem and the solutions discussed. Then state which solution you prefer and why.

Prepare: 20 seconds

Answer: 60 seconds

You can read a sample response in the speaking transcript at the end of this section.

Speaking Task 6: Lecture

In the sixth and final task of the Speaking section, you must summarize an academic lecture. After listening to the lecture, you have 20 seconds to organize your notes and 60 seconds to speak.

Now, you can review strategies for completing task 6 of the Speaking section. The first strategies below discuss listening to lectures, which you've already studied in the Listening section of this book. As noted below, you may benefit from reviewing General Listening Strategies: Lectures.

Use the following strategies to help prepare you for this task.

Strategy 1: Listen actively.
You must understand the main idea, purpose, and key points of the lecture. This usually requires an understanding of new concepts, definitions and explanations. Moreover, you need to recognize and record key supporting points in order to summarize the lecture in your answer. Therefore, you must think about what you hear.

Strategy 2: Be familiar with the basic organization of a lecture.
In order to understand a lecture and take good notes, you must be aware of how it is organized. You can use your knowledge of reading passages as a guide to the organization of paragraphs: introduction, body, and conclusion. Of course, you don't see a transcript of the lecture, but knowing the basic structure can help you to follow the sequence of ideas and focus on the right ideas.

Lectures (like reading passages) have a main idea, which includes the speaker's topic and what the speaker says about it. The main idea is mentioned in the introduction as a thesis statement, which gives the main idea.

Listen for the main idea early in the lecture. Although lectures can have hooks, short ones often do not. The lecture in this part of the Speaking section is relatively short compared with those in the Listening section, so the first or second sentence often gives the thesis. It may forecast supporting points with parallel structure, but you may have to infer supporting points yourself and then listen for them in the body of the lecture.

Each subsequent paragraph discusses one part of the main idea. Without a transcript, you have to recognize the supporting points as elements of the main idea: people, events, causes, effects, problems, solutions. Each supporting point is followed by details: examples, reasons, descriptions (see also General Listening Strategies: Lectures).

Strategy 3: Use the introduction and organization to anticipate key points and details.
Anticipation is an important part of active listening: You should always think about what might or could come next. Once you know the main topic, you can think about possible supporting points, and as you hear more of the lecture, the organization can help you anticipate the flow of ideas. In other words, the more you hear, the more you can anticipate. Look at the following transcript of a lecture on schizophrenia:

> What is now known as schizophrenia is a mental disorder that has been witnessed in human societies for many centuries. One of the symptoms of schizophrenia—hearing voices—has not always been perceived as a curse. On the contrary, in some ancient civilizations it was believed that such voices came from the gods. Research partnerships between neuroscientists and classical historians are finding indications that the ancient Greek oracle speakers of several centuries BCE might have been schizophrenic. People who heard voices were revered as priests, as mouthpieces of the gods. Other types of mental disease were considered to be the invasion of other, more often evil than good, spirits.

> In western civilization, "madness" was feared and most sufferers were left to wander the countryside or were committed to institutions. During the age of Enlightenment in the 18th and 19th centuries, the public's awareness of such conditions grew, and improvements in care and treatment began to appear. Institutions became more humane, and treatments increasingly targeted. Doctors initially sought cures using herbal remedies. With the discovery of electricity came many treatments for psychosis including symptoms of schizophrenia. One doctor claimed that a certain type of shock treatment could cure schizophrenia. These claims were highly questionable even at the time and are now believed to have done more harm than good, in fact some 40% of patients undergoing this treatment ended up with fractured spines!

> Nowadays schizophrenia is treated with an array of medications and psychotherapeutic interventions. The drugs can help control the symptoms of the disorder while education and psychotherapy can help patients and families learn to manage it more effectively. Virtually every rigorous comparison of medical approaches and social rehabilitation has shown that medication combined with social rehabilitation leads to a better outcome than either approach alone.

First, notice that the main topic is mentioned immediately (schizophrenia) in the first sentence, and there's no hook. So the first sentence is the thesis, but you can't know immediately what the speaker will say about schizophrenia. However, since the topic is described as a mental disorder, you can anticipate that the speaker might give a definition/description, and discuss a famous scientist, its discovery, diagnosis, case histories, causes, effects, and/or treatment.

Second, although there is no forecasting of supporting points, they are implied in the phrase *for many centuries*. From this, you can anticipate that the lecture may present a historical overview of schizophrenia. This is a good clue to guide you through the lecture. In fact, the lecture begins with a discussion of the ancient world, which is an excellent clue that the supporting points will be chronological. Once you hear the discussion of the 18th and 19th centuries, it should be clear that the lecture is organized chronologically: ancient civilizations, Europe during the Enlightenment, and the present day.

Third, the speaker never defines schizophrenia, but you can gradually understand it by the related vocabulary: mental disorder, hearing voices, other types of mental disease, madness, most sufferers. Without a definition, you must recognize these key words in order to understand the topic and explain it in your answer.

Finally, the speaker doesn't explicitly state a main idea by saying something like "I want to discuss X, Y and Z about schizophrenia" and you shouldn't rely on this kind of statement. Instead, you must understand it based on the supporting points. The first paragraph discusses the ancient understanding of the disease, which didn't involve treatment because ancient people believed the symptoms were influenced by gods. However, the vocabulary related to the Enlightenment and present day goes beyond diagnosis: improvements, care, treatment, targeted, cure, an array of medications and interventions, help control the symptoms of the disorder, learn to manage it, medical approaches, and so on.

Strategy 4: Take notes on the main idea, purpose, and key points.
Write notes in fragments, abbreviations, symbols, and abbreviations. Also, try to paraphrase the ideas in your notes. Don't always try to copy the speaker's exact word choice or expression. If you know a simpler or shorter way to say something, write that. Notes on the above lecture could look like the following:

> schizo = ment ill
> 1 sym = voices
>
> anc civil
> greeks — act of gods + evil
> west - alone, no treat
>
> 18/19 = mad
> ↑ # instit, ↑ care
> herbs, shock = 40% X spines
>
> today
> ↑ # new meds/treat/therap
> drugs + edu
> study = drug + social → best

Strategy 5: Organize your summary as a paragraph.

After you listen to the lecture, you have 20 seconds to organize your ideas before you speak. You should prepare to give one paragraph that summarizes the lecture. If possible, follow the same organization as the original lecture since this is the easiest option. Organize it as follows:

> Paraphrase the main idea of the lecture
> First key point
> Second key point
> Third key point

Your summary should be between six and 10 sentences long. Look at the following summary of the lecture on schizophrenia:

> The professor talks about how the mental disorder schizophrenia was regarded by ancient Greeks, by nineteenth century medical workers and by modern psychiatry. The ancient Greeks didn't think it was a disease. They thought it was a way for the gods to talk to people. When modern medicine was developing, one or two hundred years ago, doctors tried to find cures for various types of madness including schizophrenia. They tried herbal medications and electrical treatments, and they locked people up in mental hospitals. Nowadays, doctors are finding more successful ways of treating this mental illness. Usually treatment includes medication and psychotherapy.

Practice

Now, let's summarize a lecture similar to the one on the TOEFL. So that you can review your answer and monitor your progress, record your answer on a cassette player or computer.

 Play CD 2, Track 42 and practice the following:

Narrator:	In this question you will listen to part of a lecture. You will then be asked to summarize important information from the lecture. After you hear the question, you will have 20 seconds to prepare your response and 60 seconds to speak.
Narrator:	Listen to part of a talk in a United States history class.
Narrator:	Explain the link between the introduction of machinery and the concept of mass production. Use details and examples from the lecture to support your explanation.

Prepare: 20 seconds

Answer: 60 seconds

You can read a sample response in the speaking transcript at the end of this section.

IMPROVE-YOUR-SCORE STRATEGIES

To improve your speaking ability before test day, you should practice a variety of activities on your own and with a group. This section reviews some of the activities and issues that you should pay attention to as you prepare. Before you move on, however, you must understand how sounds, particularly vowel sounds, are discussed below.

Strategy 1: Work on your Pronunciation

Pronunciation is often discussed using the phonetic alphabet, which is a unique system of symbols for representing vowel and consonant sounds. Unlike the familiar written English alphabet, the phonetic alphabet uses one symbol for each vowel and consonant sound. Some symbols resemble the traditional letters, and others don't. For example, the words *to*, *too*, and *two* are all spelled the same way phonetically: *tu*. Although the phonetic symbol for the consonant sound resembles the traditional letter, the vowel's phonetic symbol doesn't. The phonetic alphabet is used in dictionaries to explain proper pronunciation, so it is worth learning. Pronunciation is an important, but often overlooked, part of learning new vocabulary.

Many students focus on the content and length of their answers when they prepare for the TOEFL. However, content is only part of your score. Your ability to produce the sounds of English words accurately is also a large part of your mark. Unfortunately, many students study pronunciation when they first learn English and then stop practicing it in order to focus on vocabulary and sentence structure. The problem with this decision is that it leaves spoken errors uncorrected and allows them to become bad habits. These uncorrected bad habits often lower a student's mark as much as, or even more than, a lack of vocabulary or structure. For example, some students know very advanced vocabulary and can write complex sentences, but they have little or no experience actually saying them out loud.

In General Speaking Strategies, you reviewed some of the basic elements of English speech: stress on syllables and words, reduced vowels, linking, gliding, and pausing. These are all topics that you should review in greater detail with a good exercise book, a knowledgeable teacher, and an interactive group or class. You should take the time to have someone identify your weaknesses in pronunciation, pace, and rhythm. Kaplan's instructors are trained to identify and improve weaknesses in spoken English, and Kaplan's classes are structured to be interactive and student-centered. Therefore, look for a Kaplan center in your area. Some of the key issues are reviewed below.

Strategy 1 A: Review vowel sounds.

There are very subtle differences in vowel sounds in English. For example, the words listed below do not have the same vowel sound. Each of the following vowel sounds are pronounced differently:

see	soak
sit	soot (rhymes with *foot*)
say	soup
said	sir
sat	sigh
saw	sound
sun	soil
beer	bore
bear	buyer
bar	bower (rhymes with *hour*)

If you have difficulty pronouncing certain vowel sounds, you need to adjust your jaw (widen or contract), the position of your tongue (high or low, front or back) and/or the shape of your mouth (smiling, relaxed, circular) accordingly. Practice pronouncing difficult sounds while you look in the mirror.

Strategy 1 B: Practice consonants.
One important part of consonant sounds is the use of the voice, or the vibration of the vocal chords. Consonants can be voiced (vibration) or voiceless (no vibration). Voiced consonants include the first sounds of *buy, my, vie, day, go, low, no, row, zap, though,* and *Joe.* Voiceless consonants include the first sounds of *fee, key, tea, thigh, chew, shy,* and *sigh.*

Besides the proper pronunciation of the consonants themselves, voiced and unvoiced consonants are important for the length of vowels. Each vowel can be short or long, depending on the final consonant. The need to voice a final consonant lengthens the preceding vowel. This lengthening is mostly unavoidable, so you needn't overdo it, but you should be careful not to lengthen a vowel before voiceless consonants, which is unnecessary.

Short Vowel	Long Vowel
Neat	Need
Hit	Hid
Wake	Wade
Debt	Den
Shop	Shawl
Etc…	Etc…

Strategy 1 C: Make sure you stress the right syllables.
Stress is a vital part of pronouncing a word correctly. The stress is a slight emphasis on a certain syllable, which is a set of two or more consonants and vowels. Each word is stressed on a particular syllable. Although there are exceptions to every rule, English stress follows particular patterns.

Learning these patterns can help you when you learn new vocabulary or encounter unfamiliar words on the TOEFL, especially academic terminology. Also, reviewing the patterns is an excellent way to polish your pronunciation technique. To help you understand the stress patterns of English words, review prefixes, suffixes, roots, compound nouns, and phrasal verbs.

As in many reference books, stress in this book is indicated with a ' before the stressed syllable. For example, the word table is stressed on the first syllable: 'table. *Computer* is stressed on the second syllable: com'puter. *Information* is stressed on the third syllable: infor'mation. The following table reviews some of the basic stress patterns in English, but you should always look out for the many exceptions.

Syllable Stress

Category	Stress Pattern	Examples
One-syllable words	Stress the only syllable	'go, 'sun, 'force
Two-syllable nouns	Stress the first syllable	'window, 'season
Two-syllable verbs, adjectives, adverbs, prepositions	Stress the root	de'flect, con'vince, 'open ob'scure, con'fused, 'varied a'round, be'tween, 'safely
Compound nouns	Stress the first part/word	'newspaper, 'bus driver
Phrasal verbs	Stress the preposition	turn 'up, set 'off, sink 'in
Reflexive nouns	Stress self or selves	my'self, them'selves
Other words with two or more syllables	Stress according to the suffix: 1. Stress suffix 2. Stress first syllable before suffix 3. Stress second syllable before the suffix. 4. Maintain the stress of original form.	1. Japan'ese, pirou'ette, crit'ique, mountain'eer, ob'tain (just verbs with -ain) 2. super'ficial, mu'sician, astro'nomical, infor'mation, in'formative, de'fficient, 'furious, hu'mility, e'lastic, 'pacify, 'fortitude, psy'chology, oceo'nography 3. 'moderate, 'modernize, 'mortuary 4. 'comfortable ('comfort), 'plentiful ('plenty), en'joyment (en'joy), ex'pression (ex'press), des'troyed (des'troy), 'capitalism ('capital), for'gotten (for'get), 'cellist ('cello), im'provement (im'prove), 'smartest ('smart), per'former (per'form), ex'citing (ex'cite), 'priceless ('price)…etc.

Strategy 1 D: Practice stress in sentences; get the rhythm.

As you learned in General Speaking Strategies, certain words in a sentence receive extra emphasis on their stressed syllable. This extra emphasis creates part of the rhythm of spoken English. As you learned in General Listening Strategies, content words are emphasized more than function words.

You should slow down for content words: nouns, adjectives, adverbs, verbs, negatives, etc. Content words should be pronounced clearly (but with properly reduced vowels) because they give most of the meaning to a sentence. However, you should not emphasize function words: articles, prepositions, conjunctions, modals, and so on. These can be connected through linking and gliding.

The emphasis of content words creates the beat or rhythm of spoken English. When people say that someone speaks English naturally, they usually mean that the person speaks with a regular and familiar rhythm. In order to have a good rhythm, you must emphasize the content words in your sentence, and combine and/or minimize the function words. Look at the following set of sentences:

> 'John 'takes 'trips.
>
> 'John 'took a 'trip.
>
> 'John 'went on a 'trip.
>
> 'John has 'taken a 'trip.
>
> 'John has 'taken his 'trip.
>
> 'John will be 'taking his 'trip.
>
> 'John will be 'going on his 'trip.

Strategy 1 E: Link words into groups, or chunks.

Linking is a common way of maintaining the appropriate rhythm, or beats, in a sentence. When one word ends in a consonant and the next begins with a vowel, both words can be joined and spoken without a pause between them. Prepositions and articles are often joined to a preceding word in this way. For example, the second sentence above should contain two linked words (took + a):

> 'John 'took-a 'trip.

The article a can be linked, or joined, to the verb *took*, and both words should be spoken as one. Likewise, the third sentence from the list above should contain three linked words (went + on + a):

> 'John 'went-on-a 'trip.

The preposition on and article a should be joined to the verb *went*, and the three words ought to be spoken as one word without a pause between them.

Strategy 1 F: Contract auxiliary verbs.

Although contractions are not mandatory, they save time and improve the rhythm of your speech. This is why they are so common in spoken English. Native speakers almost always use contractions whenever possible. Practice them and use them frequently. Some examples are:

I am = I'm

He is = He's

They will graduate = They'll graduate

They have gone = They've gone

She must have left = She must've left

The man could have taken = The man could've taken

The student should have said = The student should've said

Strategy 2: Use transcripts and read them out loud.

You've read that it isn't a good idea to practice the Speaking tasks by reading transcripts. This remains true. To prepare to speak on the TOEFL, you should practice speaking from notes and within the time limits. You can read out loud from transcripts, however, in order to improve your pronunciation and rhythm.

Improving your speech habits requires concentration and repeated practice. This is difficult to do while you speak spontaneously from notes within a time limit. Instead, use the reading passages and listening transcripts from this book and other textbooks, including the sample responses, as well as your own written essays. Stress the content words on the appropriate syllable and identify how many beats each sentence has. Group words into phrases and clauses, and mark pauses between them whenever necessary. Try to link and glide between pauses to maintain the rhythm. Mark any reductions in pronouns (he, his, etc.) and auxiliary verbs (will, has, etc.) by crossing out the reduced letters. Try to pronounce reduced vowels appropriately, and try to avoid over-pronouncing. Get into the habit of using contractions more frequently even if the verb is spelled out completely on the page.

Using the transcripts of lectures gives you practice with academic vocabulary. A big problem for students is that they have never actually spoken some difficult but familiar words before the actual exam. You should try to practice academic vocabulary and new words before the exam.

Your own essays are also useful because they include sentence structures and vocabulary that you can already generate. You should make sure that you can say them out loud with the proper reductions, linking, gliding, grouping, and pausing.

Strategy 3: Focus on grouping words and pausing.

Pausing is unavoidable since you have to pause to breathe and think, but grouping is a difficult skill for many students to master. It is common for students to pause at the wrong place in a sentence, and this is because they haven't practiced grouping enough. English is made up of phrases and clauses, so it's important to follow that organization when you speak. Mark pauses on the transcripts that you practice with, and alternate the location of certain pauses to see which pause sounds more natural. If the transcript comes from a listening activity, use the spoken recording as a guide, and try to recreate the same pauses as the speaker.

Strategy 4: Speak in English at least as often as you write, read and listen.

Like all language abilities, speaking requires regular practice. However, speaking is usually the skill that students practice the least. It is all too common for TOEFL students to prepare for the exam by mostly reading and writing, often in silence, and communicating mostly in their native language(s). This study habit might be sufficient for some fields, but it is insufficient for language learning, especially speaking.

All language abilities are related, so practicing one skill does reinforce the others through vocabulary and structure, among other things. Reading, for example, is the best way to learn new words in context. However, many students write much more easily and fluently than they speak, so they focus on writing because it is more familiar and comfortable. It's natural to prefer to do what we are good at, but you need to speak as often as possible.

First, many ESL students are better writers than speakers because they've done far more academic writing in English than speaking. If you write better than you speak in English, then this is proof that practice works! It's important to give yourself as much speaking practice as possible. With more practice, you can become a better speaker.

Also, writing allows for more thought and self-correction than speaking does. Therefore, even though you may know many ways of expressing something, you may be most accustomed to generating the vocabulary and structure slowly and rewriting it often. However, unlike writing, speech must be connected and continuous. You can't begin a sentence, then go back, say it a few different ways, and finally finish it when you've found the best way to say it. This is why practice is so important for speaking.

You must gain the ability to turn the vocabulary and structure you know into speech. You have to train yourself to be able to speak in fluid, connected sentences without starting and stopping too often. That takes practice.

Organize or join a conversation club or class, and practice as often as possible. Make sure the group or class is not too focused on lectures. In other words, make sure the teacher or moderator doesn't speak too much. Find a communicative and interactive group, so you will have many opportunities to speak and get feedback. For example, look for a Kaplan class in your area.

Strategy 5: Practice speaking under test-like conditions.

Don't rely on transcripts and written material too much. It is fine for pronunciation practice, but you must be able to speak with limited time and little preparation in order to perform well on the TOEFL. Therefore, you should practice under the same conditions. When you practice with a partner, group or tutor, don't give yourself too much time to generate ideas and speak.

You need to become more spontaneous in your speaking abilities. By giving themselves too much time to think before speaking, students often get a false idea of their skill level. It is harder to think of something to say in a few seconds compared to a few minutes.

Also, students don't always read, take notes, and speak within a set time limit when they practice. Therefore, when they do the actual exam, these students get nervous and confused because they aren't used to the speed and pressure of an actual exam.

Strategy 6: Practice speaking from fragments and words.
Using only notes when you speak is one way for you to practice speaking under conditions similar to those of the actual exam. One of the most important skills for the Speaking section is the ability to speak in complete sentences from only notes. Remember to avoid speaking from transcripts when you practice the six tasks.

Practice by thinking of a familiar noun, argument, or concept, writing a few key words down on a piece of paper, and speaking from just those notes.

> Park—parents' house
> beautiful, convenience, enjoy
> Flowers, trees, grassy, open field
> Close, don't pay
> Activities—soccer, riding

Strategy 7: Practice writing notes that are easy to speak from.
A common problem for students is that they write notes in many different parts of speech. If the notes are overly mixed, it requires more mental effort to put them together in sentences. It also makes it difficult to use parallel structure in sentences like the thesis. For example, the list of words of above begins with notes for a thesis: Park - beautiful, convenience, enjoy. However, the key points are all different parts of speech (adjective, noun, verb), so they can't easily be used together in one sentence. Parallel structure requires the same part of speech. This is something that confuses many students when they start to give their answer. Instead, the list above could be rewritten as follows:

> Park—parent's house
> beautiful, convenient, enjoyable/fun
> Flowers, trees, grass, open fields
> Close, free
> Activities—soccer, bicycles

Of course, not all your sentences have to contain parallel structure, but you can be more concise by using it and concision saves time and allows you to say more.

Strategy 8: Practice expressing ideas in a variety of ways.
A student who can speak well from notes is usually able to say the same thing in a variety of ways. In other words, the student is good at paraphrasing. This requires an ability to change the part of speech (noun, verb, adjective, adverb, etc.) and put it into an appropriate sentence structure. If you can do this well, then it doesn't matter what part of speech you write down, read, or hear. With a wide knowledge of word forms and structures, you can use any part of speech in a sentence. For example, look at the following two options for rewriting the notes for the thesis above:

> Park—parents' house
>
> beauty, convenience, fun / beautiful, convenient, fun
>
> With practice, you should be able to quickly put the above sets of words into an appropriate thesis statement:
>
> One important place for me is a park near my parents' house because it is very beautiful, convenient, and enjoyable.
>
> There is a really beautiful, convenient, and enjoyable park in my parents' area, and it's one of my favorite places.
>
> One significant place for me is a park close to my parents' house because of its beauty, convenience, and fun.
>
> I really enjoy the beauty, convenience, and fun of a park in my parents' neighborhood.
>
> I love the beauty, convenience, and fun of a park down the street from my parents' house.

Strategy 9: Note any weaknesses in vocabulary, especially verb choice.
One thing students notice is that it is easier for them to combine some ideas than to combine others. Usually, this is because they don't know an appropriate word or phrase to complete a sentence. A word could be illogical or awkward compared to the preferred style of native speakers. For example, if you wanted to use park as a subject, and *beauty, convenience,* and *fun* as objects, you would need one or more verbs:

> Park + verb + beauty, convenience, and fun.
>
> Park + verb + beauty and + verb + convenience and fun.
>
> Park + verb + beauty, verb + convenience, and + verb + fun.

Since there is no verb that would make sense with all three nouns, the second option is best. Has or possess is appropriate for the beauty of a park but not the convenience of one. *Offer* or *provide* is good for a park's convenience. One could say:

> I like the park near my parents' home since it possesses a lot of beauty and offers convenience and fun.

The appropriateness or awkwardness of word choice in English often seems arbitrary to ESL students, and it is one of the most challenging aspects of language learning. You can greatly improve your performance on the Speaking (and Writing) sections of the TOEFL by adding the correct combination of words to your study of vocabulary. Whenever you learn a new word, record or find the most common words that are used with it, especially verbs.

Strategy 10: Be well prepared for the independent tasks.
Before the actual exam, you should practice describing at least one important person, place, event, class, and possession from your life. Likewise, you should practice describing the qualities of a good parent, neighbor, teacher, scientist, and politician. Although you can't memorize your entire answer, you can practice a set of key points for each possibility. Generate ideas for a specific person, place, event and class, and so on, and then practice speaking from those notes. Be sure to practice a variety of word forms and sentence structures. Don't keep trying to say the same answer.

Also, practice supporting or opposing a wide variety of projects and issues: construction projects (roads, highways, shopping malls, theaters, etc.), living arrangements (country/ city, roommates, co-ed, multigenerational, stay at home until marriage, etc.), education philosophies (school size, home schooling, no homework, no grades), and government involvement (surveillance, censorship, etc.). Use your imagination and challenge yourself and your study partners to come up with similar topics and debate them.

Strategy 11: Speak about academic topics.
One reason that some parts of the Speaking section are so difficult is that some students have never discussed academic topics in English before. Students who regularly talk about art, history, and science with their friends have a huge advantage because they have experience defining, explaining, and exemplifying.

You should give yourself as much of that practice as possible. As mentioned above, it is a good idea to join or start a conversation group. However, a possible problem with such groups is that they are too casual in topics. Discussing the latest Hollywood marriages will not prepare you for the TOEFL. Find people who are willing to have informed discussions about current events, new discoveries, new theories, and so on.

As you've already learned, all language skills are related, and one way that reading and listening can improve speaking ability is through information. The more you read and hear, the more you can talk about. Therefore, read and listen widely and encourage friends or partners to do the same. Then, you can discuss what you've watched or read, explaining what others don't understand and debating what others don't agree with.

Strategy 12: Learn more by focusing your study and practice.
You will learn more vocabulary if you concentrate your reading, listening, and speaking in a few areas, or fields. Often, students read and listen in too many unrelated fields, such as volcanoes, Egyptian pyramids, and colonizing Mars. Although variety is good, too much

can be futile and confusing. The problem is the lack of repetition, which reinforces learning, and the lack of connections between the vocabulary and concepts of one field and those of another.

You should try to read, listen, and speak about related topics so that you can expand and build on what you've already learned. Look for interesting topics that appear regularly in the newspapers and on television, such as global warming, cancer treatment, and poverty, among many others. You need to encounter ideas more than once and use them repeatedly in order to learn well.

SPEAKING TRANSCRIPTS

CD2, Track 37

Independent Task 1: Description

Narrator: Choose a person from your childhood. State why that person was significant to you. Give details and examples to support your choice.

Sample Response

My high school music teacher was a very important person while I was growing up. His name was Steve Vernon, and he was extremely encouraging, supportive, and helpful. I doubted myself a lot when I was young. Because I didn't have a lot of self-confidence, I didn't take a lot of risks. That's why I played the triangle at first. I chose the simplest instrument. However, Mr. Vernon challenged me to try new things. So, I tried to play the violin. I wasn't very good, but he helped me a lot by giving me instructions, books, and tutoring me after school. I eventually played violin in the school recital at the end of the school year, and my parents were so proud of me.

CD2, Track 38

Independent Task 2: Opinion

Narrator: In many countries, the government practices censorship of television programs. Do you support this practice or not? Give specific reasons and examples to support your answer.

Sample Response

I definitely don't support state censorship of television because that's the job of the public and also because free speech is necessary to control government. Bad programs occasionally are broadcasted, such as shows with too much violence or bad language. However, when this happens, people complain to the stations and to the government. Citizens threaten to boycott the stations, and the offensive shows are canceled. The government doesn't need to get involved. Also, government officials sometimes make mistakes or even commit crimes.

For example, an official might give large amounts of money to friends or family members. Journalists should report this behavior, but government censors might not allow them to report it. For those reasons, I think the government should leave television alone.

CD2, Track 39

Integrated Task 3: Reading and Listening

Narrator:	For this task you will read a short passage about a campus situation and then listen to a talk on the same topic. You will then answer a question using information from both the reading passage and the talk. After you hear the question, you will have 30 seconds to prepare your response and 60 seconds to speak.
Narrator:	The University newspaper is looking for students to join its staff. Read the advertisement from the faculty advisor. You will have 45 seconds to read the advertisement. Begin reading now.
Narrator:	Listen to two students as they discuss the announcement.
Student 1 (female):	You should apply, Mark. You'd be great on the paper.
Student 2 (male):	English isn't exactly my best subject.
Student 1:	There's plenty of stuff you can do on the paper besides writing. You're a whiz on the computer. You'd be great with design. Or you could sell advertising. I'll bet you could persuade lots of people to place an ad.
Student 2:	Working on the paper would take a lot of time. Now that we're in junior year, I really have to concentrate on my schoolwork.
Student 1:	Now that we're in junior year, what you really need to concentrate on is your résumé. A position on the school paper would really impress a future employer. It would give you something to talk about during the interview.
Student 2:	I just really don't think I should spend that much time on extracurricular activities right now.
Student 1:	But you're always cooped up in your room, studying. Especially ever since you broke up with Anna....
Student 2:	There you go again, trying to set me up.
Student 1:	I'm not trying to set you up at all. It's just that working on the paper is a great way to meet people. And you never know who you might meet.
Narrator:	The woman expresses her opinion of the announcement. State her opinion and explain the reasons she gives for holding that opinion.

Sample Response

The Campus Voice, which is the school paper, is hiring students for credit. Applicants need good marks but they don't need experience, and there are many different positions available. The woman sounded very excited about the opportunity, and she really encouraged the man to join the paper. At first, she emphasized the variety of positions, such as design and advertising, and she said that the man would be good at many things. But, the man said he preferred to focus on school work. Second, the woman suggested that the man was too focused on work and should meet new people, but the man seemed annoyed by the suggestion and said that she was always telling him to do that.

CD2, Track 40

Integrated Task 4: Academic Reading and Listening

Narrator:	Now read the passage about Impressionism. You have 45 seconds to read the passage. Begin reading now.
Narrator:	Now you will listen to part of a lecture on the topic you just read about. Take notes while you listen.
Professor:	One of the reasons the Impressionists caused such a scandal was because their colors were so bright. And one of the reasons these colors were bright was because the Impressionists were able to use synthetic pigments that had never existed before.
	Along with all the scientific achievements of the Industrial Revolution came major advances in the science of chemistry. Chemists were hired by industry, and with all the new elements that were being discovered—things like cadmium, chrome, cobalt, and so on—almost as quickly as an element would be discovered, someone would be exploring it for its commercial potential as a pigment or dye. And so, dozens of colors were invented in the early 1800s. And these colors were bright, and they were stable—for 600 years there had never been a yellow, for example, that didn't fade in the light. And most important, these new colors were cheap, so even the poorest artists could afford to buy them.
	Another invention helped make the Impressionist revolution possible, and that was the invention of the metal paint tube. Paint became portable. The Impressionists could leave their studios and capture the flickering light they saw in the world outdoors. So the Impressionists represented a new sensibility, for sure, but it was a sensibility made possible by the new tools they had at their disposal.
Narrator:	Summarize the points made in the lecture you just heard, explaining how the lecture supports ideas mentioned in the reading.

Sample Response

The reading defines Impressionism, which was a movement in painting during the late 19th century. According to the writer, the Impressionists were interested in how a viewer perceived light at different times, and they wanted to represent this effect in their paintings. The lecture discusses some of the reasons that the Impressionists were able to depict light in a unique way. First, chemists had developed brighter colors by that time, such as a more permanent yellow. Also, more painters could afford the new paints because they were cheaper. Finally, the paints were packaged in smaller containers. Because of this new portability, the Impressionists could paint outside and record light more effectively.

CD2, Track 41

Integrated Task 5: Conversation and Opinion

Narrator:	For this task you will listen to a conversation. You will then be asked to talk about the information in the conversation and to give your opinion about the ideas presented. After you hear the question, you will have 20 seconds to prepare your response and 60 seconds to speak.
Narrator (female):	Now listen to a conversation between two students.
Student 1 (male):	Hey Patty, could I ask your advice about something?
Student 2 (female):	Shoot.
Student 1:	Well the thing is, my girlfriend is playing in the finals of the intramural volleyball tournament tomorrow, and she really wants me to be there.
Student 2:	(*filling in the rest of Joe's thoughts*) But . . . you have something else going on at the same time . . .
Student 1:	Well, yeah. Ages ago, I don't know, maybe three months ago. I signed up to help clean the dance-theater—you know, scrape gum off the bottom of seats and stuff.
Student 2:	Oh that sounds like fun.
Student 1:	Anyway, I'd completely forgotten about it. But then Sam, the stage manager, do you know him?
Student 2:	Who doesn't know him?
Student 1:	Okay. So, anyway, I bumped into him at the gym and he reminded me about it. I didn't know what to say, so I just pretended that I hadn't forgotten.
Student 2:	Just tell him your girlfriend's in this tournament.
Student 1:	Just tell him, huh? I feel bad though. I don't like not following through on a commitment. Especially with Sam…you'll never hear the end of it.

Student:	(surprised) Really?
Student 1:	Well, he doesn't get angry. He just jokes about it. But, you know, he's sensitive about things like that.
Student 2:	Just offer to help out another time. I'm sure he'll understand.
Student 1:	Hmm, maybe . . . that might work. But, oh, well, jeez (*sounding flustered*) . . . actually I think this is the end of the season. So . . .
Student 2:	Well, if he's going to be doing a major end-of-season cleaning job on that theater, he'll probably still be at it after the tournament ends. Why don't you go over and help him after watching your girlfriend's game?
Student 1:	No, that won't work. I'm late on an assignment for Russian history. You know . . . the one that was due last Thursday? I was able to extend the deadline. But I'm sure he'll penalize me if I don't get it in tomorrow.
Student 2:	Well, the only other alternative is to tell your girlfriend that you can't be at her game.
Student 1:	Oh, no way. I value this relationship. That would cause a real problem.
Student 2:	Well in that case, it seems pretty clear to me. If I were you, I would let Sam know as soon as possible, so he has time to find someone else.
Student 1:	Yeah, that's a good point. Actually . . . Stephanie might be able to help out. In fact . . . hmm . . . you know, I think Stephanie kind of likes Sam anyway. If they get together to work on the theater . . .
Student 2:	Whoa, don't even go there, Joe. You just take care of you. Your life is complicated enough without trying to play matchmaker!
Narrator:	The students discuss a problem faced by the man. Summarize the problem and the solutions discussed. Then state which solution you prefer and why.

Sample Response

The man's problem is that he has too many things to do and can't decide what he should miss. First, the man had promised Sam a while ago that he would help him clean the school's theater. But, the man had forgotten about this until he saw Sam recently. Also, the man said that he had to attend his girlfriend's volleyball game, which was at the same time that he had to help Sam. The woman first suggested that the man just tell Sam, but the man didn't want to upset Sam. The woman's other advice was to help Sam after the game, but the man said he had homework and couldn't get another extension. Finally, the woman recommended that the man tell Sam right away and find someone else, and I think this is the best advice. He should care about his girlfriend more than someone that he rarely sees, so he must apologize for his mistake, and hope that Stephanie can replace him.

CD2, Track 42

Integrated Task 6: Academic Lecture

Narrator: In this question you will listen to part of a lecture. You will then be asked to summarize important information from the lecture. After you hear the question, you will have 20 seconds to prepare your response and 60 seconds to speak.

Narrator: Listen to part of a talk in a United States history class.

Professor (female): Mass production—the production of standardized goods for a mass market—was first introduced in the early 20th century. Henry Ford was the first industrialist to really make full use of this system in a major way. As a result, the system became known as "Fordism." So we can say that Fordism is the mass production of standardized goods using dedicated machines and moving assembly lines and employing unskilled and semi skilled labor in fragmented jobs, with tight labor discipline in a large factory setting.

What does all this mean in practical terms? Henry Ford is, of course, known as a car manufacturer, as the maker of the Model T. But in a more general sense, his greatest innovation—his greatest contribution to manufacturing—was this concept of Fordism, this idea that the most efficient way to make something is to break down the process into hundreds of small tasks. So picture a factory and picture a production line, and there are perhaps 50 people along the production line… There's a conveyor belt constantly moving the parts along from one person to the next… Each person is assigned a certain very specific job, and that specific job is all that person does all day for his or her whole shift. . . . So it is Henry Ford whom we can thank for introducing the concept of what is now commonly known as a "complex division of labor."

It was Ford who first realized that this process would make it possible for him to lower his costs and increase his profits. And indeed, when the Model T was first built, it took 14 hours to assemble one car. Ford was eventually able to reduce the assembly time to 93 minutes. During this same period, between 1908 and 1913, the price of the Model T fell from $1,000 to $360. Very soon, this process was adopted by other industries.

Narrator: Explain the link between the introduction of machinery and the concept of mass production. Use details and examples from the lecture to support your explanation.

Sample Response

According to the lecture, machines put assembly lines into factories and made manufacturing faster and cheaper. The lecturer discusses the assembly line and the concept of Fordism. This term is based on Henry Ford, who first used assembly lines when he started building cars, such as the Model T Ford. Ford's innovation was to take a big project, like building a car, and turn it into a series of small jobs. He put many people in a long line and gave them the same job to do repeatedly. A conveyer belt was used to move cars down the assembly line, and each worker did the same task as each car passed down the line. The speaker defines this as a complex division of labor, and it revolutionized manufacturing. According to the lecturer, Ford's process reduced time, costs and price. The time went from hours to minutes, and the price fell from $1,000 to several hundred. For this reason, other industries also used assembly lines.

Chapter 4: **Writing**

The Writing section of the TOEFL is the fourth and final section. There are two parts to this section: First you will have an integrated task, in which you must read an academic passage, then listen to a related academic lecture, and finally write a summary of both. Second you will have an independent persuasive essay similar to one you may have seen in the old computer-based TOEFL test.

When the Writing section begins, you should still be wearing your headphones. You should keep the headphones on because the first task involves listening. Moreover, the headphones block out noise and reduce distractions, so it would be a good idea to keep them on for the entire section.

Some of the strategies to be used here are similar to those used for the Speaking section, since the two Writing tasks resemble the second and fourth Speaking tasks. Of course, the Writing section involves written English, not spoken, so the strategies below have been adjusted accordingly. Also, both Writing tasks are longer than those in the Speaking section, which allows for more planning and review.

GENERAL WRITING STRATEGIES: INDEPENDENT AND INTEGRATED

There are some strategies that you should keep in mind for both parts of the Writing section. Generally, these strategies emphasize organization, familiar sentence structure, variety, and editing.

Strategy 1: Organize your thoughts first.
Because of the time limits, some students feel pressured to begin writing right away. In order to write a complete essay spontaneously, you must be able to organize your ideas in your mind as you write. It's possible, but risky. The risk is that you may get lost or confused while you write. As a result, you will have to rewrite and edit more than you should, which will waste time and cause you to make more errors. This is especially true for the first task, which involves unfamiliar concepts and vocabulary.

Instead, organize your ideas briefly in note form. As you should already know, the basic organization of an essay is the paragraph. Each paragraph discusses a supporting point and begins with a topic sentence that mentions that point. The rest of the paragraph contains detailed explanation, argument, examples, description, comparison, contrast, and so on.

Do not spend more than two or three minutes on an outline before writing since you won't be marked on your outline. You will be marked only on the final essay you type onto the computer screen at the testing center. Decide quickly how you will organize your essay. Ask yourself, "What is the topic of each paragraph?" and "What detail will I discuss in each paragraph?" You can record these decisions in note form and use them as a guide while you write.

Strategy 2: Use familiar vocabulary.
Because this section involves written English, students feel that they must demonstrate a high level of vocabulary. They try to use long, sophisticated vocabulary from previous work or from the integrated texts. Although both tasks, especially the first, will involve academic terms and concepts, you shouldn't try and use long, obscure words when you don't have to.

Be yourself and use the vocabulary that you would normally use. When you want to express something, don't waste time thinking of a longer word. If you do, your writing may sound awkward, and you will make more errors. That will lower, not raise, your mark. For example, look at the following:

> The reading passage discusses Greenpeace and its presentational methodology of political protestation. According to the writer, Greenpeace desires global perceptiveness and has used theatricality in its eventfulness to increase the attentiveness of the population.

The writer of the above passage wants to use big words (*methodology, protestation*), but these words are either incorrect (*perceptiveness*) or unnecessary and awkward (*political protestation*). There are far more common word choices and forms to express the same ideas, such as the following:

> The reading passage discusses Greenpeace and how it organizes political protests. According to the writer, Greenpeace wants to be seen by people around the world, and it uses theatrical events to get their attention.

Strategy 3: Use familiar and natural sentence structure.
Like your spoken answers in the Speaking section, your written answers in this section should reflect your usual sentence structure. However, some students feel that they have to demonstrate complex grammatical structure in order to get higher marks. As a result, they try to write longer sentences with more dependent clauses than the sentences they would normally write. Like the unnecessary word choice discussed above, this is risky. You may write awkward, confusing sentences, and you will almost certainly make more grammatical errors. For example, read the following sentences from a paragraph in a persuasive essay for the question, "Do you agree that zoos are beneficial because they educate the public?"

> The first reason that I disagree that zoos are beneficial and educate the public is that they don't provide education, which can come from schools that can teach people more. People who visit zoos can't learn very much about the animals because the animals, which are kept in cages and small areas, don't do a lot while the people, who can't touch the animals, must walk around and can't see very much because the animals can't do anything. The fact that the people and animals are separated, which was boring for me when I went to a zoo, is because people must be protected, but animals are kept in small cages and spaces, which is unfair and maybe makes them angry and more dangerous to people.

The writer uses long sentences with many adjective and adverb clauses. Students who want to gain extra marks for grammar often overuse adjective and adverb clauses. The paragraph above is an example of this overuse. Although the grammar is technically correct, the style is very awkward. Because the sentences are very long, the style could be called long-winded or rambling.

It is not only difficult and risky to write extremely long sentences due to the possibility of error, but it is also confusing for the reader. Notice how you have probably forgotten the beginning of the second sentence by the time you finally finish it. Long sentences contain many ideas, and a reader has trouble identifying and remembering your main point. Avoid this kind of style.

The actual TOEFL exam is not the appropriate place to use tricky, unfamiliar sentence structure. Use sentence structure that you have used in the past, and keep the sentences short. You will not only make fewer errors this way, but you will also be more confident. This will allow you to summarize more thoroughly and argue more persuasively. Look at the following rewrite of the paragraph above:

> First, I disagree with the educational benefits of zoos because they don't provide education. People learn more at school. However, visitors to a zoo can't learn very much because the animals are kept in small cages and areas. Therefore, they can't do very much, and people don't see anything. People can't touch the animals since the people and animals are separated to protect the visitors. When I visited a zoo, I was bored. However, maybe the animals are angry because they are unfairly kept in cages.

Notice that the sentences are shorter. Like spoken English, written English is easier to follow when the sentences have two or at most three clauses. The first sentence in the rambling paragraph has five clauses (five unparallel subjects and verbs). The first sentence of the rewritten paragraph has two clauses. All of the sentences in the rewritten paragraph are shorter than those in the first rambling one, and you should recognize how much easier the rewritten sentences are to follow and understand. Therefore, limit your sentences to a similar, manageable length.

Strategy 4: Write as you speak but with better style and more care.
Generally, good writers write in the same voice, or style, in which they speak. This does not mean, however, that good writers use exactly the same words and structure in both writing and speaking. It just means that their writing doesn't differ radically from their speech. Most

academic writing is more formal than speech, but the difference usually comes from the greater amount of time one has to choose words and edit written English as opposed to the shorter time one has when speaking.

Because of the extra time to write and edit, your writing should have your best vocabulary and sentence structure. At the same time, however, it shouldn't sound like it was written by someone else. You should still use the vocabulary and structure you're most comfortable with, but you should choose the best from that group. For example, compare the two following paragraphs from separate arguments about the same question; the first is a spoken answer, and the second is a written response:

Do you agree or disagree with the following statement? *All education should be free.*

Spoken Answer

First, I think education should be free because it will help society. More people can get an education if they don't have to pay for it. And, because of that, people will create more companies, and this will give a lot of people better jobs, like high-tech jobs in computers. Also, these people will make more money, and they can buy bigger homes and nicer cars. This will help other industries, like construction.

Written Answer

First, I support free education because of the social benefits. More students can get an education if they don't have to pay tuition. The higher number of educated citizens will lead to more companies. Therefore, many people can have better careers. For example, more graduates can become computer programmers, engineers, or doctors. Moreover, these workers will have higher salaries, so they can afford bigger homes and nicer cars. The increased spending will help other industries, such as construction.

The two answers are not extremely different. However, the first, spoken answer uses more pronouns (that, this) and coordinating conjunctions (and, so), which is common in speech. The second answer doesn't include longer sentences; it includes fewer pronouns (for example, *that = higher number of educated people*), better vocabulary (tuition, careers, salaries), and more transitional adverbs (therefore, moreover). If both answers came from the same person, the difference would come from the extra time that a writer can use to think and edit.

Strategy 5: Vary your vocabulary and sentence structure.

Like speakers, writers sometimes repeat the same word or sentence structure. This is natural because people develop habits over time, so they get used to expressing themselves in the same way. People often don't even recognize that they do this, so it takes conscious effort to break the habit. As you learned in the Speaking section, repetition may not be grammatically incorrect, but too much repetition is boring and suggests a limited knowledge of the language. Therefore, to get the highest mark possible, you should try to be versatile in your use of sentence structure and choice of words.

Fortunately, it is easier to add variety to your writing because of the extra time for editing and rewriting. Therefore, readers, especially those who mark your essays, expect more variety in your writing. For example, look at the two Sample Responses above for Strategy 4. The first answer repeats some words (people), but the second response uses synonyms appropriate to the context (students, workers, citizens).

As you learned in the Speaking section, variety depends on your knowledge of word form (*noun, adjective, adverb, verb, infinitive*), function (*subject, object*), and the possible sentence structures. Some, but not all, of the possible structures are the following:

Subject + verb + subject complement: Golf is my hobby; golf looks difficult

Subject + verb + object: I'm studying chemical engineering; I like to skate

Subject + verb + direct object + indirect object: I bought my brother a wedding gift

Subject + verb + object + object complement: The weather makes me happy; my boss made me a manager

You need to know not only how to change the form of a word but also how to use it in a correct sentence. For example, the verb *persist* means *continue an intention, goal, action or state, usually despite some resistance.* But *persist* can also be a noun (persistence), an adjective (persistent), or an adverb (persistently), among other forms:

Despite our antidrug laws, drug dealers *persist* in selling narcotics illegally.

Although we have antidrug laws, drug dealers show *persistence* in selling narcotics illegally.

There are antidrug laws, but drug dealers are *persistent* in selling narcotics illegally.

In spite of antidrug laws, drug dealers *persistently* sell narcotics illegally.

Variety also comes from using different phrases and clauses. For example, in the four sentences above, the existence of antidrug laws is contrasted with the continued sale of narcotics. Notice how each sentence uses a different method of contrast:

Prepositional phrase (preposition + object): despite our antidrug laws / in spite of our antidrug laws

Dependent/adverb clause (conjunction + subject + verb): although we have antidrug laws

Independent clause + but/yet (subject + verb + but/yet): There are antidrug laws, but…

As long as you are able to use many different word forms in a variety of sentences, you can alter your writing (and speech) appropriately, paraphrase accurately, and improve your score.

Strategy 6: Avoid redundancy.
Redundancy means unnecessary repetition. When a sentence contains a redundancy, two or more different words, phrases, or clauses have the same meaning. Therefore, only one is

necessary and the other(s) should be removed. Often, sentences are redundant because a student tries to write in a formal style and uses too many words with the same meaning.

All of the following words mean addition or continuation: *also, too, as well, furthermore,* and *moreover,* and so it follows that the following sentences contain redundancies. Only one word in italics is necessary.

> *Moreover,* I *also* feel that the new highway will be very noisy *as well.*
> *Furthermore,* the construction of the expressway will *also* be extremely dangerous *too.*

All of the following words or phrases can be used to explain cause and effect, or reason: *therefore, as a result, because, since,* and *due to,* and so it follows that the following sentences contain redundancies. Only one word in italics is necessary.

> The new highway project will require huge machinery that is loud. *Therefore, because of that,* local residents will be disturbed.
> The proposed highway will require a lot of traffic, carrying heavy loads. *As a result, due to this,* pedestrians could be in danger, especially children.

Students often overuse language of argument or opinion: In my opinion; It is my belief that…; I feel that…etc. For example, the following sentences overuse this kind of language:

> In my opinion, I believe that the new highway will be too noisy.
> My belief is that I feel the new highway will result in a lot of noise pollution.
> It is my belief that I think the new highway will create too much noise.
> As I see it, I argue that the noise from the new highway will be a problem.

Avoiding redundancy is the reason that *should* is not used with recommendations or instructions. Therefore, *should* must be removed from the following sentences:

> I recommend that the highway should not be built in my area. (*Recommend* means *should do X.*)
> I insist that the new highway should be canceled. (*Insist* means *should do X.*)
> It is my recommendation that the highway should be moved to another area. (*Recommendation* means *X should be done.*)

Students sometimes repeat themselves when they refer to a reading or listening passage. For example, the following sentences contain redundancies:

> According to the writer, the writer feels that global warming is a grave threat to the planet.
> Based on the reading passage, the writer argues that global warming threatens the planet.

Strategy 7: Use cohesive devices effectively.

Cohesive devices unify a paragraph and a whole passage by showing connections and relationships between clauses, sentences, and paragraphs. The most common cohesive devices are personal pronouns (*he, him, his*, etc.), demonstrative pronouns (*this, these*, etc.), adjectives, articles (*a/an, the*), synonyms (*keep, hold, retain*), transitions (*therefore, moreover*), and word form (*persuade, persuasion, persuasive, persuasively*).

First, cohesive devices help you to avoid repetition by adding variety to your writing. For example, the following paragraph from a summary of a lecture about ozone depletion contains too much repetition:

> First, the speaker explains that pollution is depleting the ozone layer. Pollution comes from vehicles, farms, and factories. Vehicles, farms, and factories put chemicals into the air. Chemicals, such as Bromine and Chlorine, rise into the atmosphere, and the chemicals react with the ozone layer. When the chemicals react with the ozone layer, the chemicals split the ozone molecules by combining with the ozone atoms in each molecule. After the ozone atoms combine with other atoms, the ozone atoms never recombine to form ozone again. Because the ozone atoms never recombine to form ozone again, the ozone's level goes down over time.

The writer here repeats many words and clauses unnecessarily. He repeats some words and clauses to connect the ideas of one sentence with those of the surrounding sentences. However, it is not necessary to repeat the same words and structures all the time. That repetition lengthens your writing and bores the reader. Cohesive devices would help to shorten the passage and make it more interesting. Look at the following rewrite of the passage above:

> First, the speaker explains that pollution is depleting the ozone layer. The pollution comes from vehicles, farms, and factories. They put chemicals into the air. The chemicals, such as Bromine and Chlorine, rise into the atmosphere, and they react with the ozone layer. During this reaction, the chemicals split the ozone molecules by combining with their atoms. After this combination, the ozone atoms never recombine to form ozone again. Therefore, the ozone's level goes down over time.

Now the writer uses several cohesive devices: pronouns, word forms, and transitions. The pronoun *they* replaces many nouns: vehicles, farms, factories, chemicals, and molecules. The use of the article *the* (the pollution, the chemicals) comes after the first mention of the nouns. This is because the nouns are known after they are mentioned once, so they can be specified with *the*. Likewise, the use of the demonstrative pronoun *this* (this reaction, this combination) refers back to a preceding idea. When *this* is used, it modifies a known/specified noun because it has already been mentioned. *This* modifies nouns (reaction, combination) that are noun forms of verbs in previous sentences (react, combine). In the final sentence, *therefore* indicates that the entire preceding sentence is the reason for the next; the writer does this instead of repeating the entire previous clause. See also Reading Strategies, Listening Strategies, and Speaking Strategies.

Strategy 8: Know the meaning of transitions.

Transitions are also called transitional adverbs or transition signals. They are words or phrases that writers (and speakers) use to move from one sentence to the next or from one paragraph to the next. The transitions indicate, or signal, the rhetorical structure of your writing (or speech). In other words, the transitions tell the reader when you are explaining, defining, exemplifying, and so on. They refer to the relationship between the sentences and paragraphs, and they help your reader or listener to follow your argument and explanation.

Although they are useful in speech, transitions are far more common in writing. In the TOEFL, the written answers are much longer than the spoken ones, so transitions are more useful and necessary in the Writing section of the test. The various transitions are grouped below according to meaning and function:

Meaning and Function	List of Transitions
Indicate sequence or show the order of ideas or facts	First, second, third, next, then, finally, lastly
Show similarity or to add information	Also, too, in addition, furthermore, moreover
Show contrast or make a concession	However, but, on the other hand, conversely, yet, though, in contrast, nevertheless
Introduce examples	For example, to illustrate, for instance
Explain, define, or elaborate on an idea	That is, in other words
Emphasize an idea or add a surprising/tangential anecdote	In fact, indeed
Refer to common knowledge	Certainly, clearly, of course
Generalize or discuss habits/customs	Generally speaking, in general, overall, for the most part, usually, typically
Show cause and effect or result	Therefore, thus, subsequently, as a result, as a consequence, hence
Show personal attitude	Positive: fortunately, happily, thankfully Negative: unfortunately, sadly, regrettably

Conclude or end an argument or discussion	To conclude, in conclusion, to wrap up, to summarize, to sum up
	(These transitions require summary as well as final comment/conclusion)

Transitions are also called *transitional adverbs* because they modify an entire sentence or independent clause. That is why they usually come at the beginning of a sentence or paragraph.

Look at the following body paragraph from an argument in favor of exercise for all students:

Question: Some North American colleges and universities don't require students to take gym classes, or classes that involve physical exercise. Other schools require that students take at least one. Do you agree that students should be required to take one organized gym class involving physical exercise during each semester? Use specific examples and reasons in your answer.

> First, I support mandatory gym classes because they will help students learn more. Generally speaking, most students feel different levels of stress due to various factors. For example, some students must live away from home while others have to work to support themselves. Other students might have relationship problems, for instance a difficult roommate. Moreover, the stress can lower concentration and cause sadness and even depression. Regular exercise reduces stress by giving students an outlet for their frustrations. In other words, exercise gives students a way to release or let go of stress. Exercise releases or removes the stress through vigorous physical activity. For instance, a student could lose her frustrations during a swimming lesson or a weight-lifting class. Therefore, she would feel more relaxed and focused, and she could concentrate more. Of course, extreme emotional or psychological problems, such as depression, require therapy and treatment. However, a regular gym class can relieve everyday stress, improve concentration, and help students learn.

Notice how each transition/transitional adverb begins the sentence in which it appears. The paragraph begins with *first* since the paragraph is the first one in the body of the essay. Sentence Two begins with *generally speaking* to indicate that the sentence contains a generalization, or a fact that relates to an entire group or class. This fact is one of the supporting reasons for the topic sentence. *For example* begins Sentence Three because that sentence gives examples of *factors*. The adverb *moreover* begins Sentence Five because that sentence gives another supporting reason; it continues or adds to the explanation of how exercise helps learning. The writer uses the phrase *in other words* to indicate that the sentence gives a definition, synonym or explanation; *an outlet* is defined as *a way to release or let go of something*.

Strategy 9: Vary transitions, conjunctions, and prepositions, but avoid confusion.
The transitions in the chart above are not the only ways to indicate rhetorical structure, such as explanation, examples, and cause and effect. Conjunctions and prepositions can also be used, though they have different grammatical functions.

Students often confuse the functions of words like the adverb *therefore*, the dependent conjunction *because*, and the preposition *due to*. Transitions indicate relationships between separate sentences or paragraphs, but they do not connect them grammatically. Conjunctions join clauses together grammatically into one sentence. Prepositions join phrases to a sentence. The transition *therefore* shows the cause and effect relationship between two independent sentences from the paragraph above:

> For instance, a student could lose her frustrations during a swimming lesson or a weight-lifting class. Therefore, she would feel more relaxed and focused, and she could concentrate more.

Therefore shows a cause and effect relationship, but the word doesn't join the two sentences into one sentence. There is a period after *class* because everything between *for instance* and *class* is a separate sentence. Likewise, everything between *therefore* and *more* is a separate sentence: For instance, + sentence. Therefore, + sentence.

Although a transition such as *therefore* adds meaning to a sentence, it is not grammatically necessary or required. In other words, *therefore* could be removed without damaging the grammatical correctness of the sentence. Of course, some meaning would be lost and the reader would have to infer the cause and effect relationship, but the sentences would still be grammatically correct.

This is not the case with dependent conjunctions, or subordinate conjunctions (*subordinate* is a synonym for *dependent*). The various dependent conjunctions that form adverb clauses are listed below:

Meaning	Dependent/Adverb Conjunctions
Time	When, while, since, as, once *While I attended college, I delivered pizzas.*
Place	Where *A huge tree now stands where I planted a small seed.*
Reason/Cause	Because, since *Since I get sunburns easily, I don't suntan.*
Contrast/Concession	While, even though, although *While my childhood was calm, my adulthood is very hectic and stressful.*
Manner	Similarity: just as, like Explanation: in that *I prepare for a test like an athlete trains for the Olympics.*
Condition	If, unless, as long as *If it rains, I don't jog.*
Intention/Purpose	So that, in order that *My school cancelled classes after the snowstorm so that students could be safe.*

Unlike a transition, a dependent conjunction cannot be used alone and cannot be followed immediately by a comma. A dependent conjunction must be followed by at least one subject and verb, or more if they are parallel. The conjunction joins clauses into one sentence:

First, I support mandatory gym classes because they will help students learn more.

Because is a dependent conjunction that forms a dependent clause: *because they will help students learn more.* Since a dependent clause cannot be a sentence alone, it must be connected to an independent clause, which can be a sentence alone. *I support mandatory gyms* is an independent clause. The conjunction *because* joins the independent clause *I support mandatory gym classes* to the dependent/adverb clause *they will help students learn more.* Together everything is one sentence: First, + independent clause + conjunction + dependent clause.

It is also important to note that a dependent conjunction is vital to the grammatical correctness of the sentence. Removing *because* in the example above would create a run-on sentence, which is a grammatical error. There is no connection between the two subjects and verbs: *First, I support mandatory gym classes they will help students learn more.*

A preposition forms a phrase, not a clause: preposition + object = prepositional phrase. Therefore, a prepositional phrase cannot be a sentence alone. The following sentence ends with a prepositional phrase:

> Generally speaking, most students feel different levels of stress due to various factors.

Of course, the prepositional phrase *due to various factors* cannot be a sentence alone. Like all phrases, it must be joined to either a dependent or independent clause. In the example above, the prepositional phrase *due to various factors* is joined to an independent clause (*most students feel different levels of stress* can be a sentence on its own).

You should be familiar with all the possible prepositions, but there are too many to list here. However, the prepositions with the same meaning as the adverb conjunctions and transitions above are the following:

Meaning	Prepositions
Examples	Such as, for example, for instance *I play many sports, such as soccer.*
Reason/Cause	Because of, due to, as a result of *Due to my long legs, I'm a good runner.*
Contrast	Despite, in spite of, regardless of *I always try new things despite my fears.*
Manner	As *I dressed as a clown for Halloween.*

A preposition has an essential grammatical function. Unlike a transition, it cannot be removed from a sentence without creating a grammatical error. In the phrase *due to various factors*, the preposition *due to* joins the object *various factors* to the rest of the sentence. The sentence *first, most students feel different levels of stress various factors* is wrong because *various factors* has no connection to the rest of the sentence. The preposition provides that connection.

As you may have already noticed, several words can be different parts of speech. *For example* and *for instance* can be prepositions or transitions, but *such as* can only be a preposition. In the first sentence below, *for example* is a transitional adverb, but in the second sentence *for instance* is a preposition. The preposition *such as* does not have this versatility.

> For example, some students must live away from home while others have to work to support themselves. Other students might have relationship problems, for instance a difficult roommate.

Also, pay careful attention to the fact that no transitions are also dependent conjunctions. Students very often use one instead of the other. Look at the following incorrect sentence based on the paragraph above:

> Exercise releases or removes the stress through vigorous physical activity. Such as, a student could lose her frustrations during a swimming lesson or a weight lifting class. Because, she would feel more relaxed and focused, and she could concentrate more.

The writer incorrectly uses *such as*, which cannot function as a transition relating two separate sentences. The preposition *such as* can only function in a prepositional phrase: such as + object. Instead, the writer should use *for instance* or *for example*, both of which can be prepositions or conjunctions. Moreover, *because* is incorrectly used as a transition. *Because* is a dependent conjunction that joins a dependent clause to an independent one to form a complete sentence. The paragraph above could be correct as follows:

> Exercise releases or removes the stress through vigorous physical activity. For instance, a student could lose her frustrations during a swimming lesson or a weight lifting class. Therefore, she would feel more relaxed and focused, and she could concentrate more.

It's important to remember the sequence of ideas when you use either a dependent conjunction like *because* or a transition like *therefore. Because* is always followed by the cause, and *therefore* is always followed by the result:

Sequence of Ideas and Structure	Examples
One sentence with an adverb conjunction: 1. Subject + verb (result) + because + subject + verb (cause) 2. Because + subject + verb (cause), subject + verb (result)	1. Increasing taxes can reduce investment because many companies don't invest in countries with very high taxes. 2. Because many companies don't invest in countries with very high taxes, increasing taxes can reduce investment.
One sentence with a coordinating conjunction: Subject + verb (cause), so + subject + verb + (result).	Many companies don't invest in countries with very high taxes, so increasing taxes can reduce investment because
Two separate sentences with a transition: Subject + verb (cause). Therefore, subject + verb (result).	Many companies don't invest in countries with very high taxes. Therefore, increasing taxes can slow the economy.

Strategy 9: Use transitions appropriately.

Do not overuse transitions in writing, just as you should not in speaking. Too many of them waste time. More important, though, they could confuse the reader if they are used incorrectly, which will lower your mark. Generally speaking, you should not begin every sentence with a transition. At least several sentences in each paragraph should begin without them. See also General Reading Strategies, Reading Strategies: Referent Questions and Coherence Questions, General Speaking Strategies.

Strategy 10: Use a variety of sentence structures but also be concise.

As you must do in the Speaking section, you must save time when you write on the TOEFL. You have only 20 minutes to organize and write your summary for the first task and only 30 minutes to write your independent essay for the second task. At the same time, you need to include a lot of detail in both answers, such as examples, descriptions, definitions, reasons, and steps. The best way to say more in a short amount of time is to be concise.

Concision is useful and advantageous in both your paraphrased summaries and your independent essay. However, this does not mean that you should write a short, brief passage. Don't think of concision in relation to the length of the entire passage. Instead, think of each sentence. Your sentences should be efficient; in other words, each sentence should refer to as much information as possible. Although your complete written response might still be long, you will say much more if each sentence is concise.

A concise answer avoids unnecessary repetition and unnecessary sentence structure. There are always many ways of expressing something, but the best style is always the one that uses the fewest words to express the most. As you've already learned, cohesive devices and parallel structure are important ways to be concise. An adjective clause is another.

Strategy 10 A: Use adjective clauses effectively.

An adjective clause is a type of dependent clause, which cannot be a sentence alone. An adjective clause can modify a noun, phrase, or whole clause.

Part of Speech Modified by the Adjective Clause	Examples
Noun (books, exam, John)	I enjoy writing books that motivate people.
	I failed the exam, which started at 9 A.M.
	I had lunch with John, who paid for the meal.
Phrase (writing books)	I enjoy writing books, which I can do at home.
Clause (I had lunch with John; I failed the exam)	I had lunch with John, which made Bill jealous.
	(The fact that I had lunch with John made Bill jealous.)
	I failed the exam, which upset my parents.
	(The exam didn't upset my parents. The fact that I failed it upset them.)

All of the adjective clauses in the chart above add some information about a noun, phrase, or clause, but notice how each adjective clause adds information related to an action (motivate, start, paid, upset, etc.). Since the adjective clauses add that kind of information into one sentence, they add information concisely. Therefore, the adjective clauses are useful for the sentences.

Strategy 10 B: Avoid unnecessary adjective clauses.

If you want to add information related to appearance or emotional effect, however, then an adjective clause is not always necessary. Whenever possible, put a one-word adjective before the noun rather than use an adjective clause. For example, all the sentences below are correct, but those on the right are more concise; compare the adjective clauses on the left with the prepositional phrases on the right:

Noun + conjunction + be + adjective	Adjective + noun
1. I took a course that was interesting.	1. I took an interesting course.
2. A teacher who was helpful taught me study skills.	2. A helpful teacher taught me study skills.
3. I live in a country that is very crowded.	3. I live in a very crowded country.
4. *I live in Japan, that is very crowded.	
*A proper noun cannot be modified by an adjective.	

When describing the possessions or characteristics of a noun, use a prepositional phrase (with + object) instead of an adjective clause. Compare the adjective clauses on the left with the prepositional phrases on the right:

Noun + conjunction + have + object	Noun + with + object
1. I have a dog that has three legs.	1. I have a dog with three legs.
2. A man who has blue eyes and black hair is rare.	2. A man with blue eyes and black hair is rare.
3. A car that has two doors is a coupe.	3. A car with two doors is a coupe.
4. Parents who have children need money.	4. Parents with children need money.
5. A plan that has focus will succeed.	5. A plan with focus will succeed.
6. I read a novel that has six hundred pages.	6. I read a novel with six hundred pages.

Strategy 11: Write in the active voice.

In English, there are two voices: the active and the passive voice. Voices are different from tenses. Both voices can be written in any tense, depending on the verb (some verbs can't be continuous, etc.). The difference between the active and the passive voice depends on the relationship between the subject and the action or state.

It's a good idea to use the active voice as often as possible. This is true for both writing and speaking.

In the active voice, the subject performs the action or experiences the state of the verb. The actor precedes, or comes before, the active verb and the verb is immediately followed by any objects: subject + verb + direct object. Because a sentence in the active voice begins with the actor, it is more direct and easier for a reader (and listener) to follow. Look at the following examples in the active voice:

> Subject + verb + direct object: China manufactures many products.

> Subject + verb + object + prepositional phrase: I will study economics in the United States.

> Subject + verb + prepositional phrase: Many citizens live below the poverty line.

In the passive voice, on the other hand, the subject receives the action of the verb, and the passive verb is usually not followed by an object (states cannot be passive). Therefore, to rewrite an active verb in the passive voice, the active verb must have an object. Only two of the examples above can be written in the passive voice:

> Subject + verb + prepositional phrase: Many products are manufactured in China.

> Subject + verb + prepositional phrase + prepositional phrase: Economics will be studied by me in the United States.

Several ideas are important here. First, notice how the objects *many products* and *economics* in the active sentences becomes the subjects in the passive sentences. Second, while the sentence about manufacturing in China sounds fine, the sentence about economics sounds awkward. This is because the writer in the second sentence is discussing himself or herself. Therefore, the active sentence sounds more natural, and the passive voice seems unnecessary. Finally, the third active sentence (Many citizens live below the poverty line) cannot be passive because the verb *live* has no object. Verbs that require or can take an object are called transitive verbs. *Manufacture* and *study* are transitive verbs. However, *live* is not a transitive verb in the third sentence. *Live* is intransitive, so it cannot be written in the passive voice.

Strategy 12: Use the passive voice when it is necessary.
As you've already learned, you can't use the active voice all the time. Although the active voice is better, the passive voice must be used sometimes, depending on the structure, meaning, or emphasis of the sentence. The four main reasons for using the passive voice are listed below:

Reason for the passive voice	Examples (passive verb in *italics*)
1. The actor is unknown or unimportant. 2. The actor is assumed or already mentioned. 3. *The emphasis is on the receiver, not actor. 4. The paragraph needs variety and style. *The third reason is often a result of either the first and/or second. The emphasis is sometimes placed on the receiver because the actor is unknown or assumed. In the example, the receivers are contrasted, so the main idea requires the passive voice.	1. If an object ***is dropped*** from a great height, it will reach a speed of 100 km/hr in less than three seconds. 2. In my city, the mail ***is delivered*** every day between Monday and Friday. 3. Although the active voice is better, the passive voice ***must be used*** sometimes… 4. The president can veto any law that Congress writes. Congress can reintroduce the law as often as it wants. However, if a law ***is vetoed***, Congress usually edits or alters it before introducing it again.

In the first example, the actor is irrelevant to the main idea of the sentence. The sentence could be rewritten in the active voice with a subject such as someone or a person (if a person drops an object from a great height…), but that wouldn't improve the sentence's clarity, so it would be unnecessary. In the second example, the unmentioned actor *mail carrier* can be assumed based on the subject *mail* and the verb. People who perform very specific tasks are commonly assumed for the same reason, such as fire fighters, bus drivers, police officers, and doctors/surgeons. In the third example, the two possible voices in English are compared, so using the passive is more direct since it keeps the focus, or emphasis, on the voices. Rewriting the sentence actively is possible, but it would require more words and more repetition, for example *writers* and *use*: *Although it is better for writers to use the active voice, writers must use the passive voice sometimes.* Finally, the fourth example uses the passive voice in order to avoid repeating the word order of the first sentence: *The President + veto + law.*

Strategy 13: Edit your work.
Students often make the mistake of writing right until the time ends without editing their work. This means that few errors have been corrected and potentially many errors remain, which will lower the score. Students often focus too much on the number of words, or they rush at the end to include key points because they didn't plan enough at the beginning. If you take good notes, organize your ideas, and focus on writing well, not just a lot, then you should be able to finish before you reach the time limit.

Leave at least a few minutes to read over your writing at least once before you submit it. Thirty seconds is not long enough to edit anything. You need more than a minute. All writers, including the ones who wrote this book, make mistakes, so don't expect or plan to write flawlessly. Many students make mistakes that they would normally not make simply because of the pressure and stress of the test and time limit. If you plan to review your work, and you know what kind of errors to look for, then you can improve your mark significantly.

First, you need to establish the habit of reviewing your work within the time limit. Therefore, you should leave time for editing even when you practice in class or with a tutor. If you don't, then you will find it difficult to do during the actual test. When people feel pressure, they usually rely on habit, which is why it is so important to establish good habits when you practice. The stress and pressure of a test is not a normal situation, and it is very difficult to do new things under abnormal conditions.

Second, you should recognize what kinds of errors you make most often. Usually, writers make certain errors that are particular to them. Of course, not all mistakes are unique. For example, any writer could occasionally use an incorrect verb tense or pronoun, but people's errors also have patterns because people have difficulty with different parts of language. One person could have difficulty with adjective clauses while another may not really understand the past perfect. During practice with your class, study group, or tutor, you should identify any patterns in your writing. This will not only help you avoid repeating these errors, but it will also give you a set of errors to look for when you write the actual test.

Strategy 14: Edit quickly, but don't skim or read too passively.
Editing is not skimming or active reading. Active reading is a search for meaning and understanding. You want to know what is happening, who or what is doing it, who or what is affected, and how it relates to what came before and what comes after. Reading actively is analytical because you are thinking while you read, and it involves an understanding of grammar. However, reading is more passive than editing because reading doesn't involve correction. When you read a published passage, you can trust that it has been corrected already, so you trust the grammar.

Editing, however, is a search for errors and inconsistency in the structure or use of the language. Therefore, it is also analytical but more active. You have to think of the rules of English grammar and use them to find mistakes. You are changing the language, not just following it. Moreover, you must edit your own writing, so it is self-analysis, which is a unique challenge.

A common mistake for students who are editing their own work is that they skim, or read quickly and too passively, and they hope that errors will reveal themselves. Skimming is reading quickly for basic understanding; it is faster and less focused than active reading. Often students just skim the passage, hoping to find errors. This is a mistake because editing is more conscious and purposeful than reading. You have to think about particular rules and apply them to each phrase and clause in the passage. However, you have to do this quickly and efficiently. Therefore, the best way to do this is to focus your attention on groups and to limit your analysis to a few rules.

Strategy 15: Edit for specific errors, and edit in chunks.

Based on your knowledge of your own writing, choose a set of rules to apply and analyze your writing for mistakes related to those rules. You can't edit for every possible error because you don't have the time. You should devote the majority of your time to writing, so you have only a few minutes to read over your text. However, these few minutes can greatly improve your score if you can catch some basic errors. Also, it is too complicated to try to remember and apply many rules at once, so apply only two or three rules, such as verb tense, fragments, and relative pronouns in adjective clauses.

Likewise, you can't edit well if you read the entire passage from beginning to end all at once. Instead, edit one paragraph at a time. Analyze each sentence according to the rules you've chosen. Read each sentence carefully and apply one rule at a time to one paragraph before you move on to the next. This strategy will help you to apply the rules correctly and also help you to work efficiently. You need to edit your entire essay in a short amount of time, so focus on a few rules and look at your essay in sections.

Strategy 16: Understand some common errors well, and focus on them.

In order to edit successfully for certain errors, you need to understand the grammar rule associated with that error thoroughly. You can't edit well if you have to figure out the rule first and then use it. Therefore, use your practice time to identify any particular patterns to your mistakes and learn as much as you can about how to fix them.

Also, there are some common errors for ESL students, and you would be a stronger writer and editor if you reviewed them thoroughly. The following list does not include all the important rules of English grammar because there are too many to list here. You should consult a good grammar reference and exercise book for a more complete review (also see Improve-Your-Score Strategies for more suggestions). However, the following list identifies some of the most likely errors that you might find in your writing:

Strategy 16 A: Check word form and function.

Each part of speech, or word form, has specific functions in English. Students often write the wrong form even though they know the correct one because they're writing quickly and they forget to edit.

Common Error #1

The verbs *look*, *feel*, *smell*, *sound*, *taste* can be states or actions:

Subject + state + adjective (adjective describes the subject)

Subject + action + adverb (adverb describes the verb)

Errors (in *italics*)	Corrections (in *italics*)
1. The people felt *angrily* about the new rules.	1. The people felt *angry* about the new rules.
2. Someone who can't sleep always looks *tiredly*.	2. Someone who can't sleep always looks *tired*.
3. I looked *quick* out the window to check the weather.	3. I looked *quickly* out the window to check the weather.
4. Silk feels *smoothly*, so many people enjoy it.	4. Silk feels *smooth*, so many people enjoy it.

Common Error #2

An adjective can describe a noun, but only an adverb can modify a verb, adverb, infinitive, or another adjective.

Errors	Corrections
1. The writer gives more specifically examples of the problem.	1. The writer gives more specific examples of the problem.
2. The writer argues more specific about the problem.	2. The writer argues more specifically about the problem.
3. I plan to travel extensive and frequent.	3. I plan to travel extensively and frequently.
4. I plan to go on widely and frequently trips.	4. I plan to go on extensive and frequent trips.

Common Error #3

A verb form cannot be a subject or object. Although some verbs and nouns are spelled identically, such as *view* or *search*, not all are the same.

Errors	Corrections
1. The speaker disagrees with spend all the time.	1. The speaker disagrees with spending all the time.
2. Read a book is more fun for me than watch a movie.	2. Reading a book is more fun for me than watching a movie.
	To read a book is more fun for me than watching a movie.

Strategy 16 B: Verify that only actions are in the continuous tense.

A verb can be either an action or a state. Generally, an action can be perceived with one or more of the five senses, such as the verbs run, sing, or write. As always, there are exceptions; some actions can't be perceived easily, such as relax and decide. Actions can be expressed in all tenses, including the continuous.

Stative verbs, or states, cannot usually be perceived directly; they refer to the emotions, the senses, reactions, beliefs, among other ideas, and they are rarely in the continuous tenses (present continuous, past continuous, present perfect continuous, etc.). Stative verbs are very common, so review them carefully.

Type of Stative Verb	Examples
The emotions	Like, love, hate, prefer, concern, impress, please, satisfy, surprise
The senses	See, seem, smell, sound, taste
Beliefs/opinions	Believe, doubt, know, *think, realize, understand, imagine, mean, feel, agree, disagree, support, oppose *Wrong: I am thinking that the plan is illogical. *Correct: I am thinking of a solution.
Miscellaneous	Be, belong, involve, include, contain, need, have/possess/own, etc.

This rule can be very confusing for ESL students because native speakers break this rule often, especially in speech. Moreover, as you learned in Strategy 16 A, some of the verbs above have multiple meanings, and those other meanings could be actions in the continuous. For example, the sense verbs *smell*, *taste*, and *sound* are stative when they describe the condition of the subject: *The flowers smell wonderful*. However, the verbs are actions when they refer to what the subject does: *I was smelling the flowers*. Study the multiple meanings and uses of verbs, and be careful about how you use them.

Errors	Corrections
1. Although I still have some difficulties in English, I am understanding more now.	1. Although I still have some difficulties in English, I understand more now.
2. I am supporting the plan to build a new shopping center for the economic benefits.	2. I support the plan to build a new shopping center for the economic benefits.
3. Due to the environmental impact and the extreme cost, I am thinking that the new shopping mall is a bad idea.	3. Due to the environmental impact and the extreme cost, I think that the new shopping mall is a bad idea.
4. All education should be free because many poor students are needing help.	4. All education should be free because many poor students need help.

Strategy 16 C: Make sure all facts are in the present simple, not the continuous.
ESL students frequently overuse the present continuous. The present continuous should only be used for actions that are in progress or happening right now. Moreover, you must use the present simple, not the present continuous, for proven facts in nature or in academic fields (science, medicine, economics, the arts, etc.). Proven facts are actions or states that are always true, so they are expressed in the present simple tense. Read the following sentences:

China needs fossil fuels and raw materials because its economy is growing rapidly. Therefore, the country is importing huge amounts of oil as well as steel and other metals. This has raised the value of energy stocks in many stock markets around the world.

In Sentence One, the main verb *need* is stative, so it can't be present continuous. The second, dependent verb *is growing* can be continuous because it refers to an action that is in progress right now, like the verb *is importing*. Also, both actions are not facts since they have not always been true and they may not be true forever. China's economy has not always been growing rapidly and may not continue growing rapidly forever. In contrast, the verb *has raised* is in the present perfect because that action occurred at some unspecified time in the past, and it is not currently in progress according to the writer.

Errors	Corrections
1. Planting trees is a good idea because trees are producing oxygen.	1. Planting trees is a good idea because trees produce oxygen.
2. Students should always work hard because universities are accepting the best students.	2. Students should always work hard because universities accept the best students.
3. Clouds are forming partly from evaporated water.	3. Clouds form partly from evaporated water.
4. Volcanoes are shooting lava into the air when they erupt.	4. Volcanoes shoot lava into the air when they erupt.

Strategy 16 D: Ensure that the past simple has a known, specified, or implied time.
Students often misuse the past simple. Use the past simple when you specify a particular time in the past with a time, date, prepositional phrase, such as *in high school*, or clause, such as *when I was young*. Also, you can continue to use the past simple or continuous in a paragraph once you specify the time period. After a time has been mentioned, it becomes known or assumed and doesn't have to be repeated, so you don't have to keep repeating the same date. Moreover, a specific time is unnecessary if the time is implied or assumed based on the vocabulary, which is common in history:

> The veterans of World War I returned with many physical and emotional scars. (WWI is a known period from 1914–1918)
>
> The first colonies of the United States were founded by settlers from England and other European countries. (The country and verb refer to a specific, known period of history —the early 17th Century)

On the other hand, when you want to refer to the general, unspecified, or unknown past, use the present perfect. Among its many uses, the present perfect refers to past experience, which is a past action or state at an unknown or unspecified time.

Errors	Corrections
1. I'd prefer to live in the mountains because I never lived there.	1. I'd prefer to live in the mountains because I never lived there when I was young.
	I'd prefer to live in the mountains because I have never lived there.
2. I studied many instruments, but my favorite is the trombone.	2. I studied many instruments in high school, but my favorite is the trombone
	I've studied many instruments, but my favorite is the trombone.

Strategy 16 E: Correct any sentence fragments, or incomplete sentences.
A complete sentence must have at least one independent subject and verb. Depending on the sentence, there may be more subjects and verbs if there is parallel structure, but one of each is the minimum. An independent subject and verb can be a sentence alone. The following sample sentences are complete and correct:

> I agree with the plan to build a new theater.
>
> I support the new theater project.
>
> I am in favor of the planned theater.
>
> I agree with the opinion.

A subject and verb cannot be independent when they follow a conjunction. Therefore, the following clauses cannot be sentences; they are fragments, or incomplete sentences, and they must be joined to an independent clause, such as those above:

> For it will improve the economy and bring entertainment
>
> Because it will improve the economy and bring entertainment
>
> Which will improve the economy and bring entertainment
>
> That it will improve the economy and bring entertainment

Errors	Corrections
1. I agree with the plan to build a new theater. Because it will improve the economy and bring entertainment.	1. I agree with the plan to build a new theater because it will improve the economy and bring entertainment.
2. I support the new theater project. Which will improve the economy and bring entertainment.	2. I support the new theater project, which will improve the economy and bring entertainment.
3. I agree with the opinion. That it will improve the economy and bring entertainment.	3. I agree with the opinion that it will improve the economy and bring entertainment.

Unlike clauses, phrases have no subject or verb. Therefore, they cannot be sentences, such as the following:

> For example economic improvement and entertainment
>
> To improve the economy and entertain people
>
> For economic improvement and entertainment

KAPLAN

Errors	Corrections
1. I think that the new theater will bring many benefits. For example, economic improvement and entertainment.	1. I think that the new theater will bring many benefits, for example economic improvement and entertainment.
2. I support the new theater project. To improve the economy and entertain people.	I think that the new theater will bring many benefits. For example, it will create economic improvement and entertainment.
	2. I support the new theater project to improve the economy and entertain people.
	To improve the economy and entertain people, I support the new theater project.

Strategy 16 F: Look for run-on sentences, or missing conjunctions.

A run-on sentence occurs when a sentence does not have enough conjunctions or periods. In other words, it is really two or more sentences that are not separated by a period or not connected by one or more conjunctions. A run-on can be corrected by adding a period or the appropriate conjunction and punctuation.

Error	Corrections
I support the new theater project it will improve the economy and bring entertainment.	1. I support the new theater project. It will improve the economy and bring entertainment.
	2. I support the new theater project, for it will improve the economy and bring entertainment.
	3. I support the new theater project because it will improve the economy and bring entertainment.
	4. I support the new theater project, which will improve the economy and bring entertainment.

In the incorrect sentence above, there are two nonparallel subjects and verbs. One subject and verb must be the independent clause, and the other subject and verb must be a new sentence or connected by a conjunction. The first correction adds a period and leaves both as separate sentences. The second joins the subjects and verbs with the coordinating conjunction *for*. The third correction uses an adverb clause, and the fourth correction uses an adjective clause.

Strategy 16 G: Eliminate any unnecessary or extra conjunctions.
Coordinating conjunctions (and, but, yet) can join clauses, and they can join parallel words and phrases. In fact, the previous sentence is an excellent example of that. The conjunction *and* joins the nouns *words* and *phrases*, which are parallel, and it joins the first clause (coordinating conjunctions…can join clauses) to the second (they can join parallel words).

Dependent conjunctions (although, what, which) join clauses but not words. In other words, they do not create parallel structure. They form dependent clauses: adverb clauses, adjective clauses, or noun clauses.

Students can sometimes add too many conjunctions because they lose track of how many clauses they've written. You should edit your work to make sure that your sentences have the correct number of conjunctions in the right places. Review the following examples:

> Example 1: I think that the new theater which will improve the economy and bring entertainment.
>
> Correction: I think that the new theater will improve the economy and bring entertainment.

The first example includes the unnecessary dependent conjunction *which*. Because of the conjunction *which*, the dependent subject *new theater* of the noun clause has no verb: *which will improve the economy and bring entertainment* is an adjective clause modifying *theater*, so the noun clause is incomplete. Removing *which* also eliminates the adjective clause and creates the noun clause *the new theater will improve the economy and bring entertainment*. The noun clause is the direct object of the independent subject and verb *I think*.

> Example 2: Because the new theater will improve the economy and bring entertainment, so I like the project.
>
> Correction: Because the new theater will improve the economy and bring entertainment, I like the project.
>
> Correction: The new theater will improve the economy and bring entertainment, so I like the project.

The second example has only two clauses but also two conjunctions: *because* and *so*. The conjunction *and* makes the two verbs *will improve* and *bring* parallel with the subject *new theater*. Both verbs are parallel, so they are part of the same clause: *the new theater will improve the economy and bring entertainment*. Since one clause in the whole sentence must be the independent clause, one clause must have no conjunction. Each correction eliminates a different conjunction, making one or the other clause independent and creating a complete sentence.

WRITING STRATEGIES: TASK TYPES

As you practice and prepare for each task in the Writing section, keep the general strategies in mind. Always remember to organize your ideas, vary your word choice and structure, use the active voice more than the passive, and edit your work. Now, you can review specific strategies for each task in the Writing section of the TOEFL.

Writing Task 1: Reading and Listening

The first task is integrated, which means that you must do more than write. This task involves reading, listening, and writing. First, you have three minutes to read an academic passage and take notes.

At the end of the three minutes, the reading disappears, and you listen to an academic lecture on the same topic but with some variations and differences from the reading. You must take notes on the lecture while you listen to it. After the lecture is finished, you hear a question that asks you to combine information from both passages. You have 20 minutes to plan, organize, and write a summary of both. In your summary, you must explain how the key points of the lecture relate to those from the reading. After the lecture finishes and you hear the final question, the reading reappears, so you can refer back to the reading as you write.

The first Writing task is similar to the fourth Speaking task. In both tasks, you must first read a passage, listen to a lecture, and then either speak or write about both passages. However, the two tasks are not identical. In the Writing task, the reading passage is longer and includes more than a description or definition of a concept. It includes several key points and may involve an argument, or a position for or against something.

The lecture in the first part of the Writing section is about the same length as the lecture from the fourth Speaking task. However, the lecture in the Writing section includes more contrast, contradiction, or refutation. You have to record several key points from the lecture, and also pay attention to how they differ from the points in the reading.

The Essay Question that you will hear at the end of the lecture will ask you to summarize the points in both passages and explain the connections between them. Therefore, you must compare and contrast the passages as well as demonstrate your understanding of the various concepts in both.

Since the first writing activity requires reading and listening, many of the strategies below resemble some of the strategies from the Reading, Listening, and Speaking sections of this book. By now, you should be familiar with almost all the strategies for this task. Therefore, the discussion and explanations below not only prepare you for the first task of the Writing section, but they also serve as a thorough review of many key elements of Kaplan's strategies for the TOEFL.

Strategy 1: Be familiar with the possible types of arguments and discussions.
The first writing activity on the TOEFL is both a summary of two passages and a comparison/contrast of them. Both the reading and listening passages will be on similar or related main topics, but there will be some degree of difference between them. Some of the key points in each passage could be similar, but not all.

You can't predict the exact topics of the reading and listening passages for this task because the range of possible topics is too broad. However, you can become familiar with the kinds of discussions and debates that occur in most academic fields. Both the reading passage and the lecture will be academic, and either one could be argumentative or expository. An argumentative, or persuasive, passage argues for or against some opinion or plan. An expository passage describes and explains some concept(s), phenomena, or theories. Both argumentative and expository passages may involve concepts and detail from a variety of academic fields, such as history, the arts or the sciences.

Expository Passages

Reading Passage	Listening Passage/Lecture
1. An explanation of a general principle or rule Sample Topic Attorney-client privilege is a legal concept that protects the communications between a lawyer and a client. The communications are private and cannot be used in court or published in the media.	1. An exception/unusual example of the principle Sample Topic If a lawyer does not work primarily as a legal advisor, but instead works in another capacity, such as a business advisor, then the lawyer-client privilege does not apply.
2. An explanation of a problem Sample Topic The late 19th and early 20th century was a period of huge urban expansion, industrial growth, and immigration, but there was much political corruption, illness, and drunkenness. Some people blamed alcohol.	2. An attempt to solve the problem that is partially successful and/or successful in unexpected ways, and involves unintended consequences Sample Topic The Prohibition laws began in 1920 and banned the manufacture and sale of alcohol throughout the United States in order to encourage morality, improve people's health and reduce crime. The laws destroyed small alcohol businesses, but large, well-organized criminal organizations prospered, which led to an increase in crime, corruption, and violence.

Argumentative passages

Reading Passage	Listening Passage/Lecture
1. Support for a theory, interpretation, or point of view Sample Thesis Video games make children hyper-active, distracted, and insensitive to violence and danger.	1. A complete or partial rejection/refutation of the theory, interpretation, or point of view Sample Thesis Studies on the negative effects of video games on children are inaccurate. Although video games can be an encouragement to troubled children, bad behavior is largely the result of poor supervision, bad diet, and boredom.
2. Support for a plan or action Sample Thesis Oil companies should be banned from drilling for oil in the ocean due to the possible environmental damage.	2. A complete or partial rejection/refutation of the plan or action Sample Thesis A ban on offshore drilling is unnecessary because taking oil from the ocean floor reduces the risks of transporting it and the need to import it from other countries.

Strategy 2: Read actively.

Regardless of the type of passage, you must think about what you are reading, so don't skim the passage. Read closely and carefully because you are expected to learn about and discuss an academic theory, problem, or plan. In order to understand the concepts involved, you have to think logically in abstract as well as concrete terms, and you must pay attention to subtle but important differences in opinion or explanation.

Also, you must use the time wisely. Although you will be able to see and reread the passage when you write your response, you should still use the reading time to understand as much as you can. Rereading the passage later will use valuable time that is better used for writing, not reading. You should only have to go back to passage to check a detail, such as a definition or reason, or the spelling of a word.

Finally, you need to do more than just record names and data. Your answer must be more than a list of information. A high-level answer shows understanding, which is demonstrated in your paraphrasing and in your comparison of the reading and the lecture. Therefore, you have to do various things while reading, which are discussed in more detail below: identify the type of passage, look for vocabulary clues, infer, think of ways to restate ideas, connect details to the main idea, and so on.

Strategy 3: Take notes in fragments, symbols, abbreviations, and acronyms.
Your notes are only clues to remind you of such things as the key points and attitude of the writer. Therefore, don't try to write a transcript. Your notes should be individual words, phrases, and symbols, not complete sentences. Write only complete sentences in your essay. At the same time, however, there's less pressure when you read than when you listen to a lecture, so you can use fewer symbols and write more complete words when you take notes from a reading passage.

Read the following passage that is similar to one you could get during the actual test:

Deficit Spending in the United States

One of the most persistent topics of debate in America today is the size of the federal budget deficit. A government deficit occurs when the government spends more money than it earns, or receives. In other words, the government goes into debt by borrowing money from investors and even foreign governments through various means, such as selling government bonds. The use of that borrowed money is called *deficit spending.* Annual federal budget shortfalls in the United States regularly run in the hundreds of billions of dollars. At present, the U.S. government owes investors approximately eight and a half trillion dollars, an admittedly very large number that critics of deficit spending regularly point to as a sign that fiscal disaster is just around the corner.

But these doomsayers are overlooking some important facts. Since at least the Great Depression of the 1930s, deficit spending has been a mainstay of American fiscal policy. John Maynard Keynes, perhaps the most influential economist of the 20th century, recommended deficit spending as a means to help the United States escape from the economic quagmire that had trapped the country since the stock market crash of 1929. There were some very good reasons for this. When the government spends more money, this amounts to an increase in demand. Greater demand equals greater employment, and greater employment leads to more money in the pockets of people who will spend it, which in turn will create greater demand. So, deficit spending is an important way for the government to improve an unhealthy economy or to maintain a healthy one.

Deficits are far from a sign of imminent economic disaster. Rather, deficit spending is an essential tool to keep the nation's economy healthy and stable.

After a close, careful reading of the above passage, your notes could look like the following:

gov def = gov debt

$ from inv + other countries

Spend $ = def spend

Current debt = ↑ 100 m /year, total = about $7.5 billion

Critics = def → financial X

from 1930's until now, def spend common

JMK: def spend = solution to problems after '29 crash

Spend def → demand → more jobs → more $ → more demand, etc

Def spend = economic improvement during recession/depression

Strategy 4: Be prepared to infer meaning.

Not all ideas are stated explicitly, or directly. Sometimes, you must infer meaning that is implied, or stated indirectly. To *infer* means to understand unstated meaning based on logic and detail. To *imply* means to communicate unstated ideas through logic and detail. In this activity, the writer implies and you infer.

You may have to infer all or some of the author's main idea, purpose, attitude, or the connections among the author's key points. For example, the writer might not always use a key word, such as *because* or *therefore*, to indicate cause and effect. Likewise, examples are not always preceded by *such as*, *for example*, or *for instance*.

Moreover, inference is often necessary to comprehend fully what the writer's (or speaker's) explanation or argument. It is easier to paraphrase a passage when you have a better understanding of the main idea. For example, look at the following sentence from the passage:

> Greater demand equals greater employment, and greater employment leads to more money in the pockets of people who will spend it…

Here, you can and should infer that the *money in the pockets of people* comes from paychecks or salaries. The noun phrase *greater employment* implies more jobs, which pay workers salaries or wages. When people have more money due to higher employment, they are being paid while working at jobs. Making that inference is crucial to understanding the point and possibly paraphrasing it in your response.

Strategy 5: Identify the type of passage.

There are two main types of passage: expository and argumentative. An expository passage defines, describes, and explains ideas, as with a definition and explanation of representational government (the democratic process by which voters elect representatives). An argumentative passage defends a position for or against the main topic; an example would be a passage in which a writer states that the Electoral College (the process by which Americans elect the president) is unnecessary and should be eliminated.

Expository and argumentative passages are not entirely exclusive. In other words, they are not always separate and unique. Expositions don't include argument, but an argument may include some definition and/or explanation before the writer supports or rejects an idea or plan. The sample passage above about deficit spending mixes exposition and argument; it begins with a definition of deficits and deficit spending and a description of current deficit levels in the United Stats. In the second paragraph, it becomes argumentative as the writer defends, or supports, deficit spending despite the size of U.S. debt.

Pay close attention to the type of passage that you read in the first task. Identify if it is purely expository, argumentative, or a mix. The type of passage is important for understanding the writer's purpose and main idea correctly. For instance, if the writer never makes an argument, then you should not use any language of argument, such as the verbs *argue*, *propose*, and *support*. Also, if the passage is mixed, like the one above, then you need to be clear about which ideas are argumentative and which are expository. For instance, a student shouldn't write, "The writer argues that a government deficit is the same as government debt" because that idea is stated, mentioned, or explained, but it isn't argued. The definition and explanations in the first paragraph are presented as facts, not opinions.

Strategy 6: Identify the main idea and purpose of the passage.
You've learned in all three previous sections of this book that the main idea of a passage is a combination of the main topic, or the subject, and what the author says about the topic. Besides the main topic, such as deficit spending, the main idea includes how the author discusses it. For example, the author of the passage above argues in favor of deficit spending because it maintains a healthy economy. Therefore, the main idea of that passage includes deficit spending and the author's reasons for supporting it: it has been successful since the 1930's; it creates demand and jobs; etc.

The main idea of the reading passage fits the first type of argumentative passage outlined in Strategy 1: support for a theory, interpretation, or point of view. This support could be neatly expressed in a thesis statement, which is one sentence that expresses an author's main idea. For instance, the following sentence could be an accurate thesis statement for the passage above: Deficit spending is a good economic policy because it increases demand and employment.

Purpose is directly linked to main idea. In fact, the main purpose is simply another way of expressing the main idea. The purpose of an argument, for example, is to argue. The purpose of an explanation is to describe and to explain. Therefore, one could say that the purpose of the passage above is to support deficit spending based on the economic benefits.

Notice how you must partially infer the author's argument about the deficit spending. The writer never explicitly states, "My opinion is that deficit spending improves the economy" or "I think that deficit spending is economically beneficial." Direct expressions of opinion are possible, but in that passage, you need to infer the author's opinion based on the choice of words and attitude.

Strategy 7: Pay attention to attitude.
The author's attitude, or tone, refers to the kind of positive or negative emotions that the author expresses toward the topic. Attitude is directly linked to main idea and the type of passage. In a purely expository passage, the author's tone is usually neutral or objective since the purpose is just to explain. In an argument, however, the tone is either positive or negative to varying degrees depending on the opinion.

Use vocabulary clues to help you identify the attitude and, by extension, the main idea and purpose. In the passage above, the writer's definition and explanation of deficit spending is largely neutral, but the second paragraph is more positive based on the choice of words: a mainstay, recommended, help, very good reasons, more money, greater employment, greater demand, an important way, essential.

Strategy 8: Paraphrase while taking notes.

When you summarize and compare the reading and the lecture, you must also paraphrase the key points from both passages. Since you must restate the ideas in your own words when you write your response, it's a good idea to record any synonyms or similar sentence structures right away in your notes. Of course, if you can't think of any immediately, then just record a key word or symbol. However, some students focus on copying words out of the reading without thinking about their meaning and how to restate them. This leaves all the work of paraphrasing to the end. The paraphrasing is mandatory only in your essay, but you can save yourself some time and effort if you can do some of it while you read. For example, some of the notes above paraphrase ideas from the passage.

Original	Paraphrased Note-Form
1. In other words, the government goes into debt by borrowing money from investors and even foreign governments through various means, such as issuing government bonds.	1. $ from inv + other countries
2. Annual federal budget shortfalls in the United States regularly run in the hundreds of billions of dollars. At present, the U.S. government owes investors approximately eight and a half trillion dollars...	2. gov debt = ↑ 100 b /year, total = about $8.5 t

Strategy 9: Pay attention to any definition and explanation of key terms.

Since the reading (and the lecture) is academic, it includes some terminology and concepts that are necessary to the main idea. Usually, these ideas are defined in some way if they are not commonly or widely understood. Remember that you are expected to have a basic understanding of science, art, and world history. Therefore, there are always many ideas that are not defined because you are expected to know them already. For example, the reading passage probably would not define a scientific term such as *gravity*, a literary term such as *poem*, or an economic term such as *exchange rate*. These are commonly known ideas that should require no definition. In the passage above, some of these widely understood terms include *investors*, *Great Depression*, and *stock market*, among many others.

The fact that you are expected to have some basic understanding of the world and particularly the United States is one of the main reasons that reading and listening widely is so important. The more you read about on your own, the more informed you will be for the test. For example, the term *budget* is a key term in the passage above. Although it may not be as familiar to you as *stock market*, it isn't defined in the passage. A budget is a financial plan that identifies future income, costs, and savings. Just as a person might create a budget, a government or company can do the same. This is true for most modern countries, so government financial planning should not come as a complete surprise even if you hadn't studied international finance. At the same time, there are advanced or specialized concepts that you can't be expected to know.

Look for word clues to help you identify definitions in the reading. You should remember from the Reading and Listening sections of this book that the key words for definitions include the transitional phrase *in other words* and the verbs *be, refer to, include, involve, be called, be defined as, be referred to as, be related to*, among others. For instance, the passage above defines *government deficit* using the verb *occurs* and the transition *in other words*, and it defines *deficit spending* with the verb *be called*. See General Reading Strategies, General Listening Strategies for more.

Strategy 10: Use context clues to understand unfamiliar vocabulary.
Of course, it's impossible to learn all vocabulary that might appear on the TOEFL, so you won't know all the terms and concepts that are undefined in the reading passage (or the lecture as well). Therefore, you have to use the context to guess the meaning of such terms. In other words, you must use the surrounding people, nouns, actions, and states, to understand the meaning of words that you don't already know. This can involve using cohesive devices, repetition, related vocabulary, and structure clues.

The passage above includes several unusual words whose meaning is directly connected to what is said before or after them. The adjective *federal* is used a number of times—first, in Sentence One. Sentence Two offers a clue to its meaning:

> One of the most persistent topics of debate in America today is the size of the federal budget deficit. A government deficit occurs when the government spends more money than it earns, or receives.

Deficit is repeated in the second sentence, so a federal deficit must be related or similar to a government deficit. Without any other details to contradict your assumption, you should assume that *federal* refers to government in the United States. In fact, *federal* is one of the adjectives that refers to a level of government in the United States.

Repetition of and connections among vocabulary words can also help you guess the meaning of the noun *shortfall*, which appears in the first paragraph:

> Annual federal budget shortfalls in the United States regularly run in the hundreds of billions of dollars.

Based on the repetition of the adjective *federal* and the noun *budget, shortfall* must be a synonym for *deficit*. This is a safe assumption because there is no clue in the surrounding sentences that the writer has shifted or changed topics. Repetition, however, is not the only context clue.

Context also includes similarities or connections among places, times, identities, causes, effects, steps, intentions, and so forth. These connections can be made using cohesive devices, such as pronouns and adjectives. Moreover, you should try to use your own knowledge to guide you as well. A combination of background knowledge, cohesive devices, vocabulary clues, and time clues can help a reader understand the nouns *doomsayers* and *quagmire* in the second paragraph:

> But these doomsayers are overlooking some important facts. Since at least the Great Depression of the 1930s, deficit spending has been a mainstay of American fiscal policy. John Maynard Keynes, perhaps the most influential economist of the 20th century, recommended deficit spending as a means to help the United States escape from the economic quagmire that had trapped the country since the stock market crash of 1929.

The demonstrative pronoun *these* is a clue that the doomsayers have already been mentioned. An attentive student should also recognize a connection between *doom*, which has a negative tone and means *destruction* or *death*, and the noun *disaster* mentioned in the earlier sentence. Therefore, the reader could assume that the doomsayers are the critics mentioned in the last sentence of the previous paragraph. Likewise, the noun *quagmire* follows the adjective *economic* and is connected to the period immediately after a stock market crash in 1929. That date is close in time to the phrase *Great Depression of the 1930s* in the previous sentence. Even if you aren't aware that two are connected, you can safely assume that they are, and that *quagmire* must relate to a period of economic difficulty, hardship, and decline during the 1930s.

Strategy 11: Don't focus too much on unfamiliar vocabulary.

Vocabulary is an important part of reading comprehension, but not all vocabulary words are necessary to understand the main idea. Therefore, you shouldn't spend too much of your limited reading time on guessing the meaning of every word in the passage. Remember that you have only three minutes to read and take notes. For example, the noun *bonds* might be unfamiliar. Based on the fact that the noun *bonds* is an example of the means, or ways, that a government borrows money, you may be able to guess that a bond is an investment, which is true. However, notice that it is not necessary to understand exactly how a bond works, how much it costs, or anything else about it. In fact, all you need to understand is that bonds are a way for a government to borrow money. The sentence structure makes that clear, and any other meaning is unnecessary because the noun is not repeated or used in any other sentences. Therefore, it is a minor detail.

In order to save time, you must judge whether a detail is key or minor. Ask yourself questions like the following: Is it defined? Is it repeated? How many times is it mentioned? Does it define or help me understand any important concept(s)? Is it compared or contrasted with anything? Is it related to any reasons for the author's opinion? An idea is a key point if it supports the author's main idea in some fundamental way. Based on these criteria, the noun *government bonds* is minor. Mentioning the government bonds in your response would not be incorrect, but it is also not mandatory. The best way to determine if an idea is essential is to identify the idea's relationship to the main idea.

Strategy 12: Connect details to the main idea.

Make sure you understand how an idea relates to the author's main idea when you record information, such as names and data. When you mention that information in your response, you have to show that you understand its relevance by identifying it as a reason, example, cause, effect, and so on. Therefore, you need to record that relationship in your notes so that you don't have to waste time rereading the passage too much after you finish listening to the lecture. For example, the passage on deficit spending mentions the annual and total government debt in the U.S. (several hundred million dollars annually and over $8.5 trillion total). These figures are important details because they are reasons that critics predict economic problems in the future.

The connection in the passage between a detail and the main idea not only tells you which details to include, but the connection also becomes a key point in your summary. When you refer to the size of the U.S. deficit, you also must say that the size is a reason that people oppose deficit spending but the writer is not concerned with it. This is the fact that the author mentions at the end of the first paragraph and the beginning of the second. It isn't enough to just refer to the data. You have to explain how they relate to the author's overall point, or main idea.

Strategy 13: Use the key points from the reading as a guide for the lecture.

The main idea of the lecture that follows the reading passage will be related to that of the reading according to the chart outlined in Strategy 1. Therefore, based on that expected contrast, you can use the main idea and key points in the reading as clues to what the speaker might mention. For example, the writer supports deficit spending, so the speaker might totally disagree with deficit spending or might disagree with certain elements of it. However, there is no guarantee about the speaker's particular key points, such as reasons and examples.

Don't assume too much. You can't be overly certain about the main topic and main idea of the lecture based solely on the reading. For instance, if the writer defines and explains a scientific principle, such as the evolutionary advantage of size in males who must compete for females, you know that the lecture will present one or more examples that challenge or contradict that principle in some way. However, you don't know exactly what the example will be or how it fits the principle.

Nevertheless, keep the writer's topic, purpose, attitude, and key points in mind while you listen to the lecture. It is easier to listen to a speaker when you can anticipate some general ideas and you can recognize similarities and contrasts with a passage from another writer or speaker.

Strategy 14: Listen actively.
Although the reading contains clues, the clues aren't definite or obvious. Therefore, you must still think about what you are hearing. Like the reading passage, the lecture also contains abstract theories and concrete illustrations and examples. You must not only understand the lecturer's explanation or argument but also identify similarities and differences with the writer's.

Of course, unlike the reading passage, the lecture will not be printed on the screen. Since you will never see a transcript of the lecture, you will only have your memory and notes of the lecture when you write your response. Therefore, your ability to concentrate, listen closely, and take good notes is crucial.

Strategy 15: Take notes in fragments, symbols, abbreviations, and acronyms.
You must take notes during the lecture, but remember that you shouldn't try to make a transcript of the lecture. Moreover, you must avoid writing out complete sentences and even complete words. As always, your notes can remind you of what you hear, but they shouldn't replace active listening and memory. For example, look at the following transcript of a lecture about deficit spending (Remember that you will only hear, not read, the lecture on the actual TOEFL test).

> Let's talk about the issue of deficit spending. As you may know, defenders of the federal government's steadily increasing budget gaps are, uh, fond of claiming that such Keynesian spending is merely a tool to keep the national economy on an even course. They claim that greater government spending is an effective means of maintaining stability and energy in the economy. Well, in part, they are correct: when private demand is insufficient to keep the economy running at capacity, deficits can provide stimulation to a stagnant economic environment. But if this policy is used carelessly, deficit spending can have a negative impact on the national economy.
>
> In the first place, if the government runs deficits during a healthy economy, the government is, uh, competing with private companies and investors for limited amounts of credit. Remember that deficit spending means that the government is spending borrowed money. As the government and private businesses try to borrow money at the same time, this will drive up interest rates and reduce the amount of credit available to private investors, resulting in underinvestment.
>
> Second, if the economy's production capacity is already maximized, increased government demand will compete with private-sector demand, and price inflation will result. In fact, this is exactly what happened with the Kennedy tax cuts of the early 1960s: the tax cuts took effect during a time of economic health, encouraging higher prices at the same time that the government borrowed and spent more money because of the Vietnam War. The

result was the stagnant economy of the 1970s, characterized by high unemployment, high interest rates, and high inflation.

So, because it can easily damage the country's economy rather than help it, deficit spending is a dangerous, double-edged sword that must be kept under control.

After listening actively to the lecture above, your notes could resemble the following:

ds = tool, stable + energy

only ½ right: ds ↑ slow econ, but ↓ good econ

1st gov + © + peo fight 4 $

↑ int rates + ↓ credit → ↓ invest

2nd ↑ demand → ↑ prices

ex: ↓ tax 60's, good econ + ↑ gov spend → ↓ jobs, ↑ int rates, ↑ prices

Conl: ds = risky

Since you must write faster notes during a lecture, your notes for the lecture should have more symbols and abbreviations and almost no complete words. For example, since the lecture argues about the same topic as the reading does, then deficit spending can become *ds*. Also, notice the sign ©, which means copyright, as a sign for corporations. Since both passages involve economics, fluctuation (rising and falling) is common, which is why arrows are so useful. Finally, since invest/investment/investor and interest are all key ideas, the notes never use just *in*, but *inv* or *int* to distinguish the two.

Strategy 16: Be prepared to infer.
Inference is an important part of active listening for many reasons. Like the writer, the lecturer probably will not explicitly state his or her purpose, attitude, or opinion. In other words, don't listen for obvious, direct statements, such as "My purpose in this lecture is…" or "Unlike the writer, I think that…" As you did in the reading passage, you will have to infer at least some, and possibly all, of that information.

Moreover, you must recognize the differences and similarities between the two passages on your own as well. The lecturer will not mention the writer or the reading passage specifically. The speaker might refer to general, unidentified people, such as opponents, critics, or supporters of an opposing view. However, you must recognize the connection to the reading passage on your own. For example, the speaker in the lecture above refers to *defenders*, which includes the writer of the previous reading passage.

In order to get the best understanding of the lecture, you should be prepared to make inferences about the key points and details in the lecture. Your ability to infer ideas will certainly improve your ability to paraphrase and summarize the lecture as well as compare it to the reading. For example, the following excerpt from the lecture is a key example that illustrates when deficit spending is dangerous:

Second, if the economy's production capacity is already maximized, increased government demand will compete with private-sector demand, and price inflation will result. In fact, this is exactly what happened with the Kennedy tax cuts of the early 1960s: the tax cuts took effect during a time of economic health, encouraging higher prices at the same time that the government borrowed and spent more money because of the Vietnam War.

According to the excerpt, President Kennedy reduced taxes when the economy was maximized and the government was spending a lot of money also. All this eventually led to inflation, or higher prices, and economic decline. Understanding this detail requires some basic knowledge of economics and an ability to infer the significance of a healthy economy and tax cuts, or tax reductions.

As you should know, demand refers to people's need and desire for goods and services, and inflation refers to a dramatic rise in costs and prices. Also, a basic principle of economics is that when demand goes up, prices rise as well. According to the excerpt, inflation occurred as a result of too much demand. However, the passage only mentions the government's increased spending due to the Vietnam War. You must infer where the other demand came from.

First, the tax cuts came during a healthy economy. Based on the adjective *healthy*, you can infer that people were already manufacturing, buying, and selling as much as they could, so demand was high. Second, the government reduced taxes during this healthy economy. If the government cuts, or lowers, taxes, people have more money to spend. So, you can infer that people spent even more after the taxes were lowered. Therefore, the lower taxes increased demand even though it was already high in the first place. Added to the government deficit spending for the war, all this spending created too much demand, which led to inflation and eventually economic decline.

Strategy 17: Identify the type of passage.
As you did with the reading passage, you must identify whether the lecture is expository, argumentative, or a mix of the two. If the reading is argumentative and in favor of deficit spending, that should suggest that the lecture will present an opposing argument. In fact, that is exactly what the lecturer does. The lecture is argumentative, and some of the clue words include the noun *defenders*, the verb *claim*, and the adjective *correct*.

Strategy 18: Identify the main idea and purpose, using attitude as a clue.
As you listen, identify the speaker's main point, or main idea, about the main topic. As you know, this relates directly to the speaker's purpose, or what the speaker wants to accomplish in giving the lecture. The fact that the writer supported deficit spending was an excellent clue that the speaker would oppose it, but that doesn't tell you exactly how or why the speaker opposes it.

Regardless of any hints from the reading, you must still identify the speaker's key points. In the example above, the speaker opposes the use of deficit spending when the economy is doing well because it could limit investment and lead to inflation. Based on the main idea, you can say that the speaker's purpose is to reject deficit spending during prosperous economic periods because of the risk of inflation, high interest rates, and lower investment.

As always, a speaker's (and writer's) choice of words can indicate a certain tone, or attitude, in an opinionated passage. In the case of the lecture above, the speaker's use of the noun *damage* and the adjectives *negative* and *dangerous* indicates a negative, critical, and unsupportive tone.

Strategy 19: Pay attention to the specific degree of contrast or contradiction.
The reading and the listening passages will contain some similarity as well as difference. They will not differ completely. The TOEFL tests your ability to recognize these subtle differences as well as more obvious ones. One needs a clearer understanding and more sophisticated language to express specific differences. Therefore, for argumentative passages, don't assume that the lecture completely contradicts the opinion in the reading. Likewise, if the passages are expository, don't assume that the lecture's examples and descriptions are entirely different from those in the reading.

Listen carefully for varying degrees, or levels, of contrast and contradiction. For example, the lecturer in the example above doesn't completely oppose deficit spending, so the speaker doesn't completely disagree with the writer. In fact, after the speaker refers to the opinion of the defenders of deficit spending, the speaker explicitly states that the opinion is only partially true by using the prepositional phrase *in part*:

> Well, in part, they are correct: when private demand is insufficient to keep the economy running at capacity, deficits can provide stimulation to a stagnant economic environment. But if this policy is used carelessly, deficit spending can have a negative impact on the national economy.

After indicating that the refutation is only partial, the speaker specifies that the problem with deficit spending is the economic condition or climate. Deficit spending is only necessary and beneficial when the economy is *stagnant* (not growing). The writer, however, never makes any distinction or restriction about when deficit spending should be used. The writer refers to the use of deficit spending during the Great Depression but also states and implies that it is useful all the time. The writer implies this by describing deficit spending as *a mainstay of American fiscal policy*, which means a regular, common, and frequent part of policy. Also, the writer states it by describing deficit spending as *an important way for the government to improve an unhealthy economy or to maintain a healthy one.*

Strategy 20: Paraphrase while taking notes.
Paraphrasing while you listen and take notes may sound more difficult, but it could save you time while you listen and later as you write your answer. First, academic lectures, like academic reading passages, are usually formal. The vocabulary is advanced and the sentence structure is often complex, containing several clauses. You don't need to recreate that sentence structure or word choice. In fact, if you understand the speaker, you may be able to write a shorter word or phrase than the one used by the speaker.

Second, you have to paraphrase the main idea and key points of the speaker in your response anyway. Therefore, if you can use a synonym or similar structure in your notes, you save yourself the need to do it later. Of course, this is not mandatory because you are never marked on your notes, but it could be helpful. The following chart includes a few examples of paraphrasing in the notes above (the paraphrased ideas are in italics):

Original Lecture	Paraphrased Notes
1. Well, in part, they are correct: when private demand is insufficient to keep the economy running at capacity, deficits can provide stimulation to a stagnant economic environment. But if this policy is used carelessly, deficit spending can have a negative impact on the national economy.	1. only ½ right: ds ↑ slow econ, but not care = harm
2. Finally, in today's global economy, the U.S. government is as likely to borrow from foreign lenders as from those at home. That kind of deficit spending results in a net financial loss, because the interest paid on the borrowed funds does not reenter the national economy…	2. Last: non US loans = lost $ Int → out of US, harmful

Strategy 21: Listen for any new or altered definitions, descriptions, or explanations.
The lecture may define or explain new terminology, but the definition(s) in the lecture will not be radically, or extremely, different from those in the reading. For example, if the reading defines a scientific concept or principle, the topic of the lecture will not match the principle perfectly, but it also won't be entirely different. Likewise, if the writer discusses a problem and the speaker gives a solution to the problem, the solution may involve a new term or concept. Therefore, always listen for clue words related to definition, such as the verbs *mean*, *refer to*, *be described as*, and *be defined as*, but don't worry about hearing completely new, unfamiliar terms in the lecture. If you read the first passage carefully and closely, you can use those terms and concepts to understand what you hear in the lecture.

On the other hand, you also have to recognize any changes or alterations in the key points in the lecture. For example, the speaker in the lecture above specifies that deficit spending is bad at a particular time: a healthy economy in which people are buying and selling a lot. His argument doesn't involve a new definition of deficit spending, but his explanation of its effects involves a specific time, which is not a part of the reading. Other possible differences could involve certain actions, states, people, locations, directions, physical characteristics (size, shape, color, etc.), personalities (patient, paranoid, etc.), sequences, and origins, among many, many others.

Strategy 22: Use context clues to understand unfamiliar vocabulary.

As you did for the reading passage, you must use the surrounding sentences to understand any words or concepts that you don't recognize. Of course, listening to a lecture does not allow you to analyze sentences very carefully since you have no transcript to look at and you must pay attention to the speaker's next statements.

Despite the difficulty, an attentive listener can still use logic and repetition. For example, you should be able to recognize that the government is not part of the private economy:

> In the first place, if the government runs deficits during a healthy economy, the government is, uh, competing with private companies and investors for limited amounts of credit, which will drive up interest rates and reduce the amount of credit available to private investors, resulting in under-investment. If the economy's production capacity is already maximized, increased government demand will compete with private-sector demand, and price inflation will result.

Sentence One mentions that the government is in competition with private companies and investors. Even if you've never heard of private investors before, you can distinguish them from the government. In the excerpt, the government is not competing with itself since the self-reflexive pronoun is not used. Since the adjective *private* must refer to actors other than the government, private companies are probably owned by citizens, not the government. So when you hear the phrase *private-sector demand,* you'll know that it refers to the part of the economy separate from or outside the government. It's the third time you hear *private* and the second time you hear the verb *compete.*

Strategy 23: Connect details to the main idea.

Key points and details should not be mentioned in isolation, or alone. They must always be connected rhetorically to the main idea. In other words, you need to show that an idea is an example, a reason, a good or bad effect, and so on. It would not be enough to write the following:

> The speaker says that President Kennedy lowered taxes in the 1960s. Also, the economy was good while the U.S. government was paying for the Vietnam War…

The details here are not connected into a logical explanation. The tax cuts during the 1960s are examples of competition between government and private demand. They also were the causes of the high inflation and lower employment of the 1970s. Moreover, the Vietnam War is the reason for the government's deficit spending at that time.

Strategy 24: Pay close attention to the essay question, or prompt.

Your summary of the two passages must be more than a list of details. You must show some understanding in your comparison of them. The essay question, or prompt, specifies how you should relate the reading and listening passages. Your response must answer the question completely and accurately. For example, the essay question for the two passages on deficit spending might look like the following:

> The lecture discusses a particular economic policy. Summarize the speaker's position, and compare it to that of the writer. Use specific examples and details in your answer.

Notice that the essay question does not reveal very much about the exact main idea of either passage. This is because you are required to understand the writer's and speaker's ideas. The instruction may contain a clue about the type of passage (argumentative or expository) by using related language, such as the noun position, but you shouldn't rely on the question to tell you that. You should recognize the type of passage on your own while you read and then listen, based on factors such as the main idea, attitude, and word choice. Also, there may be some clue in the prompt about the main topic, such as economic policy, but that clue won't help you very much if you didn't pay close enough attention to both passages.

The most important aspect of the essay instructions is that they require you to relate the lecture to the reading. In the example above, the instructions use the verbs *summarize* and *compare*. Other possible vocabulary includes the verbs agree, disagree, support, contradict, and contrast, among others. Based on the sample main ideas outlined in Strategy 1, the following are other examples of possible essay questions on those topics:

> The lecture describes a particular type of professional relationship. How does that relationship relate to the principle explained in the reading? Use details and examples in your answer.

> The lecture describes an attempt to solve social problems in the United States. Summarize the solution discussed in the lecture, and explain the relative success or failure in solving the problems described in the reading. Use details and specific examples in your answer.

> The lecturer discusses a theory concerning a popular hobby and its effects on children. Summarize the speaker's point of view and explain how it agrees or disagrees with the theory discussed in the reading. Use details and specific examples in your answer.

> The speaker presents an argument on the banning of a particular practice. Summarize the speaker's position and explain how it supports or contradicts the opinion presented in the reading.

Strategy 25: Organize your summary as a comparison.

After you hear the prompt giving you the instructions for your essay, you have 20 minutes to write your essay. Take at least a few minutes to organize your notes, but don't use more than five minutes to create an outline of your essay. Spend most of your time writing and editing your work.

As you can see by the sample prompts above, you must emphasize similarities and differences, or contradictions and reinforcements, in your essay. Since your essay must be comparative regardless of the exact wording of the prompt, choose one of the two possible comparative outlines: Block format or point-by-point format.

Block Format

In the block format, a writer discusses each option, choice, or text in a separate paragraph. For this writing task, you would discuss the reading and listening passages in separate body paragraphs (II and III below). The topic of each body paragraph would be the main idea of either the reading or the listening passage, and the detail of each body paragraph would be the key points from that passage (There are only two below, but there could be more). This outline is useful when each passage consists of different numbers of key points, or when the key points can't be paired together very easily.

I. Introduction

II. Main Idea of Lecture

    A. Key point 1

    B. Key Point 2

III. Main Idea of Reading

    A. Key Point 1

    B. Key Point 2

IV. Conclusion

Point-by-Point Format

In the point-by-point format, a writer discusses both options, choices, or texts in the same paragraph, and each paragraph is organized by key point. For this writing task, you would discuss both the reading and listening passages in each body paragraph (II and III below), and the topic of each body paragraph would be the key point related to both. The point-by-point format requires a relatively equal number of key points, and the points must be easily paired or connected.

I. Introduction

II. Key Point 1

    A. Detail from the lecture

    B. Detail from the reading

III. Key Point 2

    A. Detail from the lecture

    B. Detail from the reading

IV. Conclusion

Both outlines are equally valid and acceptable. You must decide which one fits the passages better and which one is easier for you to write.

Strategy 26: Paraphrase the main ideas and key points; don't quote.
A summary includes just the key points that best explain a passage's main idea. Your summary should include those key points from both the reading and the listening that most support the main ideas of each passage. However, these ideas cannot be quoted, or repeated exactly, from the reading or listening. Summary is an exercise in paraphrasing, not quoting. To paraphrase means to restate another person's words in your own. This usually involves changing the words and sentence structure of a speaker or writer.

Of course, you can't restate everything since some words and ideas, particularly concepts related to academic fields, cannot be paraphrased easily. Proper nouns such the Great Depression or the Vietnam War don't need to be paraphrased. The concept of deficit spending should be explained in your summary, but you needn't use a synonym for it. Likewise, very common words such as *investor*, *dollars*, or *government* can only be paraphrased accurately with awkward expressions. For instance, a government is an organization of elected and appointed officials who create, enforce, and interpret the laws of a country, state, province, or city, but the noun *government* is far more concise and convenient.

The main error to avoid is repeating entire sentences. Although this would be difficult to do from the lecture, it is possible to do from the reading since you can reread it while you write your response. The best way to avoid this is to read the passage carefully and closely the first time you see it and take good notes. That way, you need only to verify details when you write your response.

Paraphrasing involves changing the vocabulary, voice, and structure of a sentence or series of sentences. The following review illustrates each element of paraphrasing individually, but most paraphrasing involves a combination of two or more of these elements:

Synonyms

A synonym is a word that has the same meaning as another word. This is the easiest method of paraphrasing for students to understand. However, it is actually difficult to replace synonyms in a sentence and keep the exact same sentence structure. To do that, you need to find synonyms that are the same part of speech. The noun *figure*, for instance, is a synonym for *number*, and each noun can be used in the same way in a sentence. The synonyms in the following sentence are in italics:

Original	Paraphrased with Synonyms Only
At present, the U.S. government owes investors *approximately* eight and a half trillion dollars, an admittedly *very large number* that *critics* of deficit spending *regularly point* to as *a sign* that *fiscal disaster is just around the corner*.	*Currently*, the U.S. government owes investors *about* eight and a half trillion dollars, an admittedly *huge figure* that *opponents* of deficit spending *often refer* to as *an indication* that *financial ruin is approaching fast*.

However, it is normally very difficult to just replace certain words in a sentence with synonyms that have the same part of speech and structure. Sometimes, it is impossible to do. Moreover, notice that the paraphrased sentence is not shorter than the original. This is a problem because a summary should be shorter than the original passage. Your best option is to avoid using synonyms alone: Combine synonyms with some of the other methods below.

Voice

Voice refers to the use of active or passive verbs. It is vital to remember that only transitive verbs can be passive. In other words, only verbs that can take an object can be in the passive voice. Since there are two transitive verbs in the sentence above, it can be rewritten in the passive voice (altered verbs are in italics):

Original	Paraphrased in the Passive (with Synonyms)
At present, the U.S. government *owes* investors approximately eight and a half trillion dollars, an admittedly very large number that critics of deficit spending regularly *point to* as a sign that fiscal disaster is just around the corner.	Currently, investors *are owed* about eight and a half trillion dollars by the U.S. government, an admittedly huge figure that *is* often *pointed to* by opponents of deficit spending as an indication that financial ruin is approaching fast.

Part of Speech and Sentence Structure

Part of speech, or word form, refers to the type of word or phrase, such as a noun, verb, adjective, adverb, infinitive, gerund, and conjunction. When you change a word's part of speech, such as changing the verb *owe* to the noun *debt*, you also have to alter the sentence structure to match the new word form. The sentence from above can be rewritten in a number of different structures:

Original	Paraphrased with New Word Forms, Structure, and Synonyms
At present, the U.S. government owes investors approximately eight and a half trillion dollars, an admittedly very large number that critics of deficit spending regularly point to as a sign that fiscal disaster is just around the corner.	Currently, the U.S. government debt is $8.5 trillion. Critics of deficit spending refer to this huge amount as an indication that economic disaster is near.
	Presently, the debt of the U.S. government totals $8.5 trillion. Opponents of deficit spending argue that the massive amount indicates that financial problems are coming soon
	Today, total U.S. debt has reached $8.5 trillion, and people who oppose deficit spending insist that the amount shows the proximity of economic catastrophe.
	Because the U.S government debt is $8.5 trillion, critics of deficit spending say that the economy could be in immediate danger.
	Critics argue that the U.S. government's debt of $8.5 trillion is a warning that the economy may be in trouble soon.
	According to critics of deficit spending, a U.S. debt of $8.5 trillion suggests economic doom in the near future.
	Pointing to the U.S. debt of $8.5 trillion, critics of deficit spending are worried about the economic future.

Strategy 27: Use the present tense to refer to the passages.

When you refer to the statements and arguments of the speaker and writer, you may use the present tense. This is common for summaries and criticisms of texts, such as books and articles, as well as movies and plays. As long as you are referring to the text, not the actions of the actual writer or speaker, you can use the present tense. Compare the two following sentences:

> In his book "Manufacturing Consent," Noam Chomsky argues that the state can influence the opinions of its citizens with the cooperation of the media.

> After he wrote his book "Manufacturing Consent," Noam Chomsky gave many popular lectures across the United States.

The first sentence discusses an author's opinion within a text, so the present tense is used even though the book was written in the past. The second sentence discusses the actions of the author. Because those actions (wrote and gave) occurred in the past, the past tense is required. One way to think about the difference is to realize that the argument in the book still exists while the actions in the second sentence are no longer happening.

Strategy 28: Use comparative and contrasting language effectively.
Comparing and contrasting ideas can be tricky because of the grammar involved. Make sure you are clear about how to compare and contrast ideas using phrases, conjunctions, and transitions. The chart below reviews some of the relevant vocabulary:

	Adverbs/ Transitions	Phrases	Verbs
Comparison	Similarly, also, likewise, In the same way	Preposition: like Adjective: alike, similar, comparable, equal	Fit, match, mirror, reflect, equal, resemble, compare to, support, reinforce
Contrast	On the other hand, in contrast, however (Conjunctions: although, even though, but, yet)	Preposition: unlike Adjective: not alike, dissimilar, different	Differ, not match, not fit, contrast with, contradict, clash

As always, you must pay attention to the phrase and clause structure of your sentences. Notice the use of commas, periods, and dependent/independent clauses in the following:

Rule	Example
1. A transition comes between separate sentences: Sentence. Transition, sentence.	The writer is for all deficit spending. However, the speaker is against it during positive economic times.
2. A dependent conjunction joins a dependent clause to an independent one, and forms a complete sentence: Conjunction + subject + verb, subject + verb… Subject + verb + conjunction + subject + verb…	Although the writer is supportive of deficit spending, the speaker is against it during a healthy economy. The speaker is against deficit spending during a healthy economy although the writer is supportive of it.
3. Coordinating conjunctions (and, but, yet) join independent clauses and form a complete sentence: Subject + verb, conjunction + subject + verb…	The writer supports all deficit spending, but the speaker opposes it during good economic times.
4. A preposition forms a phrase (preposition + object) that normally comes at the beginning of a sentence if the phrase is an adverb: Preposition + object, subject + verb… Subject + verb + preposition + object…	Unlike the writer, the speaker disapproves of deficit spending in prosperous economies.
5. An adjective can precede a noun or follow a connecting verb, such as be or seem. *Alike/not alike* only follow connecting verbs.	The writer's position is different from the speaker's in that… The writer and speaker have different positions in that… *The writer's and speaker's position are not alike in that…

Strategy 29: Identify the source of the ideas, but don't over-identify.

It is important for you to state whether an idea comes from the reading or the lecture. Use the following expressions and remember to add variety to your writing:

> The writer/speaker believes/feels/argues/thinks that…
>
> According to the writer/speaker,…
>
> The reading/lecture mentions/includes/refers to/discusses…
>
> According to the reading/lecture,…

At the same time, avoid overidentifying. Not every sentence should begin with one of the expressions above. Use one and then skip them until you shift from one text to the other.

Strategy 30: Write a brief introduction and conclusion; focus on the body.

Don't waste too much time on an overly creative or imaginative introduction and conclusion for this essay. A summary is a very practical type of essay whose purpose is to demonstrate your complete understanding of the passages and your ability to connect details to a main idea or argument.

Focus your attention instead on identifying the key points in both passages, understanding the relationship between them, and expressing that in your own words. A unique or attention-getting introduction is always beneficial, but it is more relevant to the second, independent essay. The second essay requires more subjectivity and personality.

Use the introduction to define any terminology or concepts, such as deficit spending. Then, immediately give a thesis that relates the two passages. Likewise, your conclusion can be a brief paraphrase of the thesis and a final description of the scope, breadth, degree or attitude of each passage.

Strategy 31: Don't state your personal opinion or preference.

The first writing task is a summary and comparison, not a personal opinion. You can comment on the characteristics of each passage. For instance, you could mention that the writer's argument is simplistic or that the speaker is pessimistic. However, you should not agree or disagree with any of the opinions or arguments in this task. Your opinion is required in the second task, not the first.

Although other tasks might ask you for your preference, you are not expected to choose between the reading and the listening. Therefore, don't state which argument is better. Focus on relating each discussion or argument to the other. Look at the following sample of a summary on the two passages about deficit spending. The summary is organized according to the block method:

> Deficit spending occurs when a government spends more money than it has by borrowing from investors. The writer and speaker agree that deficit spending is an effective tool, but they clash over when a government should use deficit spending, and they differ on the amount of risk or danger.

The author supports deficit spending even though the U.S. debt is about $8.5 trillion. The writer points out that deficit spending has been used successfully in the U.S. since the 1930s. The passage states that when a government spends money, it increases demand and creates jobs. Therefore, people spend more money and this leads to economic growth. So, according to the writer, deficit spending allows a government to control the economy. Unlike the speaker, the writer never mentions any risks or problems with deficit spending.

The lecturer agrees that deficit spending can be used to jumpstart a slow economy, but the speaker also warns about careless spending. According to the lecture, deficit spending is dangerous during prosperous periods. First, if everyone tries to borrow money at the same time, interest rates will go up, and private investment will go down. The writer never mentions interest rates. Second, if everyone spends more, this will raise prices. Low investment, high inflation and high interest rates can lead to fewer jobs, as in the post-Vietnam's 1970s. However, the writer doesn't mention the risk of extremely high demand. In fact, the writer implies that higher demand is always good and that increasing demand always creates more jobs.

Overall, the speaker makes a more complicated argument than the writer. The writer seems very optimistic while the speaker is more cautious.

Question Forms

At the beginning of Task 1, you will see a prompt outlining the first part of the exercise. The prompt will say something similar to the following:

> You will have three minutes to read the following passage. You may take notes. After reading the passage, you will hear a short lecture on a related topic. Again, you may take notes while you listen.

After the three minutes end, you will see a picture of earphones on the screen, indicating that you should put on your headphones. However, remember that the writing section immediately follows the Speaking section, during which you wore your headphones to speak into the microphone. If you don't find them uncomfortable, you should keep your headphones on during the writing section to block out distracting noise. Regardless, you must wear them to hear the lecture. A narrator will signal the start of the lecture:

> Now listen to part of a lecture on the topic you just read about.

Once the lecture ends, you will both hear and see an essay question asking you to relate the lecture and reading:

> Summarize the points made in the lecture and explain how they support/cast doubt on the points made in the reading.

You will also see, but not hear, directions on the screen, telling you that you the time limit and suggested length of your response. The directions will resemble the following:

> **Directions:** You have 20 minutes to plan and write your response. Your response will be judged on the basis of the quality of your writing and how well your response relates the points in the lecture to those in the reading. Typically, an effective response will be 150 to 225 words.

Practice

Now, you can practice the strategies you've just learned by summarizing a passage and lecture. Read and listen actively, take notes on both passages, and leave enough time to edit your work. Also, don't take more than three minutes to read the passage, and don't write for more than 20 minutes. This will ensure that your practice is authentic.

Sample Practice 1

You will have three minutes to read the following passage. You may take notes. After reading the passage, you will hear a short lecture on a related topic. Again, you may take notes while you listen.

Ernest Hemingway

Pulitzer Prize-winning American author Ernest Hemingway (1899–1961) received no shortage of praise and reward for his work as a writer and journalist. However, a more balanced and objective appraisal of his writing is long overdue. Although the quantity and scope of his work is admirable and rightly celebrated, Hemingway's subject matter and stylistic obsessions limit the ultimate artistic value of his work.

A basic platform for the criticism of his works is the material he chose for his fiction and nonfiction. Hemingway worked as a journalist, briefly in Kansas City, and then in Chicago, Toronto, and most famously Paris and Spain. As a journalist, he showed an admirable interest in many subjects, such as war and modern art, and a willingness to travel. However, he relied too heavily on his own experiences. All of his best novels, such as The Sun Also Rises and A Farewell to Arms, are extremely autobiographical. Moreover, Hemingway tried to paint a falsely heroic picture of himself in his reporting. As a result, his fiction is far too personal and his nonfiction seems unrealistic.

Moreover, Hemingway's rhetorical style is too simple for a supposed master. A lack of rhetorical devices makes Hemingway's works undeserving of the abundant praise they have received. Hemingway's fear of adjectives and his preference for limited, basic description shows a lack of style and imagination. Without rhetorical sophistication, a writer merely records events.

Related to this simplicity in writing style is the simplicity with which Hemingway constructed his characters. His portrayal of women is especially inaccurate. His female characters are purposefully one-dimensional due to his inherent distrust of the fairer sex. Feminists correctly find many of Hemingway's stories distasteful and complain that he had little understanding of the female mind.

Notes

_____

_____

_____

_____

Play CD 2, Track 43.

Now listen to part of a lecture on the topic you just read about.

Notes

_____

_____

_____

_____

Directions: You have 20 minutes to plan and write your response. Your response will be judged on the basis of the quality of your writing and how well your response relates the points in the lecture to those in the reading. Typically, an effective response will be 150 to 225 words.

Summarize the points made in the lecture, explaining how they relate to the points made in the reading. Use specific examples and details from both passages in your answer.

Sample Practice 2

You will have three minutes to read the following passage. You may take notes. After reading the passage, you will hear a short lecture on a related topic. Again, you may take notes while you listen.

Genetically Modified Foods

What are genetically modified (GM) foods? The term refers to plants intended for human consumption that have been modified in laboratories using molecular biology techniques. For example, a scientist may isolate a gene in a plant or animal that shows resistance to drought, and then insert that gene into a different plant. By performing this modification, the new plant becomes resistant to drought as well.

The potential benefits of genetic modification are almost limitless. Despite the fact that GM foods have only been around a little over 12 years, scientists have already produced genetically modified foods that provide resistance to cold, drought, salinity, pests, and herbicides, produce higher yields, withstand long shipping times, and improve nutritional content. Researchers are even attempting to develop foods containing edible vaccines to eradicate deadly diseases. There is also evidence that GM foods can provide environmental benefits by reducing the need for toxic pesticides and synthetic fertilizers that produce greenhouse gases.

Although no major health risks have emerged since GM foods were introduced, critics charge that GM foods have not been sufficiently tested to ensure their safety for human consumption. But failing to exploit the benefits of GM foods poses another kind of risk. The world's population promises to double over the next 50 years. Providing adequate and nutritious foods to this expanding population will be one of the great challenges of the 21st century. GM foods promise to meet this challenge using the best techniques modern science has to offer. The fact is, GM foods present no new or special dangers, but in fact may improve the lives of countless millions of people worldwide.

Notes

_____

_____

_____

_____

Play CD 2, Track 44.

Now listen to part of a lecture on the topic you just read about.

Notes

_____

_____

_____

_____

_____

Directions: You have 20 minutes to plan and write your response. Your response will be judged on the basis of the quality of your writing and on how well your response presents the points in the lecture and their relationship to the reading passage. Typically, an effective response will be 150 to 225 words.

Summarize the points made in the talk you just heard, showing how they cast doubt on the points made in the reading.

Writing Task 2: Independent Essay

Task 2 is the final exercise of the whole test. This second essay is independent, which means that you don't have to read or listen to an academic passage before you write. You must read a short essay question that gives you the basic main topic, and you must express and defend an opinion or preference based entirely on your own personal experience and knowledge.

You have a total of 30 minutes to generate ideas, organize them, and write your essay. There is a range of possible question types for this essay, which are similar to the essay questions in the old computer-based TOEFL exam that preceded the current iBT TOEFL. Due to the range of possible questions, many strategies include a variety of examples to cover each type.

The following strategies are specific to the second task in the Writing section of the TOEFL. The basic strategies explained in General Writing Strategies still apply to this task. However, in order to avoid unnecessary repetition, they haven't been rewritten below. Review the general strategies to make sure you are aware of all the relevant strategies.

Also, you may notice that some strategies below resemble those for Speaking Task 2: Independent Opinion. Of course, your response for this task must be written, not spoken, but some of the issues, such as concrete detail and logical reasons, are still important.

Strategy 1: Be familiar with basic essay organization.
Essays in English are organized into paragraphs. Regardless of whether you write an argument, a preference, or a comparison, the basic principle remains the same. The first paragraph is called the introduction, and it includes an interesting opening called a hook, background that defines or identifies any unfamiliar terms, and the thesis statement, which is one sentence that outlines your opinion or preference and your reasons. The middle paragraphs are called body paragraphs, and each one discusses one supporting point, or set of ideas, for your thesis. It's important that each body paragraph has a unique topic, and that all the detail within each paragraph relates to the topic of that paragraph. A thorough defense of any opinion or preference has three supporting reasons, or three body paragraphs.

Finally, the last paragraph is the conclusion, and it restates your thesis and adds some general comment, prediction, or suggestion to your essay.

I. Introduction

    A. 1st sentence = a hook

    B. Middle sentences = background

    C. Last sentence = your thesis

II. Body Paragraph

    A. 1st sentence = topic sentence (first key point supporting your thesis)

    B. Rest of the paragraph = detail (examples, description, experience, etc.)

III. Body Paragraph

    A. 1st sentence = topic sentence (second key point supporting your thesis)

    B. Rest of the paragraph = detail (examples, description, experience, etc.)

IV. Body Paragraph

    A. 1st sentence = topic sentence (third key point supporting your thesis)

    B. Rest of the paragraph = detail (examples, description, experience, etc.)

V. Conclusion

    A. *Paraphrase/restatement of your thesis

    B. *Your comment, description, prediction, recommendation about the whole topic.

*The exact order of ideas in the conclusion is not as strict as it is for the body paragraphs.

Strategy 2: Understand the difference between an opinion and a preference.
The question for this task may ask you either to defend an opinion or to support a preference. Although the organization for either essay is largely identical, the supporting points don't have to be. You should know that an opinion requires objective, logical support, while a preference could be objective or subjective.

An opinion is a position for or against a plan, action, or argument. For example, someone might support higher taxes on gas for cars because the taxes could be used to clean the environment and to create new energy sources. Someone else could oppose the same higher taxes on fuel because of the economic impact: reduced savings, especially for poor families, increased costs, and less spending. Both of these positions are defended logically and objectively. The reasons are objective because they are not based on the likes or dislikes of the writers. Since they are logical, the reasons could be accepted and repeated by a wide range of people.

A preference, on the other hand, is a choice based on a writer's taste (likes and dislikes) or personality. For example, a person might prefer traveling alone because she enjoys being spontaneous and taking risks, which others might not like. Another person might prefer traveling with a partner since he likes conversations and feels lonely easily, which could ruin a long trip. These reasons are subjective because they are based on what the writer likes (taking risks) or the writer's personality (feels lonely).

The only really important difference is that an opinion must be defended with mostly objective reasons. Your personal preferences and experience can be used, but you can't defend an argument well based entirely on what you like or don't like. For example, if you supported higher fuel taxes because you don't drive and are unaffected, this would be a weak argument. Of course, a preference could be explained logically and objectively. For example, you could explain your preference for traveling with a partner by saying that a partner reduces costs and is safer. This is a logical, objective explanation that many people could support.

However, you don't have to explain your preferences objectively. When you prefer one choice over another one, you aren't expected to convince, or persuade, someone else that your preference is better or more logical. Preferences don't have to be persuasive. They just have to be explained and supported, so you can support them entirely with your likes and dislikes if you want. Also, a preference essay is usually more descriptive than an argument. Although arguments require description, an argument often includes more explanation, such as cause and effect, while preferences can be purely descriptive.

The main point is that an opinion must be persuasive. In an argument, you are expected to try to convince your reader or listener that your opinion is better and more logical than someone else's. Therefore, your opinion should be supported with logical reasons that might change other people's minds. If your argument is too subjective, it could be weak and unconvincing, so try to be as objective as possible when you defend an argument. However, you needn't worry about this when you explain a preference.

Strategy 3: Be familiar with the conditional.

The conditional refers to the use of *if*, *unless*, *when*, or *as long as* in sentences where one idea is based on another one. Regardless of whether the ideas are possibilities or certain predictions, one cannot be true without the other. Therefore, there is a cause-and-effect logic to the conditional: If A, then B. For example, look at the following sentence:

> If I were the leader of my country, I would improve transportation and education.

Improving transportation and education requires the other idea of becoming leader. Becoming leader of the country makes the other actions possible. Since cause and effect are part of the conditional, it is useful for all essay topics, but imaginary topics, such as the ones above, require a clear understanding of the differences between the real and unreal conditional. Errors with the conditional are common in essays, so it is necessary to review the structure thoroughly. Look at the chart below:

Type of Conditional	Uses and Examples
Real Present If + subject + present tense, subject + present tense…	Fact, customs, and habits: If demand rises, prices also rise. I take a taxi to work if I'm late. Unless I'm late, I take the bus to work.
Real Future If + subject + present tense, subject + future tense…	Real predictions of future action or state: If it rains tomorrow, the trip will be canceled. A student can concentrate more if he exercises first.
*Unreal Present/Future If + subject + past tense, subject + modal + base form… *All modals except *will*, *can* and *may*.	Unlikely, impossible, imaginary present/future action or state: If I were the president, I would cancel school fees. I would make school free if I became president. If I were to become president, students could go to school for free.
*Unreal Past If + subject + had + past participle, subject + modal + have + past participle… *All modals except *will*, *can* and *may*.	Imaginary past action or state: If I had studied harder, I would have passed the test. I could have attended school in the US if I had received a higher TOEFL score.
*Mixed Unreal (If +Unreal Past, Unreal Present) If + subject + had + past participle, subject + modal + base… *This conditional is based on assumed logic: a different past would have led to a different present. However, this depends on the specific actions or states; not all actions or states can be mixed in this way.	If I had grown up in Japan, I would probably speak Japanese today. I could fix my own car if I had studied mechanics in school.

The *if*-clause is an adverb clause, which is one type of dependent clause, and like all adverb clauses, it can come before the independent clause or after it. So some of the examples start with *if* (adverb clause is first) and others don't (independent clause is first).

Strategy 4: Be familiar with modals.

Modals, such as *may* or *should*, are an important part of many expressions and statements, especially the conditional. They are useful to defend arguments and explain preferences. A modal is a type of auxiliary verb, or helping verb. Modals are not verbs on their own, but they are added to the base form of verbs in order to express possibility, ability, habit, desire, or incompletion among many other ideas. The basic structure is the following:

Subject + modal + base form (base form = an infinitive form minus *to*)

Forms of the Infinitive (ex: verb = take)

Form	Active	Passive
Simple	to take	to be taken
Simple Continuous	to be taking	/
Perfect	to have taken	to have been taken
Perfect Continuous	to have been taking	/

Base Forms (infinitive forms minus *to*)

Form	Active	Passive
Simple	take	be taken
Simple Continuous	be taking	/
Perfect	have taken	have been taken
Perfect Continuous	have been taking	/

A modal is added to the base forms above to form a verb. The modals *might* and *may* both mean possibility. They express indecision or uncertainty about present, future, or past actions or states. Based on the examples in the chart above, they could be added to those base forms to form the following verbs. Pay special attention to the different time references:

Form	Active/Passive Verb	Time
Simple	might take/might be taken may take/may be taken	Present/Future I may take a break now or later.
Simple Continuous	might be taking may be taking	Present/Future Since he isn't home, Jack might be taking his vacation now.
Perfect	might have taken/might have been taken may have taken/may have been taken	Past I might have taken the wrong course last semester, but I'll check with my advisor.
Perfect Continuous	might have been taking may have been taking	Past Susan wasn't home when I called. She may have been taking a rest. I'll find out when I see her later.

Notice that the simple and simple continuous base forms are used for either present or future time, and that the perfect and perfect continuous base forms refer to past time. This general pattern is repeated in all modals except *can*, which is never used with perfect base forms (the only options are *can take, can be taken,* and *can be taking*).

A complete discussion and explanation of all modals is too broad for this book, but you should be aware of the pattern explained above and be able to use modals accordingly. Moreover, there are additional meanings for some common modals in the past.

The modals *ought to, must, should, could,* and *would* have extra meanings when they refer to past time. In the past, and sometimes the simple continuous, they refer to actions or states that were not started or completed. Although all modals have multiple meanings, the chart below only examines the modal *should*, used for advice or suggestion. *Should*, like the modal *ought to*, is used when an action or state is recommended or advised. Look at the chart below:

Form	Active/Passive Verb	Time
Simple	should take/should be taken	Present/Future Students should take art classes to improve creativity.
Simple Continuous	should be taking	Present/Future + Omission I should be taking a test right now, but I'm too sick today.
Perfect	should have taken/should have been taken	Past + Omission I should have taken an art class last semester, but I didn't know it was a requirement.
Perfect Continuous	should have been taking	Past + Omission I should have been taking a rest when you called, but I was watching a movie instead.

Review the meanings and functions of all modals, especially *could*, *would*, and *should*, and keep in mind the added meaning of omission or failure in the past. *Should*, like all modals, has many meanings. Besides advice, it can also express assumption or expectation (I should pass the test if I study hard enough), requests for advice (Should I take an art class this semester?), and polite questions (Should I come back later?). Similarly, *would* can refer to real or imaginary desire (I would eliminate school fees if I were the King), past habit (I would study all night when I was in college), and polite requests (Would you like to see a movie?). Finally, *could* can mean present/future possibility (Free education could help the poor), past ability (I could read well before I was five), imaginary ability (I could feed the poor if I taught them farming), or a polite request (Could you give me directions?).

Strategy 5: Know the correct grammar for the verb prefer.
Some of the language of preference involves confusing grammar. There is a variety of words and structures, and students often mix them incorrectly. Therefore, you should review the options and choose a few that you can easily remember and use correctly.

The first grammar point to realize is that the verb *prefer* is transitive, which means that it must be followed by a direct object. An object of a verb could be a noun, gerund, or infinitive, so the options are the following:

Prefer + object	Examples
Prefer + *noun*	I prefer *the mountains.* I prefer *novels.*
Prefer + *gerund phrase*	I prefer *living* in the mountains. I prefer *reading novels.*
*Prefer + *infinitive phrase* *Most verbs can be followed by only a gerund or an infinitive, not both. *Prefer* is an exception.	I prefer *to live* in the mountains. I prefer *to read novels.*

Once you decide on the object for the verb *prefer*, you must be careful about how you contrast your preference with the other choice: I prefer A to B; I prefer A instead of B, etc. Depending on your word choice, there are certain grammar rules to follow. The first rule to follow is parallel structure; don't mix gerunds and infinitives. For example, the following sentence is incorrectly mixed: I prefer reading novels rather than to watch movies. The second rule is that an infinitive cannot follow a preposition. For instance, the sentence *I prefer to read novels instead of to watch movies* incorrectly puts the infinitive *to watch* after the preposition *of*. The chart below outlines the correct choices:

Possible Sentence Structures	Examples
Subject + prefer + gerund + to + *gerund *No infinitive after *to*	I prefer reading novels to watching movies.
Subject + prefer + gerund + rather than + ger	I prefer reading novels rather than watching movies.
Subject + prefer + gerund + instead of + *ger *No infinitive after *of*	I prefer reading novels instead of watching movies.
Subject + prefer + infinitive + rather than + base	I prefer to read novels rather than watch movies.
*Subject + would prefer… *All of the above can be rewritten with *would* to discuss unreal/imaginary choices	I would prefer living in the mountains to living near the ocean. I would prefer to live in the mountains rather than live near the ocean.
Subject + *would rather + base form + *than + base form *No gerund or infinitive after *would rather* *No *rather than*; only *than*	I would rather read novels than watch movies.

Choose two or three structures that you are familiar with or comfortable with, and use those structures often. You should use more than one in order to add variety and style to your writing, but don't try to use all of them. That is how students get confused and make mistakes.

Strategy 6: Decide if the question asks for an opinion or a preference.
There is a variety of possible questions for the second writing task. Some ask for your opinion and others ask for your preference. Strategy 2 above explained that each type requires a different degree of logic and personal likes/dislikes. The vocabulary and main idea of the prompt will tell you whether the question requires an opinion or a preference.

An argumentative question includes language of argument, such as the verbs *argue*, *think*, *feel*, *believe*, *arrgree/disagree*, *support/oppose*, and *endorse/reject*. Also, an argument question could be phrased around one statement, a choice of opinions, or a comparison. For example, look at the following sample persuasive questions:

Form of Argumentative Question	Example
One statement (opinion, plan, or action)	Do you agree or disagree with the following statement? Sports are unnecessary for college and university students. Or Do you support or oppose the following plan? The government plans to limit class sizes to a maximum of 15 students.
A choice of opinions	Some people feel that sports are necessary and vital to a college education. Others feel that sports are a waste of money in college. What do you think?
A comparison	Do you agree or disagree with the following statement? For college students, an arts course is more important than a sports activity. Or In some societies, elderly people live with their children and/or grandchildren when they cannot support themselves. In other societies, the elderly live in special residences, called retirement homes, which are operated by health professionals. Compare the two systems. Which system do you think is better?

Preference questions can have the same forms as argumentative questions. However, preference questions may include different vocabulary, such as the verbs *like/dislike, enjoy, prefer,* and *choose,* or they may use some argumentative verbs, such as *think* or *feel.* Preference questions could also include certain adjectives, such as *favorite, preferred,* or *most enjoyable.* Like argumentative questions, a preference question could be phrased around one noun, a choice of options, or a comparison. Look at the following chart:

Form of Preference Question	Examples
One person, place, or thing	What is your favorite hobby?
	Which country do you want to visit the most?
	If you could go back in time, which famous person would you most like to meet?
	If you were the leader of your country, what would you change?
A choice of options	Some people enjoy living in the mountains. Others prefer living near the ocean. Which lifestyle do you like?
A comparison	For many students, a small, rural college is much better than a big, urban college. Which do you prefer?
	Or
	Compare the advantages and disadvantages of studying at a small, rural college to those of studying at a big, urban school. Which type of school do you find preferable?

Strategy 7: Read the question carefully; pay attention to all parts of the question.

Read the question carefully and slowly, and think about it briefly. Don't skim it quickly and then begin writing immediately. It is a common mistake for students to read the question too quickly, and then to begin writing without really thinking about it very much. This is a mistake because you may miss or omit an important part of the question. Usually a student makes this mistake by focusing on one or two key words and missing another key word and possibly missing a comparison as well. Look at the following three sample questions:

Question 1

For college students, an arts course is more important than a sports activity.

First, Question 1 asks about a comparison of sports and arts, so you must discuss both choices (sports and arts) in your answer, using either the block or point-by-point method. Your answer cannot discuss just one even though you may believe that one is better than the other.

Second, the statement relates the comparison to college students, implying also a college education in general. Therefore, you should focus your explanation and examples on college life and college-level learning, not just yourself and your own needs. Although you can certainly mention your own experiences, they should support your discussion of education in general. Your own preferences or personal choices can't be the main reasons because you must be persuasive.

Other questions might include reasons as part of the opinions. For example, a question could ask the following:

Question 2

Some people feel that sports are necessary to a college education because they teach team work and other valuable skills. Others feel that sports are a waste of money in college because athletes don't learn anything useful. What do you think?

In Question 2, the opinions include reasons, but no comparison. Unlike Question 1, it doesn't compare sports specifically to any other field such as arts. Also, the difference of opinion in Question 2 is based on the practicality or usefulness of sports activities. Although you don't have to limit your argument to the stated reasons, you must still address the idea of value in your essay. For example, if you think sports are necessary, you could discuss other skills besides team work, such as leadership, concentration, and self-motivation.

Question 3

Students at a university often have a choice of places to live. They may choose to live in university dormitories, or they may choose to live in apartments in the community. Compare the advantages of living in university housing with the advantages of living in an apartment in the community. Where would you prefer to live? Give reasons for your preference.

Question 3 is not only a preference essay but also a comparison of on-campus dormitories and off-campus apartments. Therefore, you need to discuss both choices and also explain your preference for one of them. Moreover, the question specifically asks about the advantages of both choices. Advantages and disadvantages are part of all comparisons even if they aren't specified in the question. However, when the question specifies one or the other, the specified quality should become the focus of your comparison. Of course, disadvantages or problems must be included (nothing is perfect), but each body paragraph should begin with advantages or benefits. Along with certain benefits, one or more disadvantages could be part of the reason for your preference.

Strategy 8: Distinguish between real and imaginary topics.
Any question could ask you to discuss a real or an imaginary topic. A real topic is one that involves facts from the past, present, or future. Most of the questions in the chart for Strategy 3 are about real topics; they require you to discuss your opinions or ideas about real people,

places, things or events. Nonimaginary questions usually use the present or past simple, and don't require too many modals, such as *could* or *would*.

By contrast, a hypothetical or imaginary topic requires you to imagine that the past was different that it actually was or that the present is different than it actually is. Usually, an imaginary question involves impossible or highly unlikely actions or states. Compare the following three questions:

> Who is your favorite person from history?
>
> If you could change anything about your country, what would you change?
>
> If you were the leader of your country, what would you change about your country?

The first question is real and nonimaginary even though you probably never met or knew the person. Your respect for him or her is a fact, which can be expressed in the present simple (present facts are discussed in the present simple), and the question requires you to discuss your personal opinions of a real historical person. Your explanation must include this person's real actions in the past tense (simple, progressive, etc.), but certainly your actual beliefs and opinions must be in the present. Look at the following sample body paragraph from an answer to the first question:

> First, I respect Christopher Columbus a lot because he was brave. When he left Europe the first time in 1492, he was sailing into uncharted waters. This required a lot of courage. Although some Europeans, such as the Vikings, had explored across the Atlantic before him, nobody knew that in the 15th century. Therefore, he had limited maps that showed nothing about where he was going. He wanted to discover a trade route to China, and when he finally saw land, which is part of the Bahamas today, he assumed it was Oriental. Of course, he was wrong, but it was still courageous.

The second and third questions, on the other hand, are imaginary, or hypothetical. Both basically ask the same thing, but the second gives the writer an imaginary position (leader). Both questions include the unreal present/future conditional, and involve the modals *should*, *could* and *would*, which are necessary for imaginary discussions. Although this topic can involve facts in the present simple or past simple, it also requires the use of those modals when the actions or states are imaginary. For example, look at the following sample body paragraph from an answer to the second/third question:

> Right now, traffic is a big problem in my country, so I would improve the public transportation system first. A lot of people drive cars. One of the reasons is the lack of good, clean, regular buses. People can't rely on the transportation system. I would make the schedules more reliable, so that people could get to work on time. Also, many of the buses right now are dirty. I would clean them, so that people would feel more comfortable. If more people took the bus, there would be much less traffic than there is today.

In the sentence above, the problem (traffic) is discussed in the present simple since it is a fact, but the improvement is referred to with the modal since it is an imaginary action.

Moreover, you cannot use *will* in a hypothetical essay because *will* refers to real future actions or states, not imaginary ones. All actions or states related to the changes in that essay must be hypothetical or imaginary since they are based on an unlikely or impossible fact: limitless power to change anything in your country. Therefore, your answer to a hypothetical topic requires *could*, *would*, and *should* (not *can*, *will*, or *shall*).

Strategy 9: Generate ideas for both sides of the discussion or argument.
After you read the question carefully, make two columns on your scrap piece of paper, and list all the ideas you can think of for both sides of the topic. Often, students know immediately which side of an argument they support or which choice they prefer. Other times, an opinion or preference is not very obvious and requires some thought. Regardless of whether you know your side or position, you need to generate and record ideas for both sides for several reasons.

First, you should not spend too much time thinking and not writing. It is better if you can judge, or assess, your ideas on paper rather than in your mind. It is safer and easier to choose the best ideas when they are all written down first. Ideas come and go quickly. If you don't write an idea down right away, you may forget it as quickly as you thought of it while your mind moves on to the next thought. Therefore, avoid getting lost in thought by writing your ideas down first, and then editing them after.

Second, comment on both sides of a comparison, and possibly both opinions in a non-comparative argument. Of course, in a comparison, you should mention details related to both choices even though your choice might receive more detail. A comparison that discusses only one option is incomplete. However, a question about just one plan, action or opinion offers no comparison (The government plans to limit class sizes to a maximum of 15 students. Do you support or oppose the following plan?). Although you may discuss just one side, this would be a one-sided argument.

As you will learn later in this section, you may refute a point from the opposing position in an argument about just one plan or idea, such as the one above, depending on your ideas and language ability. Although you should always support your own thesis with at least two supporting points, you can also argue against a reason for the opposite position. This adds balance to your essay, and shows imagination since you can express and discuss more than just your own ideas.

Therefore, regardless of whether you must write a comparison or not, you should generate ideas for both options. For example, look at the following sample list for the following question:

> The government plans to limit class sizes to a maximum of 15 students. Do you support or oppose the plan?

Agree	Disagree
fewer distractions	Cost more
mp3, dvd, cellphones	Need more teachers
more help	Parents upset?
more concentration	Some still misbehave
important tests	Smaller activities
focus on students	
improve more	
less stressful for teacher	
more control	
discipline problems	
no time for all	

Strategy 10: Think of organization right away, especially for comparison essays.
The basic organizing principle of the essay is the paragraph. Each body paragraph must have a unique and clear topic that is expressed in the topic sentence at the beginning of the paragraph. This is true for all essays, so at some point you will have to organize your ideas into paragraph topics before you begin writing the essay. If you create a random list of ideas right away, such as the list above, then you have to identify paragraph topics soon afterward. However, you can save time if you think of the paragraph topics as you generate your ideas.

This is true for all essays, but especially for comparisons. A comparison presents special difficulties for a lot of students. Because students generate many characteristics and traits, the essay can be difficult to organize, especially when students have a list of disorganized, random ideas. Some students solve this problem by identifying advantages as the topic of one paragraph and disadvantages as that of the other. Although this is a possible solution, especially if the question mentions advantages and/or disadvantages, it is an extremely broad way of organizing an essay, and it can lead to disorganized and unnecessary writing.

All comparisons, and many other essay types, are based on good points and bad points, benefits and drawbacks, or advantages and disadvantages. People support ideas due to the benefits of their choice and despite some of its drawbacks but also due to the drawbacks of the rejected option. Advantages and disadvantages are good basic principles to help you think of ideas. However, you will have to focus your discussion around your own perspective, or point of view.

As you will review again later in this section, your ideas need to be focused on a main idea (your opinion or preference), so your supporting points shouldn't be too scattered or disconnected. Your thesis is the unifying theme of the essay. Therefore, avoid including good points and bad points in your essay that don't relate to your opinion or preference. Part of the purpose of the essay is to test how well you organize ideas to support your thesis.

For all essays, but especially comparisons, you need to organize good and bad points according to your point of view, not just make a list. However, you don't have to write your ideas randomly as in Strategy 9 above. Of course, you can if you want to organize and manipulate ideas quickly afterwards, but essays are easier to write if you organize your ideas as soon as possible. For example, since you know that you must organize your comparison according to either the block method or the point-by-point method, you should think about those options when you generate your ideas. (See Writing Task 1: Strategy 25 for a review of each method)

Depending on the ideas you generate, you may decide that one organization method is better than the other. If you can identify several points of comparison, or ideas that are related to both choices, then you can use the point-by-point method. However, if you can't think of these connecting ideas, then the block method is better since you can discuss each choice separately.

Compare the first set of listed ideas to the second set of organized ideas for the following question:

> In some societies, elderly people live with their children and/or grandchildren when they cannot support themselves. In other societies, the elderly live in special residences, called retirement homes, which are operated by health professionals. Compare the two systems. Which system do you think is better?

Family	Residence
lot of work	expensive
more caring	less work + caring
play with grandchildren	no effort = no respect
save money	bingo, cards boring
pay for drugs	far away
not too sick	strangers
Alzheimer's too difficult	inconvenience
more friendly/intimate	good for very sick
respect	

	Family	Residence
quality of life	intimate, grandchildren,	strangers, boring, cards, bingo,
respect	effort, caring = respect, but not too sick	inconvenience, shame? good for very sick
cost	cheaper, $ → drugs, need room	expensive, necessary if no room

The key difference between the two sets of ideas is that the chart has connecting ideas on the far left (quality of life, respect, cost). The first set of ideas is mixed, so you must sort them into groups according to some common idea, such as respect and cost. This could take time, and it might be very difficult, depending on the ideas. Therefore, think about how you will organize the essay right away. For example, the chart includes these connections (quality of life, respect, cost) between the two options, so each one can be a topic of a body paragraph. In order to use the point-by-point method of organization, you need to have these connecting ideas to serve as the points, or topics, for each paragraph. It saves time and effort if you can think of these ideas while you're generating ideas.

Strategy 11: Choose the opinion or option with the best support.
Base your decision about an argument or preference on the quantity and quality of your ideas. It doesn't matter whether you support or oppose an idea, and it doesn't matter which choice you prefer. However, it does matter how well you support your opinion or preference. Therefore, you should choose the best reasons for your essay.

Sometimes your best ideas don't follow your first reaction to the question. For example, you might initially feel supportive about an idea, but after you generate your ideas, you could have better reasons for rejecting the plan. This is actually a common occurrence. Some students might decide to support the plan regardless of their poor reasons, but this is a mistake. The smarter test taker will choose the side with the best support, so even if you initially want to support a plan, you should oppose it if your ideas against it are better.

The best support for an opinion is the most logical and the most persuasive. Generally, people are persuaded most by ideas that affect them personally or that affect the greatest number of people in the most ways. Advantages and disadvantages are more persuasive when they apply to many people rather than just one person, such as you. That is why arguments should be objective, not subjective.

Of course, preferences don't have to be persuasive. For example, you don't have to convince the reader to enjoy the same hobby that you enjoy or the same music. Therefore, in this case the best ideas are those ideas about which you can write the most. When you generate ideas about a preference, choose the option that you can support with the most examples, descriptions, and anecdotes. Since the amount of your ideas depends on your state of mind, stress, and many other factors, you may find that you have more ideas for something that you might not normally prefer. It is important to realize that you aren't marked on your honesty but rather on the quality of your writing. Therefore, your preference in your essay needn't be the option you would always choose. It just needs to be an option that you can discuss well during the test.

Strategy 12: Organize your notes into supporting points and details.
Before you start writing your essay in complete sentences and paragraphs, you need to know how you will organize your essay. Some students can begin writing immediately because they can create and hold the organization in their mind as they write. This is an excellent

but rare skill; more important, it is still risky. Even experienced writers can get confused or forget where they are going in an essay. If this happens, you could lose valuable time trying to reorganize your ideas.

Instead of taking risks, be safe and decide ahead of time what you will write about in each paragraph. Choose from your list of ideas and decide which ideas can be paragraph topics (general) and which ideas are details (specific examples, reasons, etc.). This is really a form of categorization since you must group your detail under more general supporting points. Depending on how organized your list of ideas is, you may have to rewrite, add, or ignore certain ideas if they don't match the other points or if they are too general or too specific. For example, look at the following reorganization of the random list of ideas generated for an argument in favor of smaller classes. Contrast the outline below with the original list in Strategy 9, and notice how some have been altered to fit the outline:

distraction

-mp3, dvd

-discipline

no individual help

-teach lesson

-too many students

competition

-tests

-need skills

Don't spend a huge amount of time on an outline. It can be a rough sketch over the list you've already written, for example numbered circles for supporting points and dots for details, or it can be an outline that you rewrite, such as the one above. It's important to know what you will say, but you don't get marked for your outline. Use an outline as a rough plan only.

Strategy 13: Write your thesis on note paper or the screen right away.
Your thesis statement is a vital part of your essay and your mark. It is one of the key elements that markers will look for when they mark your paper (along with organization, grammar level, vocabulary, logic, etc.). Also, it is the basis of your argument or preference. All your supporting points must relate to it.

Once you've chosen and organized your supporting points and detail, you should be able to write a thesis that expresses your opinion or preference, and forecasts the points you will mention in the body of your essay. In fact, a well-written thesis statement can become a good guide for the rest of the essay if the forecasting is clear. Make sure your thesis statement does all the following (discussed below in greater detail):

- includes key ideas from the question
- states a clear opinion or preference
- does not just repeat a fact
- forecasts your supporting points (recommended but optional)

Strategy 14: Paraphrase the question in your thesis.

It is important to include some of the vocabulary from the question in your thesis and the rest of your essay. The question gives you the main topic, and your preference or opinion adds the main idea to the essay. However, you shouldn't copy the question exactly.

First, you must occasionally change certain vocabulary, such as pronouns, when you write your thesis. Look at the following Essay Question:

> The government plans to build a highway in your neighborhood. Do you support or oppose the plan?

If you copy the question carelessly, you might forget to change the pronoun *your* to *my*, as in the following thesis statement: *I don't think a highway should be built in your neighborhood.* Read the question carefully and think before you write. This could be a correct thesis: *I don't think that a highway should be built in my neighborhood for several reasons.*

Second, you should demonstrate your understanding of the question and show your language skills. Remember that the purpose of the essay, and the purpose of the whole test, is to measure your language ability. Your actual position or preference is not important.

Therefore, use your thesis as an opportunity to demonstrate your vocabulary and grammar knowledge. Paraphrasing the question could improve your score, so try to use a synonym, change the voice, or alter the sentence structure of the question slightly. For instance, look at the following examples of thesis statements in support of smaller classes:

> The government plans to limit class sizes to a maximum of 15 students. Do you support or oppose the plan?
>
> I agree that classes should be limited to 15 students.
>
> I agree with smaller class sizes.
>
> I support the reduction in class size.
>
> I think the government should reduce class sizes.
>
> I am for the new limit on the size of classes.
>
> I am in favor of limiting classes to 15 students each.

Strategy 15: Distinguish between facts and opinions in arguments.

Facts and opinions are important parts of all essays, especially arguments. Although there are many ways of expressing them, a *that*-noun clause is a common structure: subject + verb + that + subject + verb. Look at the following examples:

The writer states that the sky is blue due to the absorption of light in the atmosphere.

I think that government censorship is wrong.

The first sentence expresses a scientific fact, and the second expresses an opinion. Although it is difficult to tell from one sentence, a fact can be recognized by explanation and proof as well as by how debatable the idea seems. Most people would recognize the explanation for the color of the sky as a fact based on the acceptable logic. An opinion, on the other hand, may have explanation but little or no proof, and an opinion is debatable. One can recognize that the second statement is an opinion based on the language, such as *wrong*, and the fact that the idea is debatable. It's possible to develop reasons to support the opposite.

You must be able to distinguish facts from opinions, particularly when you write a persuasive essay about a proposal or plan. For example, look at the following sample question:

The city plans to build a highway to downtown beside your neighborhood. Do you agree or disagree with the proposal?

Many students confuse a fact with an opinion when they write a thesis for a question like the one above. Look at the following two sentences; one of them is incorrect:

I agree that the city plans to construct a highway close to my area.

I agree that the city should build a highway near my neighborhood.

Although both sentences paraphrase the question, the first does not express an opinion. Because the noun clause (that the city plans to construct a highway close to my area) refers only to the plan, it is a fact, not an opinion. There is no doubt about whether the city's plan is true or real. The city has already made a decision, and the fact that the city plans to build a highway is not under debate. The essay question requires some comment about the plan: support or opposition based on its location, cost, benefits, convenience, dangers, etc.

Only the second sentence expresses an opinion because the writer has added the modal *should*, expressing agreement and support. Therefore, the second sentence expresses more than just a fact about the plan because the writer implies that the plan is a good idea. Of course, the modal *should* is not the only way to express a positive or supportive opinion. A writer could use many kinds of sentence structures and vocabulary to express agreement or disagreement. The important idea is to express some opinion, not just a fact.

Strategy 16: State a clear argument or a definite preference.
In your argument or preference essay, choose and defend one side without any doubt or indecision. An essay with a definite, obvious, unambiguous decision is easier to write than an essay that is only partially decided. When your opinion is strong, you can support your position more clearly and reject the opposing argument more easily because your explanations are simpler, or less complicated.

Often, people don't always have an opinion that is completely for or against a topic. Since the TOEFL topics are purposefully not controversial or upsetting, students sometimes don't feel very strongly about them. However, a smart student will be more decided or opinionated than he or she really is. The purpose of the essay is to determine how well you can defend a position or explain a choice. Therefore, the first step is to make the choice. If you spend too much time explaining why you can't really decide, then you are missing the point of the essay.

Similarly, people's preferences are not always very strong or pronounced, especially when the topics are very common and mundane. Like the argumentative essay topics, the preference topics are purposefully not very unique or exciting. The test writers want the test to be accessible and understandable to a wide range of people, so they must choose topics that are relevant to many different people and cultures. This, unfortunately, also makes the topics a little boring. Some students react very honestly but incorrectly to this fact by trying to explain why they don't really care either way or why the options don't really relate to them. This is a mistake since the whole purpose of a preference essay is to measure how clearly and effectively you can explain what you like or don't like. If you avoid making a choice, you miss the whole purpose of the essay.

Always take one side in an argument or prefer one option over the other. Of course this does not mean that you must entirely ignore the other choice or position, but you need to add balance to your discussion. Your opinion or preference should be stated clearly in the thesis statement of your introduction.

Strategy 17: Write a complete but brief introduction.
Unlike the first writing task, the second essay requires a more thorough introduction. The first task was a summary of two passages, which is a less creative exercise than the independent essay in Task 2. Your argument or preference requires a complete introduction, but don't forget that your introduction is still a short paragraph (3–5 sentences) and that the body paragraphs are still the focus of your essay. A good introductory paragraph should include the following four parts:

- a hook, to attract the reader's attention
- background information, to set the stage for your discussion
- a thesis statement, which clearly states your opinion or recommendation
- forecasting, a brief listing of all the main supporting points in your essay

First, a good introduction should begin with an opening that not only attracts the reader's attention but also encourages the reader to continue reading. A creative and imaginative opening is called a *hook*, which is the first one or two sentences of your introduction. A good hook can take many forms: an anecdote, a provocative rhetorical question (a question directed to the reader), an interesting fact, an unusual or surprising comparison/analogy, or a contradiction or reversal of an accepted belief or assumption. Any of these strategies, if carefully applied, will make an effective hook. See also Listening Strategies: Main Idea Questions for more detail on hooks.

Second, the introduction should provide useful background information, but this depends on the topic and detail of your essay. You must decide if you will discuss information that your reader may not recognize. For example, if you discuss a hobby that is unknown or unfamiliar to most North Americans, you could briefly explain that fact in the introduction. Likewise, if you explain your admiration for a historical person that is unknown in the West, you should briefly explain who he or she is. However, you must choose concise information because the introduction is not the place for lengthy explanation and description. Later, in the body of your essay, you can more fully explain your opinion or preference.

Third, your introduction should state clearly your opinion or preference in a thesis statement. Generally, this will be the last sentence of the introduction. Besides a clear opinion or preference, your thesis may include the main supporting points, or reasons, that you intend to mention in the essay. This is called *forecasting* and requires some degree of parallel structure if you want to include the points in the same sentence. Depending on your vocabulary, you may use parallel infinitives, prepositions, verbs, adverb clauses, nouns, and adjectives. The following examples in the chart are some, but not all, of the possibilities:

Model	Example
I think that + (noun clause) + infinitive + infinitive + infinitive…	I think that families should care for their elderly relatives at home in order to save money, to respect them, and to improve their quality of life.
I believe that + (noun clause) + prep + object, + object, and +object….	I believe that the elderly belong at home with family due to the cost, respect for the elderly, and quality of life.
I feel that + (noun clause) + because + subject + verb, + verb, and + verb…	I feel that caring for the elderly at home is better than sending them to a facility because it saves money, shows them respect, and improves their quality of life.

List these ideas in the same order you will be presenting them in the body, for this shows organization and improves your score. This forecasting will help your reader anticipate how your essay is organized and how your explanation will be constructed. Forecasting can be done either with parallel structure in the thesis statement or in another sentence immediately following the thesis. For example, look at the following introduction for the comparison essay on caring for the elderly:

What should you do if you see an old person with heavy bags to carry? You should offer to help. This is also the right thing to do when family members can't take care of themselves. In my opinion, living with family is much better for older people than living in facilities because of the cost, respect for the elderly, and quality of life.

Notice how the hook (a rhetorical question) mentions an idea related to the main topic, but the question doesn't ask directly about a choice of homes for old people. Instead, the writer introduces the essay by discussing the ethics of helping the needy. Then, the writer connects that idea to the main topic (this is the right thing…) and clearly states an opinion.

Also, notice the forecasting in the thesis statement. The nouns *cost*, *respect*, and *quality* are clues to the supporting points in the body of the essay. That kind of forecasting improves your score, but it is not essential. Although the thesis statement itself is essential, a writer could write one without forecasting: *In my opinion, living with family is much better for older people than living in facilities for several reasons.*

Strategy 18: Don't focus on the hook too much.
A good hook, or opening, requires some imagination as well as some experience and even background knowledge for hooks based on history or science. In order to write a strong hook, you need to think of a variety of ideas related to the main topic, such as people, things, places, actions, events, and states. In the example above, the writer uses a familiar situation (an elderly person requires help) that is related to the main topic through a common idea (helping older people is the right thing to do).

Despite the benefits of a creative hook, don't spend more than a few minutes on it. A hook gives the reader a good impression of your skills as a writer, but it is not as essential as your thesis and the supporting points in the body of your essay. Therefore, don't waste your time on it if you can't think of anything very inventive or original.

If you have trouble with the start of your introduction, just make a common, general comment on the main topic. There are always simple facts about the topic that can be used to begin an essay. Don't be overly specific. For example, all of the TOEFL topics are based on the fact that people have different opinions about them and that societies have different customs, traditions, and beliefs. Therefore, you can almost always use those ideas in an opening. Review the following examples, and use some variation on them if you get stuck:

People differ on how they X.

People have different opinions/beliefs/customs about X.

Many people have different ideas about X

There are many different ways of X in the world.

Societies around the world have various ideas about X.

X is different in many parts of the world.

Strategy 19: Support each topic sentence with concrete detail.

Once you've finished the introduction, you need to work on the body of the essay. As you've already learned, a body paragraph generally has two parts: the topic sentence and supporting details. The topic sentence identifies the controlling idea of the paragraph, and it is the paragraph's first sentence. Since it the most general sentence of the paragraph, it will contain few or no specific details about the topic.

The topic sentence identifies one supporting reason for an argument or preference. If the essay is comparative, the topic sentence may identify one option as the topic (block method) or one aspect of both options (point-by-point method). Regardless of the essay type, the basic concept doesn't change:

> Body paragraph
>
> Sentence 1 = Topic sentence
>
> Rest of the paragraph = Detail

Supporting details comprise the rest of the paragraph, and they develop, or expand, the general point introduced in the topic sentence. These details may include examples, statistics, logical explanation, or personal anecdotes. The paragraph should not include information that does not support the paragraph's controlling idea as identified in the topic sentence.

In all essay types, supporting details may describe, define, explain, or exemplify. Description, definition, and explanation are examples of rhetorical structures. As you've learned from both the Reading and Listening sections of this book, rhetoric is a technique writers use to inform and influence their readers. Overall, arguments include more explanation since they must be persuasive while preferences include more description since they are mostly informative, or expository. However, most body paragraphs include a mix of the following rhetorical devices:

Description

Use description when you want to answer the question, "What is it like?" The purpose of description is to provide concrete information about appearance, sound, taste, touch, and smell as well as traits like number, amount, direction, speed, efficiency, cost, and personality. Remember that when you are writing a description, you should choose your information carefully. It isn't necessary to provide every detail you can think of—doing so will only take more of your time and tire your reader. Instead, use only those details that best support your main idea.

Moreover, don't focus too much on adjectives. Although adjectives are useful, they aren't the only descriptive words:

Part of Speech	Example (part of speech in *italics*)
1. Adjective	Life with one's family is *more enjoyable*. One has a *more enjoyable* life with one's family.
2. Adverb	One lives *more enjoyably* with one's own family.
3. Verb	One *enjoys* life more with one's own family.
4. Noun	One gets more *enjoyment* in life with one's own family. Life with one's own family offers more *enjoyment*.

However, always add specific, concrete imagery; adjectives like *enjoyable, fun, good, bad,* and *great* are too subjective and vague. Include details that a reader can picture, or imagine in their minds. Similes and metaphors are useful when you can't think of a specific word, and examples are an excellent way to make description more concrete. See also Speaking Strategies: Independent Description for more detail. Here is an example of description from a body paragraph on the elderly and retirement homes (description in italics). Notice how the example (playing with grandkids) makes the adjectives *intimate* and *livelier* more concrete:

> Third, quality of life needs to be taken into consideration. Retirement homes can be *lonely* places, especially if the home is *far from family*. Also, the activities in retirement homes are *too repetitive and impersonal*. Playing cards or bingo with strangers is *not exciting*. In contrast, family homes have *more intimate* company and *livelier* activities. For example, grandparents can help with things like childcare in many homes. The elderly can play with their grandkids, and the children can learn from the experience of their elders.

Definition

Choose definition when you need to answer the questions "What is it?" or "What does it mean?" Sometimes you will need to introduce a term or idea that is unfamiliar to your reader, or to show exactly what you think that term or idea means. When that is the case, you should provide a definition to help your reader understand your discussion. A definition is a specific form of description; a definition describes the qualities, characteristics, or traits that make one noun or idea unique from others like it. Definitions give the meaning of a term or idea, using verbs such as *be, refer to, mean, be called, include, involve,* and *be referred to as.* The following paragraph from an essay on the elderly includes a definition of Alzheimer's disease:

Second, caring for elderly relatives at home is more respectful than putting them in retirement homes, but this is not always possible. Retirement homes are necessary for people who need a lot of attention, such as Alzheimer's patients. Alzheimer's is a disease that first affects a person's memory, and then eventually damages language and motor skills. However, if they are healthy enough, the elderly deserve to stay with family. They took care of younger family members at one time in their lives, and so their children and grandchildren should take care of them when they need it. People that allow their healthy older members to live separately from their families don't put a high value on the elderly.

Explanation

An explanation gives the reason(s), cause(s), method(s), sequence(s), purpose(s), intention(s), and effect(s) of a fact, an action, an event, or a concept. Grammatically, an explanation requires transitions (*therefore*, *as a result*), conjunctions (*because*, *so*, *for*), infinitives, and/or prepositions (*due to*, *thanks to*, *despite*). Use explanation in your topic sentences to support your thesis and in your body paragraphs to explain your supporting points. Again, like all rhetorical devices, explanation can involve other types of detail, such as description or examples. In fact, examples and description will always help explain your meaning and add strength to your logic. Here is an example of explanation from an essay supporting smaller class sizes. Notice how the paragraph includes explanation (*therefore*, *due to*), description (*50, 30, 20*) and examples (*such as*):

> Second, with so many students and problems in classes, teachers no longer have enough time to work with each student individually. In a 50-minute class with 30 or more students, a teacher must spend about 20 minutes on the lesson. Therefore, the teacher will have just one minute to spend with each student. There is a lot for a young person to learn, such as good reasoning, from teachers. However, due to the crowding, individual attention just isn't possible in today's classrooms despite the competition and high demands of education.

Exemplification

Examples are more specific types or kinds of general nouns. They are usually used to support each of the other three strategies. The best examples help the reader understand your point by providing a reference to something they already know or are familiar with. However, more examples are not necessarily better; one or two examples are usually enough. Also, make sure that your examples actually demonstrate the point you are trying to illustrate; otherwise, you will confuse your reader instead of helping her. An example can be used with a transition (*for example*, *for instance*) or a prepositional phrase (*for example*, *for instance*, *such as*).

Look at the use of *such as* in the paragraph above. On the other hand, examples don't need special vocabulary; review the first paragraph above and see how the writer gives examples of activities (playing cards and bingo) without using phrases like *such as* or *for example*. Similarly, the following body paragraph from the essay on class size also includes examples without those phrases:

Other challenges that students and teachers face are the constant distractions that surround them every day. Cell phones, music players, and even video games and DVD players are all being carried into classrooms. With the general noise and confusion in today's classrooms, it is easy to hide and even use them during class. This forces teachers to focus on discipline, which prevents the teacher from doing his or her best, and it prevents the students from getting the most out of the curriculum.

Strategy 20: Refute the opposing argument in the body of your essay.

A strong argument not only supports one opinion with logical reasons but also refutes one reason for the opposite opinion. To refute an idea means to prove that the idea is wrong and to specify how it is wrong. Although a good refutation strengthens an argument, you should only add one to your essay if you are confident in your language ability and certain about your reasoning.

A refutation is argumentative, but it is not a part of comparisons since a comparison must include key points and details from both sides or choices. Therefore, if your question includes comparative language, such as *compare*, *better*, or *more beneficial*, then don't worry about a refutation. Whether you choose the block method or the point-by-point method, you will discuss both sides of the topic. However, for an argument essay that doesn't require you to compare anything, you should spend most of the essay supporting your thesis and opinion. For example, the following argumentative essay involves no comparison:

Do you support or oppose the following plan?

The government plans to limit class sizes to a maximum of 15 students.

In that argument, you don't need to compare two choices or options. Therefore, you can use all of the essay to support your opinion. In other words, each body paragraph presents one supporting point that defends your opinion. An essay without a refutation might look like the following:

Have you ever tried to study in a shopping mall? I assume you haven't. Probably that's because of the number of people and distractions and the amount of noise. Unfortunately, classroom size is turning some classrooms into loud, distracting places. This is an important issue that must be dealt with immediately. I support the reduction in class size so that teachers can do more and students can concentrate better, get more help, and succeed.

First, students and teachers face constant distractions that surround them every day in the classroom. Cell phones, music players, and even video games and DVD players are all being carried into classrooms. With the general noise and confusion in today's classrooms, it is easy to hide and even use them during class. This forces teachers to focus on discipline, which prevents the teacher from doing his or her best, and it prevents the students from getting the most out of the curriculum.

Second, with so many students and problems in classes, teachers no longer have enough time to work with each student individually. In a 50-minute class with 30 or more students, a teacher must spend about 20 minutes on the lesson. Therefore, the teacher will have just one minute to spend per student. There is a lot for a young person to learn, such as good reasoning, from teachers. However, due to the crowding, individual attention just isn't possible in today's classrooms, despite the competition and high demands of education.

Finally, students are competing more and more for schools and jobs. Student performance on standardized tests is becoming increasingly crucial to the opportunities that lie before a student when they complete their studies. The scores that students receive on tests like the SAT or TOEFL are important factors in determining the college or university that a student attends in the United States. Because large class sizes prevent teachers from spending time with them, students lose some knowledge and skills they need in order to perform well on these important tests.

Reducing class size is not only a matter of relieving pressure. It will enhance what students learn and how they learn it as well as make the job easier for the person they learn it from.

Notice that each body paragraph supports the thesis with one key point (distractions, individual attention, and test scores). As an argument this is fine, but it lacks a refutation of one reason for the opposing opinion. If you only discuss reasons that you are right, your argument will be weaker than the argument of someone who also explains why the other side is wrong.

If you generated ideas for both sides of the argument, then you can choose an idea from the opposing side and see if you can prove that it's wrong in some way. This is a big advantage of writing ideas for both sides even though you only support one. Once you choose the side with the best support, you can use one idea from the opposite side in your refutation.

However, only devote one paragraph to a refutation. Usually, it is the second or third body paragraph, just before the conclusion. In the essay above, you would replace the paragraph that begins *Finally, students are competing*…with a refutation. This doesn't change the basic organization of the essay: body paragraph = topic sentence + detail. The only real change is that your refutation is the topic of the paragraph instead of another supporting point. Use one of the following expressions, or some variation, in your topic sentence:

Opponents argue/believe/think/feel that + opposing opinion…

It can be argued/believed/thought that + opposing opinion…

An argument can be made that + opposing opinion

According to opponents/others, + opinion…

After you state one reason that other people support the opposite view, identify a factual or logical problem with the opinion. Your refutation is your explanation of this problem. All refutation involves strong logic and precise detail, and there are several common ways of refuting an opposing argument. Although each method is discussed separately below, most refutations include a mix of at least two of the following methods:

Basis for Refutation	Example
1. Inaccurate/Incorrect facts or logic Often, this method focuses on incorrect causes or effects. Opponents argue that X causes Y, but you explain how it is impossible or unlikely.	Opponents of smaller class sizes argue that reducing the number of students will lower the level of discussion in the class. However, this is not true since fewer students will behave better. Therefore, they will be able to discuss and learn more, not less.
2. Incorrect/Faulty assumption This is an extension of the first method since a bad assumption is a logic problem, but it is worth discussing separately. An assumption is an unquestioned belief that is unmentioned but implied in the argument. In this refutation method, opponents believe X, but you identify an incorrect belief that is logically part of the argument. If the assumption is wrong, then the argument is wrong also.	Critics of reduced classroom size warn that parents will be upset about the higher fees for smaller classes. Although they are right that smaller classes will cost more, they incorrectly assume that most parents won't want to pay more for a better education for their child. This is a weak assumption. It is far more likely that parents will gladly pay more so that their child can receive adequate attention and higher marks.
3. Irrelevant In this refutation, opponents believe X, but you explain that although X is true, X doesn't relate to the topic or X doesn't matter. Sometimes, this can be similar to the first method above because the refutation may argue that X doesn't have the effect that opponents claim.	It can be argued that smaller classes will make games and activities less entertaining. With fewer students, the class cannot organize games with as many students as before. However, this argument is entirely irrelevant to the goal of educating children. Games are fun, but they are only secondary concerns, not primary ones. Also, fewer participants will not ruin anyone's fun significantly.
4. Too general/Missing key detail In this method, opponents make a very broad comment about an unspecified or vague issue, person, group, activity, etc. In your refutation, you show that the argument omits some key detail (type, time, place, nationality, gender, quality, etc.), and the added detail changes the argument. This is a common method because few arguments apply to everyone, everywhere and at all times. You can often refute a point by focusing on an overlooked detail.	An argument can be made that students will still misbehave in a smaller class. Although this could be true, it is necessary to look at which children. Not all kids will continue to act irresponsibly; only troubled or disturbed students will not improve. These children should be removed from the class temporarily or permanently and given help elsewhere.

Strategy 21: Relate all supporting points and details to your main idea.

In the same way that all your details (example, descriptions, etc.) relate to each supporting point (the topics of your body paragraphs), all your supporting points should relate to your main idea. You shouldn't discuss or explain an idea in the body if it doesn't support your thesis in some way. In the sample argument essay in Strategy 20, all the supporting points relate to the thesis statement. Even the refutation paragraph, which explains a problem with an opposing reason, still supports the thesis. A refutation shows a weakness or error in the opposite argument, and this strengthens the argument expressed in the thesis.

In comparisons, it is a common mistake for students to compare and contrast ideas without relating them to the main idea, or thesis. Don't compare or contrast traits that don't support your opinion or preference. Remember that you must compare choices for a reason. For example, read the sample essay below. It is organized in the point-by-point method and is based on the following question:

> In some societies, elderly people live with their children and/or grandchildren when they cannot support themselves. In other societies, the elderly live in special residences, called retirement homes, which are operated by health professionals. Compare the two systems. Which system do you is better?

> What should you do if you see an old person with heavy bags to carry? You should offer to help. This is the right thing to do for family as well. In my opinion, living with family is much better for older people than living in facilities because of the cost, respect for the elderly, and quality of life.

> The first advantage to having an elderly family member live with his or her family is an economic one. A retirement home might be necessary if a family has no room, but it is very expensive. Some families may not be able to afford the cost. At the same time, a family does not usually have to pay much more to give an available room to a grandparent. Also, the saved cost of a retirement home can pay for medical bills and other costs, such as drugs. Living with relatives is a better financial solution for some families.

> Second, caring for elderly relatives at home is more respectful than putting them in retirement homes, but this is not always possible. Retirement homes are necessary for people who need a lot of attention, such as Alzheimer's patients. Alzheimer's is a disease that first affects a person's memory, and then eventually damages language and motor skills. However, if they are healthy enough, the elderly deserve to stay with family. They took care of younger family members at one time in their lives, and so their children and grandchildren should take care of them when they need it. People that allow their healthy older members to live separately from their families don't put a high value on the elderly.

> Third, quality of life needs to be taken into consideration. Retirement homes can be lonely places, especially if it is far from family. Also, the activities in retirement homes are too repetitive and impersonal. Playing cards or bingo with strangers is not exciting. In contrast, family homes have more intimate company and activities. For example, grandparents can help with things like childcare in many homes. The elderly can play with their grandkids, and the children can learn from the experience of their elders.

To summarize, the elderly are a valuable part of society and deserve to be treated that way. They should not be put in some home somewhere unless they have to be. Instead, the elderly should be surrounded by loved ones. In fact, society would be a better place if more people cared for each other more.

Strategy 22: Remember to conclude your essay effectively.
The conclusion is an important but often overlooked part of the essay. Students correctly try to be creative in their hook and thesis, but unfortunately they don't always apply the same amount of attention to their conclusion. Due to the time limit, this can't always be corrected, but you should plan to write an effective conclusion and monitor your time accordingly. A missing or poor conclusion creates an imbalanced essay, and leaves the reader with a poor impression of your abilities and imagination.

Your concluding paragraph is the last thing your reader sees of your essay. It is important that your conclusion leaves the reader with a clear understanding of both the message conveyed by your essay (your main idea) and its relevance or importance. With a little practice, you will be able to write strong, effective conclusions that accomplish both of these goals.

Remember the two meanings of the English word *conclude*. The first, most common meaning is to come to an end: He concluded his speech on time. This is not the meaning of *conclude* you should be focusing on as you write your concluding paragraph. Instead, you should be thinking of the second meaning: to make a decision after a reasoned consideration of the facts. In other words, your concluding paragraph ought to emphasize the message or lesson that logically proceeds from the facts and arguments in the body of your essay. Your conclusion should remind your reader of your thesis and perhaps also point out the broader implications or consequences of that thesis.

First, signal or identify the conclusion with an appropriate transition, such as to summarize or to conclude. Then, recall and emphasize the essay's thesis. It is best for you to do this in a fresh way, so paraphrase your thesis; don't simply repeat or copy it. Copying requires little effort and no thought, so it generally doesn't earn you any marks. However, you needn't focus on the parallel structure as much as you did in the introduction. Unlike the introduction, the conclusion doesn't have to forecast anything, so you don't have to put all your key points in one sentence.

Second, since your TOEFL essay will be relatively short, you needn't worry about a complete and thorough review of the essay's supporting points. This is usually only necessary if the essay is particularly long, but an independent TOEFL essay is not very long. Therefore, don't over-explain or over-describe your ideas. Paraphrasing them briefly is good enough.

Third, try to show the relevance of your essay by connecting it to current events, popular culture, or social trends. Comment on the connection to the outside world, make a prediction, or suggest a course of action for the reader or others. However, do not include additional arguments or specific details, which belong in the body paragraphs. Look at the following example from the essay on retirement homes:

> To summarize, the elderly are a valuable part of society and deserve to be treated that way. They should not be put in some home somewhere unless they have to be. Instead, the elderly should be surrounded by loved ones. In fact, society would be a better place if more people cared for each other more.

Strategy 23: Don't forget to edit.
Editing is important. All writers make mistakes. Although you probably won't be able to find all of them due to the time limit, you should at least try to catch as many as you can. Remember to edit for specific grammar or punctuation errors; don't just reread the essay. Know what you are looking for ahead of time. Look for common mistakes, such as word form, verb tense, and clause structure.

Question Forms

There is only one essay question for this task. As explained in Strategy 2 and 3 above, the question may ask you for an opinion or a preference, and it may take one of the following forms:

Arguments
[A group/institution] plans to X. Do you agree or disagree with…?

Do you agree or disagree with the following statement? [A debatable opinion/comparison]

Some people believe X. Others think Y. Which side do you support?/What do you think?

Compare X and Y. Which is better…?

Preferences
Who/what is your favorite/the most important/the best…?

Some people enjoy X. Others prefer Y. Which do you prefer/like/enjoy?

If you were/could X, what would you change/say/ask/do…?

Imagine X. What would you change/say/ask/do…?

Compare X and Y. Which do you like/prefer/enjoy/want more?

Practice

Using the strategies you've just studied, write a 30-minute essay for each of the sample essay questions below. Read the question carefully, decide if it's persuasive or expository, generate ideas for both choices when necessary, organize them into supporting ideas, support you ideas concretely, and edit your work.

You have 30 minutes to plan, write, and revise your essay. Typically, an effective response will contain a minimum of 300 words.

Essay Question 1

Students at a university often have a choice of places to live. They may choose to live in university dormitories, or they may choose to live in apartments in the community. Compare the advantages of living in university housing with the advantages of living in an apartment in the community. Where would you prefer to live? Give reasons for your preference.

Essay Question 2

Imagine that you have received some land to use as you wish. How would you use this land? Use specific details to explain your answer.

Essay Question 3

Do you agree or disagree with the following statement? A zoo has no useful purpose. Use specific reasons and examples to explain your answer.

Essay Question 4

Some people prefer to spend their free time outdoors. Other people prefer to spend their leisure time indoors. Would you prefer to be outside, or would you prefer to be inside for your leisure activities? Use specific reasons and examples to explain your choice.

IMPROVE-YOUR-SCORE STRATEGIES

This section reviews extra strategies that you should use to improve your overall writing ability and prepare for the TOEFL. Good writers usually have habits that helped them improve and that help them remain talented. The most common theme in the strategies below is practice and repetition. If you begin practicing and studying long before you register for the actual TOEFL test, you will be more comfortable and focused. Your choices of vocabulary, sentence structure, and organization will become clearer and more automatic with practice. Therefore, give yourself enough time to practice the following suggestions.

Strategy 1: Practice typing.
Hand-written essays are no longer accepted at most test centers. You will probably have to use a keyboard to type your essays on the computer screen. Typing can be an awkward activity if you're not used to it. Although you don't need to learn to type as fast or well as a professional secretary, you must still become familiar with the keyboard. Learn the general location of each key: left or right side, top or bottom. With practice, your eyes will automatically go to the general area of each letter you think of as you type. However, the only way you will learn that is with practice. So, start typing!

Strategy 2: Write every day.

Like all language skills, writing requires regular practice. Most students know how difficult it is to speak or write after a long break. If you stop for a while, it is difficult at first, and you need time to return to your earlier skill level. Most students practice occasionally, but this really only keeps you at the same level. Periodic practice only maintains your current level over the short term.

If you want to improve noticeably and reasonably fast, you need to practice daily. This is true for all skills: reading, listening, speaking, and writing. This doesn't have to involve full length essays all the time. Many students keep journals, in which they record the day's events, some thoughts, and maybe some quotes from friends or even strangers.

It's a good idea to have a tutor or teacher read your writing and comment on the vocabulary, style, and so forth. However, even if you can't get everything graded and corrected, you should still try to write all the time. The more you practice, the more fluent and natural you will become.

Strategy 3: Read and listen widely and often.

All language skills are related. Reading and listening are both passive, receptive abilities, while speaking and writing are both creative, active skills. Also, as you read and hear new vocabulary and structure, you get more familiar with it. Even if you study a lot of grammar, if you've never read or heard it, it's difficult (maybe impossible) to use it correctly.

Also, the TOEFL is an integrated test. You need to become comfortable writing about what you read and hear. Moreover, the reading and listening must be academic, not comic books and music videos. So, read and listen to a variety of topics as often as possible.

Strategy 4: Summarize short pieces of writing frequently.

Along with regular writing practice, you need to summarize and paraphrase regularly. Use English newspapers, magazines, text books, and even your own essays. However, don't overload yourself by trying to summarize long articles all at once. Overeager students frustrate themselves by trying to work with long texts instead of breaking them up.

You don't have to summarize entire passages all the time. Just as you don't have to write whole essays all the time, you don't have to paraphrase long articles. Read entire pieces to understand the main idea, but focus on one paragraph at a time. Review the vocabulary thoroughly and research synonyms. Paraphrase each supporting point in several different ways to practice your sentence structure. Also, it is easier to find a tutor or teacher to correct a paragraph every day instead of a whole essay.

Strategy 5: Look for debates, contradictions, and exceptions.

As you might have realized by now, the TOEFL includes a lot of subtle differences between and among ideas in all four sections of the test. People, places, and things are similar in some ways, but unique or different in important ways. Opponents agree on some issues but generally disagree, especially on specific key points.

You should become familiar with these kinds of relationships. Look for debates, exceptional people, or surprising occurrences in English newspapers and magazines, and on English television and websites. Try to find out what makes the topic special, why two parties disagree, why a person is hard to categorize, or why an event doesn't fit the usual pattern. For example, new discoveries and theories in various fields often depend on unusual or exceptional details.

Strategy 6: Fill gaps in your background knowledge.

You are not expected to have any particular expertise or professional knowledge for the TOEFL since the test is not focused on any one field or profession. However, the test covers a wide range of basic topics in science, history, and the arts. While you aren't expected to solve any problems in these academic fields, you are expected to be able to read, listen, talk, and write about people, debates, and discoveries from many of them.

Therefore, someone with a broad education will have an easier time on the TOEFL than someone who never really paid attention in class. You know your own weaknesses better than anyone else, so focus on areas about which you know little or nothing. For example, can you talk about art, such as Impressionism and Cubism? Can you discuss the reasons for the American Revolution? Is a whale a mammal? Why or why not? Can you explain why a sodium atom and a chlorine atom bond to form a sodium chloride molecule, or salt? Understandably, students have the most difficulty with unfamiliar topics.

Strategy 7: Practice changing clauses to phrases and then back to new clauses.

Mostly, English sentences are made up of phrases and clauses. Paraphrasing requires a good vocabulary and an ability to manipulate sentence structure. The importance of phrase and clause structure cannot be overstated. For example, the chart below shows only some of the possibilities:

Dependent Clause (in *italics*)	Phrase (in *italics*)
1. *Before he graduated*, Steve got married.	1. *Before his graduation*, Steve got married.
2. *Because many politicians seem corrupt*, so too many voters don't vote.	2. *Due to corruption among politicians*, too many voters don't vote.
3. The scan shows doctors *where the tumor is*.	3. The scan shows doctors *the location of the tumor*.
4. *If you have enough money*, you can go.	4. *With enough money*, you can go.
5. Many voters argue *that stem cell research is a good idea*.	5. Many voters *support stem cell research*.

Strategy 8: Learn more verbs, and study their related forms.
A lot of students focus on nouns when they are learning new vocabulary. Although nouns are obviously necessary, verbs are extremely powerful. The reason is that they relate to so many other word forms. Infinitives, gerunds, and participles are sometimes called *verbals* because they are formed from verbs.

For example, the verb *move* is directly related to all the infinitive forms (to move, to be moved, etc.), the gerund (moving, being moved) and the participles (moving, moved). The key issue is that if a verb is stative, then so are the verbals; if the verb is transitive, so are the verbals; if a verb can't be passive, then the verbals can't be passive either.

Move can be a transitive action because it can be followed by an object: I will move the car. Therefore, the infinitive, gerund, and participle forms are also transitive actions:

Infinitive: I don't want *to move the car*.

Gerund: I enjoyed *moving the car*.

Participle: The man *moving the car* is my brother.

However, *sit* is an intransitive action because it can't take an object: I will sit on the chair. Therefore, the infinitive, gerund, and participle forms also cannot take objects:

Infinitive: I don't want *to sit* on the chair.

Gerund: I hate *sitting* on the chair.

Participle: The man *sitting* on the chair is my brother.

The more verbs you know, the more infinitives, gerunds, and participles you know. Learn how to use these related forms and you will be able to say and write a lot more in English.

WRITING TRANSCRIPTS AND SAMPLE ESSAYS

The Writing section of the TOEFL is the fourth and final section of the test. There are two parts to this section: Task 1 is an integrated task, in which you must first read an academic passage, then listen to a related academic lecture, and finally write a summary of both. Task 2 is an independent persuasive essay similar to one you may have seen in the old Computer-based TOEFL test.

Task 1

CD2, Track 43

Sample Practice 1

Narrator: Now listen to part of a lecture on the topic you just read about.

Professor: Now...I admit that I'm biased when it comes to Hemingway...I've been reading his short stories and, um, novels for years and years and I never get tired of them...but that's just praise based on my personal preference. So today I'd like to give you some more, shall I say...scholarly justification for Hemingway's greatness and how that greatness affected his works.

First and foremost, Hemingway's background as a journalist probably influenced his minimalist approach to writing which, ah, in turn influenced other great writers like, say, Raymond Carver. Though some critics say that Hemingway's rhetorical style was too simplistic, most literary pundits applaud his...well, his sheer genius. It's amazing how just by describing situations or senses very specifically, Hemingway can get quite complex ideas across to his readers. His dialogues are especially famous...they are simple, but so realistic. Reading them, you feel you are eavesdropping, spying on real conversations. It is because his rhetorical style is so direct that Hemingway is able to accomplish this.

Next, again...despite what many critics say, Hemingway was not incapable of using rhetorical "tricks." He tried double narration, when a story is told from several points of view...In fact, he loved to experiment with point of view and...oh, this brings up another point, he did try to write from a woman's perspective at times and, considering he was a man, did so quite well, I think. Many feminists might tell you differently and we can discuss this point further once you've all read more of his works.

Uhhh...anyway, getting back to my previous point...another rhetorical technique, the most frequently used, in fact, that...uh...Hemingway employed was inference. He didn't tell the reader everything, but left out some details so that the reader would have to infer, to work for the meaning. The result of this technique is that Hemingway's stories are extremely challenging for a reader, but also more satisfying because of the effort we must make to solve the puzzle.

Sample Response for Practice 1 (point-by-point Organization)

Ernest Hemingway was an American writer and journalist who lived during the first half of the 20th Century. The writer is highly critical of his work, and suggests that Hemingway has not been criticized enough. However, the speaker sounds enthusiastic about Hemingway's talent and refutes some of the points mentioned in the reading.

Both passages refer to Hemingway's experience as a journalist, but the fact is discussed differently. The writer praises Hemingway's broad and wide range of interests, such as war, but argues that his work is too autobiographical. On the other hand, the lecturer mentions that Hemingway's journalism experience was the reason that he wrote in a very simple style.

Both passages disagree on Hemingway's level of sophistication. The writer argues that Hemingway's style was too unsophisticated and unimaginative. For example, Hemingway apparently hated adjectives, and the writer criticizes his lack of rhetorical devices. However, the speaker contradicts this point by arguing that Hemingway used inference and double narration, or two perspectives. The speaker praises this simplicity as realistic, and says that Hemingway still included complicated ideas, especially in his characters' conversations.

Finally, the lecturer states that Hemingway wrote successfully from a woman's perspective, but the writer disagrees. According to the reading passage, Hemingway was suspicious of women, which is why he wrote unrealistic female characters.

CD2, Track 44

Sample Practice 2

Narrator: Now listen to part of a talk on the topic you just read about.

Professor (male): The risks associated with GM, or rather genetically modified foods, fall into three categories: environmental, health-related, and economic.

Let's talk about environmental risks first. Involuntary cross-pollination is a subject that GM proponents don't like to discuss. Example: a company genetically modifies a crop so that it's resistant to herbicides. This sounds good—you get higher crop yields, pests don't bother you, etc. But now the wind comes along and the plants transfer their herbicide-resistant genes to other plants and you get these herbicide-resistant weeds—super weeds!—that are almost impossible to eradicate! So that's one possibility, here's another. Last year, a study that found that pollen from genetically modified corn caused high mortality rates in Monarch butterflies. Okay, Monarchs feed on milkweed plants, not corn, but now we know that if pollens from GM cornfields blow onto nearby milkweed plants, it can destroy the Monarch butterfly population—and who knows what other species are at risk. Involuntary cross-pollination has already been the subject of several lawsuits, with companies that produce GM foods suing farmers for harvesting patented crops that the farmers claim were pollinated by GM crops that blew onto their fields. These are obviously not intended consequences, but these examples show how, once GM crops enter the ecosystem, there's no controlling them.

As far as safety for human consumption, the main concern has to do with GM foods introducing new allergens, or causing allergic reactions in susceptible individuals. Recently, a proposal to inject a gene from Brazil nuts into soybeans was struck down for fear of causing unexpected allergic reactions. Lots of people are allergic to peanuts—you have to figure that any plant that's had peanut genes introduced into it may carry some risk. A lot more testing needs to be done, but even proponents of GM foods admit this is a risk.

The economic argument against GM foods boils down to the fact that companies spend a lot of money developing and patenting GM crops. Many critics charge that new plant varieties will raise the price of seeds so high that small farmers and poor Third World countries won't be able to afford them. So rather than providing food for the world, GM foods may have the consequence of widening the gap between rich and poor.

Narrator: Summarize the points made in the talk you just heard, showing how they cast doubt on the points made in the reading.

Sample Response for practice 2 (point-by-point organization)

The lecture and the passage both focus on genetically-modified foods, but express opposing viewpoints. Whereas the author of the passage clearly agrees with GM foods, the lecturer strongly disagrees with them for several reasons.

The professor's main argument against GM foods is based on cross-pollination. He explains that if the wind transfers the genes of a GM plant to other plants, there can be negative effects. Super-weeds, or weeds that are resistant to herbicides, could be created. Moreover, cross-pollination can kill certain species, like the Monarch butterfly. This can lead to lawsuits between GM food companies and farmers when crops are accidentally affected by GM genes. These arguments cast doubt on the reading passage. The writer also states that GM foods are resistant to drought and pests, but argues that this is good. Unlike the speaker, the writer argues that the GM foods help the environment.

The professor also discusses the economic disadvantages of GM foods. He maintains that GM food production will lead to an increase in the price of seeds, and small farmers and third-world countries will no longer be able to afford them. This directly contradicts the author's opinion that GM foods will help feed more people.

Finally, the text and the lecture both touch on the issue of safety. The professor is certain the GM foods are dangerous since they may introduce more allergens to the world. The author, on the other hand, recognizes that although GM foods may be dangerous, there is no conclusive proof as of yet.

Task 2

Essay Question 1

Students at a university often have a choice of places to live. They may choose to live in university dormitories, or they may choose to live in apartments in the community. Compare the advantages of living in university housing with the advantages of living in an apartment in the community. Where would you prefer to live? Give reasons for your preference.

Sample Response 1

One of the things that worried me most about going to college was leaving the comfort and familiarity of my parents' home. When I received my housing packet and had to decide where to live, I was unsure of whether I'd be happier in a dorm or in an apartment. As a compromise, during my first two years of university, I lived in a dorm and in the last two years, I lived in an apartment. I have good memories from both of these experiences, but in hindsight would probably prefer the latter.

Dorms provide many advantages, but two of the most obvious are social contact and security. Living in a dorm is clearly the best way to go if you want to make friends. Dorm advisors often organize social activities and the sheer proximity of other people makes it nearly impossible to live in a dorm without getting to know others. I made two of my closest friends in my dorm, and we are friends to this day. Another advantage of dorm life is that it is extremely safe. There is usually a security desk or a main locked door in addition to the lock on your room. Dorm advisors live on site and keep track of any suspicious people. Moreover, your roommates tend to know more about you and your whereabouts since you share the same room, not just the same building. Thus, my entire dorm stay chased away me feel welcome and secure.

On the other hand, apartment living is also pleasant due to the freedom and privacy it affords. Without dorm advisors to check that you are not having a party or playing your music after a certain time of night, you can feel the freedom to have a little fun. You also are free in your choice of apartments. You can live near or far from campus, in a big or small apartment, a cheap or expensive one, with a roommate or without. In dorms you have no control over any of these choices. In addition, privacy is a major benefit of life in an apartment. You can have your own room and sometimes your own bathroom. After a shower, you don't have to walk down the hallway in a towel in front of dozens of people, sometimes of the opposite sex.

Because privacy and freedom are two of my most important values, I prefer apartment life. I cannot imagine living some place like a dorm forever. Still, dormitories are immensely beneficial for the early parts of college life and I have few regrets about living there to begin with.

Essay Question 2

Imagine that you have received some land to use as you wish. How would you use this land? Use specific details to explain your answer.

Sample Response 2

Owning property like a house or land is one of the best ways to make money. Therefore, if I were given a gift of land, it would make sense for me to sell it or keep it as an investment. Despite the wisdom of this decision, and due to my love for animals and desire to make a difference in the world, I would probably use the land to create a no-kill animal shelter for cats.

Since I was a child I can remember loving animals, especially cats. I used to collect cat pictures and stuffed animals, and I begged my parents for a cat for 6 or 7 years before I finally was allowed to have one. At that age, I would never have imagined that there were so many homeless cats in the world. As I got older, I visited a few animal shelters where hundreds of cats were crammed into tiny cages looking uncomfortable and unloved. My first instinct was to adopt them all and give them a warm home. If I had a piece of land to build on, I could do just that. I would make a large house with indoor and outdoor rooms for the cats to enjoy their lives. They could live there indefinitely without risking euthanasia, and I could employ veterinarians to spay and neuter the cats in order to reduce the number of strays in the world. Although many people would say this is a waste of money, my love for cats makes it seem worthwhile to me.

Connected to the idea of money is my philosophy on life: success is making a difference, not making money. When I die I will not be happy unless I can look back at my life and feel like I have changed the world for the better in some small way. Building a no-kill animal shelter will not end world hunger or help achieve a cure for cancer, but it will make a difference for some animals that are suffering, provide peace of mind for other cat lovers, and even create a safer environment for many housecats who are frequently attacked by strays carrying diseases. Moreover, small changes are often the first steps to bigger ones.

I'm sure any financial advisor would not approve of my plan to make an animal shelter on a piece of valuable land. My parents might not approve either, and would probably beg me to reconsider as fervently as I begged them for my first cat. However, I believe it is important for each of us to follow our hearts, not our pocketbooks, and this is why I would use the land as I've outlined above.

Essay Question 3

Do you agree or disagree with the following statement? A zoo has no useful purpose. Use specific reasons and examples to explain your answer.

Sample Response 3

Zoos may not be as useful to society as hospitals or schools. However, to say that they have absolutely no useful purpose is clearly an overstatement. Zoos have the dual function of educating children and preserving endangered species.

Zoology is a major part of science study for young children, probably due to the fact that animals are so interesting. Accordingly, most elementary schools organize field trips to the zoo in order to give children the chance to see the animals that they have been learning about, sometimes even to touch them. Many zoos now have keepers who give informative speeches about the animals they care for. I recall a short lecture about tarantulas that I heard when I was at the zoo as a child; I learned that most tarantulas are not at all dangerous to humans, despite their fierce reputation, and that they are extremely delicate and easily injured. At the end, the keeper brought a tarantula around for everyone to touch. Children often learn best in this type of "hands-on" environment, so zoos are quite motivational in their process of education.

Zoos are key organizations in promoting wildlife protection. Educating children about animal preservation is perhaps the most important aspect of a zoo. If the next generation does not learn to love and respect animals, many will become extinct. The fate of a small monkey living thousands of miles away in the Brazilian rainforest may not seem crucial to a child until he sees how cute the monkey is or learns firsthand about the monkey's unique characteristics. Moreover, breeding programs are the norm at most large zoos. With the funding they receive from the government, private donations, and entrance fees, zoos are trying to prevent many endangered animals from becoming extinct by increasing their numbers in captivity. Although this solution is not ideal, at least it ensures that certain animals will not disappear from the planet forever.

Opponents of zoos often argue that they are cruel because they keep animals in small spaces with little food and interaction. However, this is no longer true of most zoos. Although it was unfortunately the case in the past, only a handful of zoos still mistreat the animals. Animal rights organizations routinely monitor zoos and report infractions or crimes. Also, the public is no longer willing to support disrespectful organizations.

In conclusion, protecting animals and teaching people about them is the best way to make sure animal diversity is a future reality. From this standpoint, it is evident that zoos do serve an important purpose.

Essay Question 4

Some people prefer to spend their free time outdoors. Other people prefer to spend their leisure time indoors. Would you prefer to be outside, or would you prefer to be inside for your leisure activities? Use specific reasons and examples to explain your choice.

Sample Response 4

Generally speaking, weather tends to determine if I will go outside or stay indoors in my free time. If it is raining or very cold, I typically prefer to sit on my sofa and read or go to the kitchen and cook a nice meal. If it is warm and sunny, I can't stand to be in my house; I want to go outdoors and do something active. However, let's imagine that weather is not a factor. In most cases then, I would rather be outside enjoying the nature and my favorite hobby, hiking.

In my opinion, nature is truly the most beautiful gift on Earth. I can stand and stare at trees, water, and mountains for hours because I feel completely at peace when surrounded by natural beauty. Buying a picture of such scenery or looking at it from inside my house is not satisfying to me. I like to crunch the autumn leaves, hear the waves crashing on the beach, and smell the fresh mountain air. In fact, I cannot stand exercising in a gym because I want to feel surrounded by nature, not by sweaty people or beeping machines. The lure of natural beauty is therefore a major reason I am more inclined to spend time outdoors than in.

Another reason I prefer spending time outside is because hiking, a purely outdoor activity, is my favorite pastime. A friend of mine introduced me to it when I was 16 and since than I can't get enough of it. Every spring I wait for the snow to melt so that I can spend every weekend wandering on trails. Of course, the beauty of the nature attracts me to this sport, but also the challenge of reaching a specific goal: the peak of a mountain or a waterfall accessible only on foot. The feeling of accomplishment I get when I make it to the top of a steep incline after a five-hour struggle is priceless. I would give up all my other hobbies, reading, dancing, and cooking in order to keep hiking.

If the temperature in my city was always a comfortable 70 degrees and the climate was usually dry, I would not hesitate in saying that I always prefer outdoor activities to indoor ones. The beauty of the nature and the thrill of hiking would keep me constantly occupied. Unfortunately, as I live in Seattle, about nine months of my free time is spent inside.

NOTES

NOTES

NOTES

NOTES

NOTES

NOTES

NOTES

NOTES

NOTES

rocket is launched with initial velocity $\geq v_e$, it will never return to earth; hence v_e is called the escape velocity.

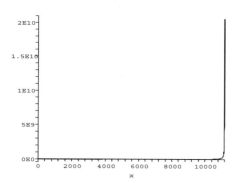

69. Suppose the degree of q is n. If we divide both $p(x)$ and $q(x)$ by x^n, then the new denominator will approach a constant while the new numerator tends to ∞, so there is no horizontal asymptote.

71. When we do long division, we get a remainder of $x + 2$, so the degree of p is one greater than the degree of q.

73. The function $q(x) = -2(x-2)(x-3)$ satisfies the given conditions.

75. True.

77. False.

79. True.

81. Vertical asymptote at $x = 2$. Horizontal asymptotes at $y = 4$ and $y = 0$.

83. For any positive constant a, $e^{-at} \to 0$ as $t \to \infty$. Since $\sin t$ oscillates between -1 and 1, $e^{-at}\sin t \to 0$ as $t \to \infty$. In the following graph, we see that suspension system A damps out at about 5 seconds, while system B takes about 18 seconds to damp out.

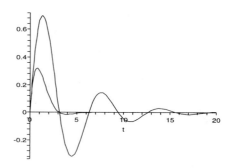

85. $g(x) = \sin x$, $h(x) = x$ at $a = 0$

87. $\lim\limits_{x \to 0^+} x^{1/(\ln x)} = e \approx 2.71828$

89. $\lim\limits_{x \to \infty} x^{1/x} = 1$

1.6 Formal Definition of the Limit

1. (a) From the graph, we determine that we can take $\delta = 0.316$, as shown below.

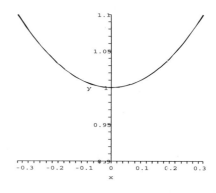

(b) From the graph, we determine that we can take $\delta = 0.223$, as shown below.

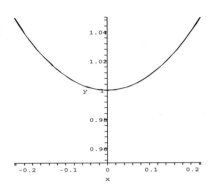

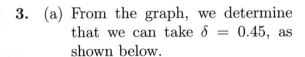

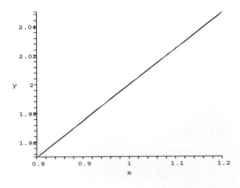

(b) From the graph, we determine that we can take $\delta = 0.2$, as shown below.

3. (a) From the graph, we determine that we can take $\delta = 0.45$, as shown below.

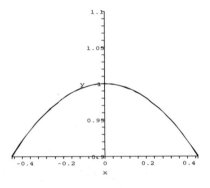

(b) From the graph, we determine that we can take $\delta = 0.315$, as shown below.

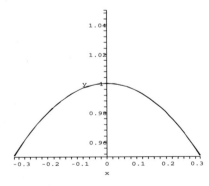

7. (a) From the graph, we determine that we can take $\delta = 0.02$, as shown below.

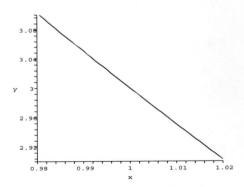

5. (a) From the graph, we determine that we can take $\delta = 0.38$, as shown below.

(b) From the graph, we determine that we can take $\delta = 0.01$, as shown below.